BRYOPHYTORUM BIBLIOTHECA

BAND

A morphogenetic re-evaluation of *Haplomitrium* Nees (Hepatophyta)

by

SHARON ELAINE BARTHOLOMEW-BEGAN

With 508 figures and 11 tables

J. CRAMER
in der Gebrüder Borntraeger Verlagsbuchhandlung
BERLIN · STUTTGART 1991

Editors

Dr. J.-P. Frahm, Duisburg and Dr. S.R. Gradstein, Utrecht

Author's address:

Sharon Elaine Bartholomew-Began
Department of Plant Biology
Southern Illinois University
Carbonale, Illinois 62901
U.S.A.

Present address:

Department of Biology
West Chester University
West Chester, PA 19383
U.S.A.

Printed in Germany by strauss offsetdruck gmbh, 6945 Hirschberg 2
ISBN 3-443-62013-2

ABSTRACT

The hepatic genus Haplomitrium has traditionally been of great phylogenetic interest due to its wide array of presumed ancestral characters, some of which are unique and others of which are reminiscent of Takakia and other bryophyte groups. However, the evolutionary affinities, taxonomy, and underlying morphogenetic patterns of Haplomitrium remain uncertain. Of the 13 described species, only 5 have been studied, in part, for developmental patterns, mostly at the light microscope level. New discoveries of both unique and moss-like ultrastructural characters in Takakia, a postulated close relative of Haplomitrium, and recent suggestions that Haplomitrium is more closely related to the Metzgeriales than to the Jungermanniales makes renewed investigations, including ultrastructural analyses of Haplomitrium appropriate. In this study, the ontogenetic patterns responsible for morphological features of the genus were investigated using experimental culture techniques, light and electron microscopy. Morphogenetic potentials were examined to gain insight into both infrageneric affinities and the phylogenetic relationships with other bryophyte groups, Takakia and the tracheophytes. Of the 13 species studied, 7 are retained and 6 new synonyms are proposed. Likewise, the only other extant member of the Haplomitriales, Steereomitrium, was synonymized with Haplomitrium. The taxa are grouped into two subgenera: subg. Haplomitrium (with two sections: sect. Haplomitrium and sect. Archibryum) and subg. Calobryum. It is concluded that, the genus is a highly plastic, plesiomorphic taxon, probably of Gondwanian origin. Haplomitrium's assemblage of derived traits negate it as the progenitor of either the Metzgeriales or the Jungermanniales. Rather, it appears to have a shared ancestry with the Metzgeriales, and is consequently best classified as a coordinate order in the newly described subclass Metzgeriidae. The shared ancestry of the Haplomitriales and the Metzgeriales suggests that primal hepatics were cylindrical forms with the potential to give rise independently to both dorsiventral and erect, radial gametophytes in response to selective pressures.

ACKNOWLEDGEMENTS

I wish to express my sincere appreciation to Dr. Barbara Crandall-Stotler for her inspiration and guidance throughout this study and for critical evaluation of this manuscript. I am also grateful to Dr. Joseph Beatty, Dr. John Bozzola, Dr. Lawrence Matten, and Dr. Raymond Stotler for their assistance and analyses of this work. Special thanks go to Dr. John Bozzola and the staff at the Center for Electron Microscopy at Southern Illinois University for technical advice and the use of facilities; Dr. Raymond Stotler, Dr. Patricia Geissler, Dr. Pekka Isoviita, and Dr. Jiří Váňa for taxonomic advice; Dr. Ella Campbell, Dr. Alan Fife, Dr. Zennoske Iwatsuki, Dr. Naofumi Kitagawa, Dr. Masami Mizutani, Dr. Malcolm Sargent, Dr. George A. M. Scott, Dr. Patricia Selkirk, and Dr. Barry Sneddon for contributing plant materials, as well as Dr. Patricia Geissler, Dr. S. R. Gradstein, and the curators of the following herbaria for their assistance in obtaining specimens: AD, BM, CHR, F, G, H, HO, L, LWU, MANCH, MUCV, NY, STR, SYD, TNS, WELT. Finally, but most importantly, I wish to express special thanks to my husband Timothy Began for his love, patience, and unselfish aid in preparing this manuscript, and to my parents Edwin and Elaine Bartholomew for their continued encouragement, love and support.

This manuscript is a revised version of a dissertation submitted to the Graduate School of Southern Illinois University at Carbondale in partial fulfillment of the requirements for the Doctor of Philosophy Degree in Botany.

TABLE OF CONTENTS

INTRODUCTION

Along with *Takakia* Hatt. & Inoue, *Haplomitrium* Nees has been regarded as one of the most ancient hepatics, mainly because of its erect, radial habit, its massive archegonia, and the similarity of early developmental stages of its archegonia and antheridia. *Takakia* was postulated to be closely related to *Haplomitrium* and was included in the Calobryales by Proskauer (1962), and later by Schuster (1967). However, recent developmental investigations have shown that *Takakia* gametophytes possess several unique characteristics which separate the taxon from all other bryophyte groups (Crandall-Stotler 1986). In contrast, preliminary examinations of the newly discovered antheridial and sporophytic material of *Takakia* support suggestions that it may be linked to the Andreaeopsida (Murray 1988; McFarland et al. 1989; Smith 1990). Thus, the taxonomic and phylogenetic relationships of *Haplomitrium* remain questionable and require re-evaluation.

Although a few workers have considered *Haplomitrium* as a very specialized group (Cavers 1911; Grolle 1969), most have considered *Haplomitrium* to be an ancient taxon (Smith 1955; Schuster 1966, 1967). The early alliance of the Calobryales with both the Jungermanniales and the Metzgeriales contributed to the hypothesis that the Calobryales were the common ancestor of the leafy and thalloid hepatics. In contrast, current authors suggest that the Jungermanniales and the Metzgeriales probably did not evolve from a recent common ancestor (Schuster 1966, 1967; Crandall-Stotler 1981). Therefore, *Haplomitrium* need not be viewed as the "missing link" between these two taxa. Also, the anacrogynous status of the genus indicates that *Haplomitrium* is not as closely related to the Jungermanniales as previously postulated. In fact, it has been suggested that *Haplomitrium* is probably more closely related to the Metzgeriales (D. H. Campbell 1920; Schuster 1966, 1967; Watson 1971) since it is anacrogynous, forms similar preprophase spore exine precursors (Brown and Lemmon 1986, 1987) and has apical derivative development and leaf ontogeny that is like that of the Metzgeriales (Leitgeb 1875; D. H. Campbell 1920; Crandall-Stotler 1981; Bartholomew 1987). On the other hand, certain sporophytic characteristics, such as the unistratose capsule wall, very long elaters, and dehiscence along a single suture, imply that Haplomitrium may be more closely related to certain members of the Marchantiopsida (D. H. Campbell 1920). In contrast, a line of evolution separate from other bryophytes is suggested for Haplomitrium by the spermatozoid anatomy of H. hookeri (Smith) Nees (Carothers and Duckett 1978, 1979; Duckett et al. 1982, 1983). The occurrence of an accessory band of microtubules overlying the osmiophilic crest at the spline anterior in mature spermatozoids is reported only in Haplomitrium and homosporous ferns, thus suggesting that Haplomitrium may be the closest living ancestor linking the bryophytes and tracheophytes (Duckett et al. 1982, 1983).

Despite the apparent phylogenetic significance of Haplomitrium, the morphogenetic patterns underlying its characters have not been previously described. According to Engel (1981), Haplomitrium consists of 13 species: H. hookeri, H. blumii (Nees) Schust., H. mnioides (Lindb.) Schust., H. andinum (Spruce) Schust., H. giganteum (Steph.) Grolle, H. gibbsiae (Steph.) Schust., H. intermedium Berie, H. indicum (Udar & Chandra) Schust., H. ovalifolium Schust., H. chilensis Schust., H. dentatum (Kumar & Udar) Engel, H. grollei Kumar & Udar, and H. monoicum Engel. Since 1981, one new species, H. kashyapii Udar & Kumar, has been described (Udar and Kumar 1982), while H. chilensis has been synonymized with H. gibbsiae (Hässel de Menéndez 1975). Some aspects of development have been studied in only five of these 13 species. Haplomitrium hookeri (Gottsche 1843; Leitgeb 1875; Lilienfeld 1911), H. gibbsiae (E. O. Campbell 1954, 1959), and H. blumii (Goebel 1891; D. H. Campbell 1920) have received the most intensive treatment. Apical growth and development were described by Leitgeb (1875) and Goebel (1891) in *H. hookeri* and *H. blumii*, respectively. Likewise, Leitgeb (1875) investigated leaf ontogeny, gametangial development, and branching in *H. hookeri*. Other studies on branching were performed by E. O. Campbell (1959) on *H. gibbsiae*, and by Crandall (1969) on *H. mnioides*.

Most of the developmental information available concerns the primitive gametangia of *Haplomitrium*. In addition to the previously mentioned work by Leitgeb (1875), archegonial and antheridial development have been studied by Gottsche (1843) and Lilienfeld (1911) in *H. hookeri*, by D. H. Campbell (1920) in *H. blumii*, by E. O. Campbell (1954, 1959) in *H. gibbsiae*, and by Mehra and Kumar (1990) in *H. indicum*. The structure of the shoot calyptra was investigated by Fulford et al. (1958) in *H. blumii* and by E. O. Campbell (1959) in *H. gibbsiae*. Also, the fine structure of the mature spermatozoid of *H. hookeri* (Carothers and Duckett 1978, 1979) and *H. gibbsiae* (Carothers and Rushing 1988) have been described.

Although the developmental sequence involved in the formation of the conducting tissue of *Haplomitrium* has not been described, several anatomical studies have been performed on the mature conducting tissue. E. O. Campbell (1959) described the anatomy in *H. gibbsiae*, while Grubb (1970) studied this organization in *H. hookeri*, *H. gibbsiae*, *H. intermedium*, and *H. mnioides*. Ultrastructural details of the conducting tissue of *H. gibbsiae* were contributed by Burr et al. (1974) and by Hébant (1977).

Developmental information regarding the sporophyte is also limited. Siessegger (1966) included *H. hookeri* in his comparative study of elaters. The spores of *H. mnioides* were observed using electron microscopy by Miyoshi (1967) while Taylor et al. (1974) published descriptions and scanning electron micrographs of *H. gibbsiae* spores and elaters. The only ontogenetic study at the ultrastructural level was performed by Brown and Lemmon (1986) who described spore wall development in H. hookeri using transmission electron microscopy.

Attempts have been made to determine the pattern of sporeling development in H. gibbsiae (E. O. Campbell 1959), H. mnioides (Nehira 1961; Yang 1966, 1967) and H. blumii (Yang et al. 1968) but the sporelings usually died early in development. Only Yang (1967) and Yang et al. (1968) were successful in growing the sporelings to maturity. However, their descriptions of the patterns are incomplete and consequently, a re-evaluation of sporeling ontogeny is necessary.

Basile (1968, 1973, 1974, 1979) has experimented with the application of the exogenous proline antagonist hydroxy-l-proline to various prostrate liverworts possessing leaves in two ranks. These treatments have induced changes in symmetry, leaf morphology, and branching patterns. Basile (1973, 1974) deemed these changes phylogenetically and systematically significant, postulating that during the course of evolution and speciation, alterations in the genes responsible for hydroxyproline protein synthesis resulted in morphological changes similar to those observed when hydroxy-l-proline is experimentally added to the plants. Thus, phenovariations resulting from the exogenous addition of proline antagonists are possible phenocopies of other taxa and reminiscent of what have been considered by Fulford (1965) and Schuster (1966) as the more ancestral forms in the evolution of leafy hepatics. Such tests utilizing proline antagonists on a plesiomorphic plant with questionable phylogenetic and systematic status such as Haplomitrium should theoretically contribute evidence useful in determining relationships and understanding evolutionary trends.

The purpose of this investigation, then, is to analyze the developmental patterns underlying the morphological features of Haplomitrium using data obtained from a combination of light microscopy, electron microscopy and experimental culture techniques. A comparison of morphogenetic potentials between Haplomitrium and presumably related taxa in the

Takakiophyta, Jungermanniopsida, and the Marchantiopsida should provide clarification of the systematic position of *Haplomitrium*, its role in hepatic evolution, and its relationship to the tracheophytes.

MATERIALS AND METHODS

Specimens Examined

Both living material and dried herbarium specimens were utilized in this investigation. Representatives of every *Haplomitrium* species except *H. ovalifolium* were examined either as living or herbarium specimens. The type of each species was seen with the exceptions of *H. chilensis*, *H. indicum*, and *H. ovalifolium*. For more detailed information on the specimens examined, refer to the taxonomic treatment.

Culture Techniques

All living specimens were maintained on moistened vermiculite in culture dishes. In addition, a portion of each live collection was axenically surface sterilized for 1 min in a 1% aqueous solution of Chlorox-Tween-20, washed three times in autoclaved, distilled water, and cultured on Hatcher's macronutrient agar, supplemented with Hatcher's micronutrient solutions B and C (Hatcher 1965). All cultures were maintained in a Percival model 135 I growth chamber equipped with 20 W cool lights under a light intensity of 1929.6 lux and a photoperiod of 8 h light:16 h dark. The plants received 0.6 $\mu W/cm^2/nm$ of light in the red spectrum and 0.72 $\mu W/cm^2/nm$ in the blue spectrum, with no detectable radiant energy in the far-red spectrum. A day/night temperature regime of 14° C/10° C was maintained.

Dried voucher specimens were made for all living collections and deposited in the Hepatic Herbarium of the American Bryological and Lichenological Society (ABSH). A small amount of each collection was also preserved in formalin-acetic acid-alcohol (FAA).

Sporeling ontogeny was investigated utilizing mature spores taken from elongated, undehisced capsules secured from live *H. mnioides* (*Doei s.n.*, *Furuki 7408*) and *H. blumii* (*Furuki 384*) plants. The spores were axenically cultured under controlled conditions according to the technique outlined by Renzaglia and Bartholomew (1985). Five types of macronutrient agar media were utilized. These consisted of Beneke (1903), Hurey-Py (1948), Bold (1949), Hatcher (1965), and Basile's F^1 (1975). Each of the macronutrient media was supplemented with Hatcher's micronutrient solutions B and C (Hatcher 1965). All culture flasks were positioned on the same shelf within the growth chamber and rotated in an orderly sequence once every two weeks to insure equivalent light intensities for each flask. The cultures were maintained under the environmental conditions previously described for living material. Observations were performed every two weeks as described by Bartholomew (1985) and vouchers were placed in FAA.

Light Microscopy Techniques

Live specimens prepared for light microscopy were fixed in FAA or Randolph's modified Navashin's fluid (Randolph 1935), dehydrated, paraffin embedded according to the procedure outlined by Johansen (1935), and sectioned using an American Optics, Spenser 820 rotary microtome. The resulting 10 μm sections were stained with Iron Alum O (Gray and Pickle 1956)- Fast Green FCC combination. Herbarium specimens of taxa unavailable as live samples were restored for paraffin sectioning using the technique of Crandall-Stotler (1976).

All observations, illustrations, and photographs involving light microscopy, including gross morphological observations, were performed with the aid of Wild Heerbrugg drawing and photographic apparati. Measurements were made using a Wild Heerbrugg ocular micrometer.

Electron Microscopy Techniques

The samples for transmission electron microscopy (TEM) were washed in 1% Triton-X-100, fixed for 6 h in 2% glutaraldehyde-2% formaldehyde in 0.1 M sodium cacodylate buffer, pH 7.0, post-fixed in 2% buffered osmium tetroxide for 24 h, then dehydrated at 1 h intervals in a graded ethanol series. After three 30 min washes in propylene oxide, the specimens were infiltrated with a 1:1 mixture of Spurr Low Viscosity and Embed 812 embedding media according to the following schedule. The specimens were first placed in a 1:1 mixture of propylene oxide and plastic resin for 24 h, then maintained for 4 days in 100% resin with changes every 24 h, and finally, embedded and polymerized at 60° C for 3 days. Thin sections were cut either on a Reichert Om U2 ultramicrotome using a glass knife, or on an LKB III ultramicrotome using a diamond knife. The resulting sections were placed on nickel or formvar grids and stained for 30 min in 2% aqueous uranyl acetate and for 7 min in basic lead citrate. Observations were made with a Hitachi H-500H transmission electron microscope.

Fresh materials for scanning electron microscopy (SEM) were removed from the previously described TEM preparation after being post-fixed in osmium tetroxide. The specimens were then washed in distilled water for 30 min with changes every 10 min and dehydrated in a graded ethanol series. Samples were critical point dried in a Tousimis Samdri-790 critical point drier using CO_2 as a transition fluid, mounted on stubs, and sputter coated with 400 Å of palladium-gold in a Technics Hummer V sputter coater using a pulse mode of deposition. Herbarium specimens were prepared for SEM according to the technique outlined by Stotler and Crandall-Stotler (1987). The material was then critical point dried, mounted on stubs, and coated with palladium-gold as previously described.

Sectioned paraffin material for SEM was prepared using the previously described light microscopy procedure, except the sections were cut at a thickness of 30 µm, and were adhered to round, glass coverslips instead of glass slides using 3% formalin-Haupts solution (Bissing 1974). After removing the paraffin with xylene, the coverslips were mounted on stubs and coated with 400 Å of palladium-gold (technique by Crandall-Stotler, personal communication). All SEM observations were made on a countertop Alpha-9 ISI scanning electron microscope.

Experimental Techniques

The experimental studies were designed to ascertain the effects of kinetin on the morphogenesis of *H. hookeri* gametophytes, as well as the effects of media composition, light quality, light intensity, and exogenous L-proline on the morphogenesis of *H. mnioides* sporelings. Adult *H. hookeri* plants were axenically cultured on Hatcher's macronutrient agar medium (supplemented with Hatcher's micronutrients B and C) containing 0.2 mg/l of 6-furfurylaminopurine kinetin. The cultures were maintained under the same environmental conditions previously described for cultured specimens. After three months, the plants were removed from culture and studied using light and scanning electron microscopy.

The effects of macronutrient media composition on the morphology, ontogeny and growth rate of *H. mnioides* sporelings were ascertained by comparing the sporelings grown on the five different media previously described. All other environmental factors (light regime, photoperiod, and temperature) were held constant.

The L-proline, light quality, and light intensity studies were performed using *H. mnioides* juveniles which were grown from spores on Hatcher's medium. This medium was chosen because the most "normal" morphology, i.e., a juvenile morphology most like that observed in adult plants, was obtained on Hatcher's medium. Juveniles were axenically transferred to 125 ml Erlenmeyer flasks each containing Hatcher's medium (Hatcher 1965), and the fungicide Nystatin. All cultures were maintained in a Percival model 135 I growth chamber under a uniform temperature ($14^{o}C/10^{o}C$, day/night) and photoperiod (8h/16h, day/night).

Light intensity differences were tested by maintaining one set of three cultures at 3002.2 lux, while another set received only 861.1 lux. The control set received 1929.6 lux.

Differences in light quality were investigated utilizing three wooden boxes each lined with aluminum foil and equipped with either a red, blue, or clear, 1/16 inch plexiglass filter. The boxes were arranged within the growth chamber so that each received an initial light intensity of 3216.0 lux before being covered with plexiglass. A set of three cultures was placed in each box.

The radiant energy of the wavelengths in each box was measured at plant level using a Plant Growth Photometer IL 150. The control box, which was fitted with clear plexiglass, received full spectrum light of 1.2 $\mu W/cm^2/nm$ in the blue spectrum and 0.8 $\mu W/cm^2/nm$ in the red spectrum. The second box was equipped with a red filter and received light of 0.69 $\mu W/cm^2/nm$ in the red spectrum and no detectable radiant energy in the blue spectrum. In contrast, the third box, fitted with a blue filter received light of 0.56 $\mu W/cm^2/nm$ in the blue spectrum and no detectable radiant energy in the red spectrum. No radiant energy was detected in the far-red spectrum in any of the above set-ups due to the cool white lights of the growth chamber. To test the effects of radiant energy in the far-red spectrum, an additional 100 W incandescent light was placed in the growth chamber to illuminate a set of three cultures. These plants received 11 $\mu W/cm^2/nm$ in the far-red spectrum, 8.5 $\mu W/cm^2/nm$ in the red spectrum, and 2.1 $\mu W/cm^2/nm$ in the blue spectrum.

The effects of L-proline were tested by adding 10^{-4} M L-proline to five culture flasks and placing them, along with three control cultures on the same shelf of the growth chamber under a light intensity of 1929.6 lux. As was the case for all the cultures involved in the experimental studies, the flasks were rotated in an orderly manner once every two weeks to maintain uniform intensities. The plants were grown for six months under their respective culture conditions and observed using light microscopy and SEM. Voucher specimens were placed in FAA.

MORPHOLOGY AND ONTOGENY

The Vegetative Gametophyte

Gross Morphology

The morphological characters of the vegetative gametophytes of the species of *Haplomitrium* are summarized in Table I. The vegetative traits which may be used to differentiate among the species of *Haplomitrium* include the habit and size of the plant, the symmetry and branching patterns of the leafy shoot, the size, shape, and insertion of the leaves, and, in a few instances, the mode of stolon production. These traits must be considered cautiously, however, since there is great plasticity within the genus. For example, there is notable variability in leaf shape among individuals of a species and often even among the leaves of a single shoot. These gradations in outline make decisive descriptions difficult and have led to the use of confusing terminology. Also, these characters are easily altered by environmental conditions (see experimental section).

As evident from Table I, *H. hookeri*, *H. ovalifolium*, *H. monoicum* Engel, and *H. intermedium* are similar in numerous vegetative characters. They are all small plants with erect aerial shoots that are usually nearly leafy throughout and exhibit great plasticity in symmetry and leaf insertion within the same species. Although most shoots are isophyllous with transverse leaf insertion, occasional shoots are slightly anisophyllous with one row of transversely inserted leaves and two rows of leaves with subtransverse insertion. Leaf shapes and sizes similarly vary within a species, although most leaves are longer than broad. Each species possesses leaves that are unistratose, but bistratose at least near the point of insertion. The sizes of the median leaf cells range from 25-80 µm long X 25-72 µm wide. The plasticity of traits exhibited by these species attests

TABLE I. MORPHOLOGICAL FEATURES OF THE VEGETATIVE LEAFY SHOOTS OF HAPLOMITRIUM SPECIES[a]

	Habit	Dimensions (LXW; mm)	Symmetry	Leaf Insertion	Leaf Shape	Basal Field Thickness
H. hookeri	erect	5.0-16.0 X 1.4- 2.6	iso, or slightly aniso	tr, or 1 row tr & 2 rows subtr	ellip, orb, rhom, ovate, ling	2-stratose
H. ovalifolium[b]	erect	8.0-16.0 X 1.2- 1.5	iso	tr	oval, ovate-lanc	2-stratose
H. monoicum	erect	4.5- 8.0 X 1.2- 1.8	iso, or slightly aniso	tr, or 1 row tr & 2 rows subtr	ellip, ovate, orb; 1-celled border	2-stratose
H. intermedium	erect	7.0-12.0 X 1.1- 2.0	iso, or slightly aniso	tr, or 1 row tr & 2 rows subtr	ovate, obovate, ellip, orb; 1-celled border	2-stratose
H. gibbsiae/ *H. chilensis*	erect	6.0-30.0 X 1.5- 2.2	iso, or aniso	tr, or 1 row tr & 2 rows suc	ovate, rhom, subro, orb	2- to 4-stratose

TABLE I. CONTINUED

	Habit	Dimensions (LXW; mm)	Symmetry	Leaf Insertion	Leaf Shape	Basal Field Thickness
H. kashyapii	erect	9.0-15.0 X 1.3- 2.8	iso, or aniso	tr, or 1 row tr & 2 rows suc	ovate, rhom, suborb	4-stratose
H. mnioides	prost	13.0-23.0 X 3.0- 6.0	aniso	ll suc, dl tr	ll subro, orb; dl ovate	2- to 3-stratose
H. blumii	erect	14.0-50.0 X 3.9- 5.9	aniso	ll suc, dl tr	ll obovate, orb, subro; dl ling, ovate	2- to 4-stratose
H. andinum	erect	20.0-55.0 X 6.0- 6.3	aniso	ll suc, dl tr	ll orb, ellip, subro; dl ling, bluntly trian	2-stratose

[a]vegetative material unknown for H. indicum, H. dentatum, H. grollei, and H. giganteum; [b]from Schuster 1971; aniso, anisophyllous; dl, dorsal leaves; ellip, ellipsoidal; LXW, length X width; lanc, lanceolate; ling, lingulate; ll, lateral leaves; orb, orbicular; prost, prostrate; rhom, rhomboidal; subtr, subtransverse; suc, succubous; suborb, suborbicular; subro, subrotund; tr, transverse; trian, triangular.

to their probable plesiomorphic status within the genus.

Haplomitrium mnioides, H. blumii, and H. andinum (Spruce) Schust. are less plastic in the expression of their vegetative traits than most of the previous species (Table I). These species are gametophytically very similar. Each is typified by fairly large, fleshy plants which possess robust, extensively branched rhizomatous systems. The leafy shoots are anisophyllous with the lateral leaves being decurrent on the stem and clearly succubously inserted, while the reduced dorsal leaves are transversely inserted. The leaf margins are entire and the median leaf cells range from 50-125 µm long X 37-80 µm wide. The basal-most leaves are distinctly reduced, few in number, and orbicular to slightly elliptical in outline.

Haplomitrium hookeri (Figures 1-8; 9-16). Numerous workers have provided descriptions of H. hookeri including Smith (1813), Hooker (1814), Gray (1821), Nees (1833, 1836), Gottsche (1843), Carrington (1874), Lindberg (1885), Cooke (1894), Stephani (1900), Pearson (1902), Müller (1906), Lilienfeld (1911), Evans (1917), Meylan (1924), Macvicar (1926), Schiffner (1941), Frye and Clark (1943), Worley (1969), and Landwehr (1980). As pointed out by Lilienfeld (1911) and Worley (1969), H. hookeri plants exhibit great phenotypic variability, depending on their habitat. The plants are bright green and often relatively small, slender, and leafy throughout (Figs. 1, 2), but may also be stocky with the leaves crowded distally (Fig. 3). The plants range from 5.0-16.0 (-27.0) mm tall and 1.4-2.6 mm wide. Most leafy shoots of H. hookeri are isophyllous with transversely inserted leaves (Fig. 2), but, a few are anisophyllous, though not conspicuously so. These individuals possess a dorsal row of leaves that are slightly reduced in size (Figs. 9a, 9b, 10a, 10b) and lateral leaves that may be subtransverse in insertion, especially near the base of the shoot. Although no other workers referred to H. hookeri as anisophyllous, many did recognize the existence in some individuals of a somewhat reduced row of leaves (Carrington 1874; Husnot 1922; Cooke 1894; Inoue 1971; Schuster 1967). Evans (1917) and Schuster (1966, 1967) also noted oblique insertions, while Gottsche (1843) described the leaves as oblique on the stem, but was unclear as to which row was "the third".

The leaves of H. hookeri may be patent (Fig. 2) or somewhat imbricate (Fig. 3). Although the leaves are mostly longer than broad, leaf morphology is extremely variable. Ovate, ovate-triangular, orbicular, broadly elliptical and sometimes even broadly lingulate, asymmetrical, or rhomboidal leaves, often with three blunt lobes are evident (Figs. 4-6, 9-15). As observed by Frye and Clark (1943), a variety of leaf shapes may occur on the same shoot. The leaf sizes range from 1.1-1.6 (-2.0) mm long X 0.5-1.5 (-2.4) mm wide X 0.5-0.9 mm from the base to the broadest portion. When anisophyllous, the dorsal leaves are somewhat reduced and more lingulate (Figs. 9a, 9b, 10a, 10b), measuring 0.7-1.0 mm long X 0.4-0.7 mm wide X 0.3-0.5 mm from the base to the widest region. The leaf margins are typically entire but may be undulate, incised, toothed, or lobed and bear unicellular slime cells (Figs. 4-6, 9-15). A review of the literature verifies this variability in leaf morphologies. For example, in the original description, Smith (1813) described the leaves as being ovate or oblong and angularly lobed. Nees (1838) echoed this description, while Gottsche (1843) also noted cordate and spear-shaped leaves. Carrington (1874) additionally reported lingulate, obtuse, and distantly lobed leaves. Pearson (1902), Schuster (1966, 1967), and Worley (1969) emphasized their polymorphism stating that the leaves are usually ovate, diamond-shaped, occasionally broadly lanceolate to lingulate, rhomboidal, often possessing 2-3 obtuse lobes, or are sometimes strongly asymmetrical. In transection, the leaves are unistratose, although a small bistratose zone exists near the point of insertion (Figs. 16a, 16b). The median leaf cells are slightly longer than wide or nearly quadrangular and measure 40.0-65.0 µm long X 32.0-42.5 µm wide.

Sometimes, small "bumps" which represent unelongated branches spot the basal region of the leafy shoot and rhizome. These seemingly dormant branches were first noted by Gottsche (1843).

The rhizomatous system is pale yellow-green to whitish and exhibits the typical Haplomitrium growth pattern (Figs. 1, 7). Grubb (1970) stated that there is no clear distinction between stolons and prostrate rhizomes. However, as is apparent in Figures 3 and 9, a prostrate axis is readily discernable. In fact, Grubb (1970, fig. 1c, p. 36) illustrated such a rhizomatous system. In some plants, the rhizome is very reduced and as pointed out by Gottsche (1843), it is sometimes difficult to distinguish stolon branches from those that will become erect and leafy due to their abbreviated growth. This is likely due to environmental conditions since the leafy axes of these plants are similarly stunted and the leaves are clustered distally as opposed to being leafy throughout.

The stolons are rather short, usually not exceeding 4.0 mm in length (Figs. 1, 7). The rhizomatous system and stolons are covered by numerous sessile and single-stalked slime cells (Fig. 8). As evident in Figure 8, reduced scales like those described by Lilienfeld (1911, fig. 6, p. 327 and fig. 7, p. 328) form near the apex of the rhizome.

Haplomitrium ovalifolium. Although H. ovalifolium was unavailable for study, Schuster (1971) indicated that this species is virtually identical to H. hookeri when sterile. The following brief description is derived from Schuster (1971). The plants are clear green and slender, 8.0-16.0 mm tall and 1.2-1.5 (-2.2) mm wide. The rhizomatous system is freely branched.

The aerial axes are typically leafy throughout with the lower leaves quickly becoming mature in size and form. The leaves are erect-patent and transversely inserted. They are usually oval to ovate-lanceolate in outline, thus always longer than wide, 0.92-1.00 mm long X 0.65-0.70 mm wide. Most possess smooth to repand-dentate margins with blunt angulations, but rare leaves may be deeply divided almost to their bases. The median leaf cells measure (20-) 25-32 (-36) X 25-38 (-60) µm and the leaves are unistratose except at the very base.

Haplomitrium monoicum (Figures 17-18; 21-25). The vegetative morphological characters observed in this investigation for H. monoicum generally agree with those published for the species by Engel (1981). The plants are the smallest of all Haplomitrium species measuring 4.5-8.0 mm tall and 1.2-1.8 mm wide. No living material was seen in this investigation, but Engel (1981) described the plants as fleshy, brittle, and pure grass green with a pale green to whitish rhizomatous system. The aerial shoots are nearly leafy throughout and as described by Engel (1981), are isophyllous or subisophyllous (Figs. 21-23). A dorsal row of leaves can sometimes be distinguished both by leaf size (Figs. 17a-c) and insertion (Fig. 23). Thus, some shoots are anisophyllous.

The leaves are slightly decurrent on the stem (Fig. 23) and are contiguous and patent (Figs. 21-23). Although leaf insertion is almost always transverse, individuals may possess one row of transversely inserted leaves while the leaves of the other two rows are subtransverse (Fig. 23). All three rows of leaves are elliptical, ovate, orbicular, or sometimes obovate (Figs. 17a-c, 21, 22). As described by Engel (1981), the leaves are longer than broad with narrowed bases and usually broadly rounded apices. They measure 0.9-1.3 mm long X 0.6-0.9 mm wide X 0.3-0.6 mm from the base to the widest region of the leaf. When distinct, the dorsal leaves are slightly smaller than the lateral leaves averaging 0.6-0.8 mm long X 0.5-0.8 mm wide X 0.5-0.7 mm from the base to the broadest portion (Figure 17a). The leaf margins are entire and possess a distinct border of 1 or sometimes 2 rows of large cells (Figs. 24, 25a). A few minute, unicellular slime cells were found scattered on the leaf margins in a few plants (Figs. 25a, 25b). Thus, the absence of slime papillae, cited by Engel (1981) as a diagnostic species character, is not valid. The leaves are unistratose with a small, median bistratose basal field (Figs. 18a, 18b). The median leaf cells measure 52.0-80.5 µm long X 50.2-73.6 µm wide and are therefore, larger than those of H. hookeri.

The prostrate, freely branched rhizome displays the growth pattern typical for Haplomitrium (Fig. 21), i.e., most leafy shoots arise from upturned rhizome apices and stolons descend from the rhizome.

Haplomitrium intermedium (Figures 19-20; 26-34). Only Berrie (1962) and Schuster (1967) have provided detailed descriptions of H. intermedium. This species is composed of very slender, delicate plants (Figs. 26, 28) which are 7.0-12.0 (-20.0) mm tall, but only 1.1-2.0 mm broad. The leafy shoots are light to almost translucent green, while the rhizomatous system is very pale green to whitish. The erect shoots are leafy throughout and the leaves may be remote to contiguous and are patent (Figs. 26, 28, 29, 31). Although Berrie (1962) described the symmetry of the leafy shoots as isophyllous, it is sometimes possible to distinguish a dorsal row of leaves through size differences (Figs. 19a-d) and, rarely, by a slight variation in leaf insertions between the rows (Fig. 28). Schuster (1967, p. 29) also noted this distinction stating in his species description, "Dorsal leaves, when distinguishable, 0.6-0.85 area of lateral leaves, often more obovate, ca. 1,000-1,100 µ wide X 1,375-1,575 µ long." By definition, then, these plants must be considered anisophyllous. Thus, both isophyllous and anisophyllous individuals comprise this species.

The lateral leaves, which are slightly decurrent on the stem (Figs. 28, 30), are longer than broad and may be slightly obovate, ovate, elliptical, nearly orbicular, or occasionally somewhat rhomboidal even on the same plant (Figs. 19a-d, 28). This variability in leaf shape is also reflected in the literature. Berrie (1962) described the leaves as rhomboidal to suborbicular with those near the base of the shoot occasionally deeply bilobed. He also reported leaves with 1-4 short, basal marginal teeth each terminated by a slime cell such as the leaf shown in Figure 29. Schuster (1967) described the leaves as ovate to elliptical to sometimes ovate-orbicular, widest in the median two-tenths and usually narrowed in the basal quarter. The leaves measure 1.1-1.7 mm long X 0.7-1.3 mm wide X 0.7-0.9 mm from the base to the broadest portion. This size range corroborates the observations of other workers (Berrie 1962; Schuster 1967). When dorsal leaves are evident, they are slightly smaller measuring 0.65-1.2 mm X 0.4-0.7 mm wide X 0.3-0.7 mm from the base to the broadest region. These reduced leaves are ovate, nearly obovate, or more lingulate in outline (Figs. 19a, 19b, 30). The leaf margins are usually entire (Figs. 19a, 19c, 28, 30), but in some cases may be undulate (Fig. 19d). A few unicellular slime cells (up to six according to Berrie [1962]) are spaced along the margin (Fig. 30). Occasionally, a 1-celled border of larger, somewhat swollen cells can be distinguished (Figs. 29, 30). The leaves are unistratose but a small 2-stratose zone exists near the point of insertion (Figs. 20a, 20b). The median leaf cells average 57.2-88.0 µm long X 49.6-69.0 µm wide, thus being similar in size to those of _H. monoicum_.

Although the majority of _Haplomitrium_ species remain unbranched in the upper regions of the leafy shoots, _H. intermedium_ frequently exhibits terminal, exogenous branching (Fig. 31). Berrie (1962) and Schuster (1967) also reported terminal branching.

Stolons may be produced as branches from the rhizome in the fashion typical for the genus (Figs. 26, 27), but, more commonly, stolon production is from the leafy shoot (Figs. 26, 32). Each leafless, geotropic stolon originates near the base of the leafy shoot, typically from a leaf axil (Fig. 32). This axillary mode of stolon production is unique and distinguishes the species from all other _Haplomitrium_ taxa.

In the protologue, Berrie (1962) reported that the axillary branching pattern is predominant in plants that are densely overshadowed by other bryophytes. In contrast, when *H. intermedium* is not crowded in such a manner, branching is confined to rhizomes, i.e., the rhizome grows obliquely through the substrate and new leafy axes form when branches ascend vertically (Berrie 1962).

Schuster (1967) emphasized the systematic importance of the axillary geotropic stolons, but did not mention the existence of stolon production from the rhizome. Similarly, Grubb (1970) emphasized the axillary stolon ontogeny and stated that no distinct rhizome is differentiated. However, he did acknowledge that a new leafy shoot may "closely approximate" the creeping rhizome typical for the genus by growing horizontally for a time prior to becoming erect. Indeed, the existence of axillary stolon production is exceptional and indicates a variation in the genetic mechanism responsible for stolon branching, thus separating *H. intermedium* from other *Haplomitrium* species. However, the existence of the typical mode of stolon and leafy shoot production from the rhizome links *H. intermedium* with the other species. The potential for the production of stolons from two points of origin reflects the plasticity of the genetic mechanism that controls branch formation in this species.

Similar plasticity is displayed in *Takakia lepidozioides* Hatt. & Inoue. Although in nature the stolons of this taxon originate from the rhizome, Crandall-Stotler and Bozzola (1988) reported stolons originating from leaf axils in the upper portions of leafy shoots in plants cultured on moistened vermiculite. However, these stolons were green, and although descending, were not necessarily subterranean.

The stolons of *H. intermedium* are also morphologically unique in being extremely slender and much longer than those of any other *Haplomitrium* species. The stolons often measure more than 1.5 cm in length, making them as long as to often longer than the leafy shoots themselves. In addition, the stolons frequently branch in a monopodial pattern (Figs. 33, 34a). No other *Haplomitrium* species is known to possess branched stolons. Also, unlike the other species, very few slime papillae exist on the rhizome and stolon axes (Fig. 34). As first noted by Berrie (1962), this virtual lack of slime cells results in the absence of a thick mucilage sheath.

Haplomitrium gibbsiae/ H. chilensis (Figures 35-39; 42-44). The vegetative gametophytes of H. gibbsiae and H. chilensis are identical and will be considered together in this description. Stephani (1924), E. O. Campbell (1959), Schuster (1967), Hässel de Menéndez (1975), and Allison and Child (1975) have published descriptions of H. gibbsiae, while only Schuster (1971) has described H. chilensis. It is apparent from these descriptions, a survey of herbarium material, and experimental studies, that the morphology of these plants varies greatly depending on environmental conditions. The plants are bright to yellowish green and often robust with the leaves crowded distally on the aerial axes (Fig. 35). In contrast, some may be slender, feeble, and nearly leafy throughout. The gametophytes range from 6.0-30.0 mm tall and 1.5-2.2 mm wide. All three rows of leaves are typically isophyllous and almost exclusively transversely inserted. However, occasional individuals occur in which the leaves of two rows (the lateral rows) may be succubous (Figs. 36, 37). Hence, the dorsal row may be identified by its transverse insertion. Schuster (1967) also reported this infrequent oblique insertion pattern.

The leaves may be widely patent or somewhat imbricate. Most descriptions agree that leaves which are subrotund or rhomboidal in outline, often with three blunt to acute lobes, are characteristic of the species (Figs. 36, 42a, 42b, 43a) (Stephani 1924; Campbell 1959; Schuster 1967; Allison and Child 1975). However, ovate to orbicular and even elliptical leaves may also be present (Figs. 38, 42c, 43b, 43c). Leaf sizes range from 2.5-3.5 mm long X 2.8-4.0 mm wide X 0.9-2.5 mm from the base to the broadest portion of the leaf. The leaf margins may be entire (Figs. 38, 39), repand, or angled (Figs. 37, 42a, 42b, 43a) with marginal slime cells (Fig. 39). Campbell (1959) reported that up to six marginal mucilage cells may be present, but these wither and are lost when exposed to dry air. The leaves are unistratose with a large polystratose basal field that ranges from 2-4 cells in thickness (Figs. 44a, 44b). The median leaf cells are longer than broad and measure 45.0-78.5 µm long X 43.0-65.8 µm wide. Schuster (1967) reported great variability in the sizes of the median leaf cells with cells ranging from 45-60 µm in length by 40-50 µm in width up to 65-85 µm long X 55-66 µm wide.

The upper portions of the aerial shoots are usually unbranched, but, sporadic, lateral branching was present in a few specimens. Stephani (1924) described frequent branching of this type. As noted by Schuster (1967), semi-elongated, "dormant" spurs are produced at or near the bases of the erect leafy shoots (Fig. 35). Although similar unelongated branches occur in other Haplomitrium species, they typically appear as small, globose "bumps" and do not consistently elongate to the extent seen in H. gibbsiae.

The branching pattern of the rhizomatous system is also somewhat variable (Campbell 1959; Grubb 1970), although, the basic pattern is that typical for the genus. Stout, prostrate rhizomes such as those described by Grubb (1970) were clearly distinct in most of the H. gibbsiae material examined (Fig. 35). However, in individuals with a lax habit and very long rhizomes a regular differentiation of stolons is not always obvious. This difficulty in distinguishing stolons is likely due to their abbreviated growth form and is similar to the condition that also occurs in H. indicum, H. dentatum (Kumar & Udar) Engel, and some individuals of H. hookeri.

The entire rhizomatous system, as well as the basal spurs of the leafy shoots is covered by numerous 1-2-celled slime papillae. The resulting mucilage sheath is very thick. According to Campbell (1959), this slime layer may be as thick as the rhizome axis itself.

Although similar to the previously described taxa in numerous aspects, *H. gibbsiae* differs in that the majority of aerial shoots possess leaves that are crowded distally as opposed to being leafy throughout, the symmetry of the leafy shoot is less variable, the leaves possess polystratose basal fields, and finger-like, lateral spurs are prominent at the bases of the leafy shoots.

Haplomitrium kashyapii (Figures 40-41). No vegetative material of any of the Indian taxa except *H. kashyapii* was available for study. Apparently, *H. grollei* (Kumar and Udar 1977) has been based exclusively on female plants, while *H. indicum* (Udar and S. Chandra 1961, 1965) and *H. dentatum* (Kumar and Udar 1976) are known from male and female plants. No vegetative plants were mentioned or illustrated by the authors for any of these latter taxa.

Although the type specimens of *H. kashyapii* were examined, the plants were tattered and in such bad condition that information was difficult to obtain. Therefore, the protologue (Udar and Kumar 1982) was relied upon heavily for the following description. According to Udar and Kumar (1982), live *H. kashyapii* plants are green to bright green. The plants are erect (Figs. 40, 41) and measure about 9.0-1.5.0 mm tall X 1.3-2.8 mm wide.

The aerial axes are reportedly leafy throughout (Udar and Kumar 1982) (Fig. 41), but at least one individual in the type collection possesses leaves that are clustered above (Fig. 40). The symmetry and insertion patterns of the leafy shoots were difficult to determine from the depauperate material at hand, but, like the previously described taxa, some individuals appeared to be isophyllous, while others were anisophyllous. This concurs with the report by Udar and Kumar (1982) that the symmetry is isophyllous to subisophyllous. Although Udar and Kumar (1982) emphasized the variable orientation and morphology of the leaves, they recognized only transverse insertion.

Some shoots of the type material also appear to possess succubously inserted leaves. The leaves may be suborbicular, ovate, rhomboidal, or slightly reniform, 2.5-3.2 mm long X 1.8-3.0 mm wide X 1.2-2.0 mm from the base to the broadest region. When anisophyllous, the dorsal leaves are slightly reduced. In the type material they measured 1.8-2.1 mm long X 2.4-2.7 mm wide X 0.9-1.0 mm from the base to the broadest portion. Udar and Kumar (1982) reported entire leaf margins. Like *H. gibbsiae* and *H. chilensis*, the unistratose leaves possess a large polystratose basal field. The median cells of the leaves measure 68.5-92.0 µm long X 24.2-39.2 µm wide.

Udar and Kumar (1982) remarked that the branching of the rhizomatous system of *H. kashyapii* lacks a definite pattern, but the plants illustrated in the protologue (Udar and Kumar 1982, figs. 1-7, p.76) possess rhizomes whose apices have become erect and leafy. The stolons, although not extemely numerous, arise from the rhizomatous axes.

Since no differences in the gross morphology of the rhizomatous systems exist between sterile and fertile plants of the other *Haplomitrium* species, it is assumed that none exist among the Indian species either. Therefore, based on the herbarium specimens examined and the illustrations provided by the authors (Udar and S. Chandra 1961, 1965; Kumar and Udar 1976, 1977; Udar and Kumar 1982), it seems that the growth patterns of the rhizomatous systems of all the Indian taxa are typical for the genus, i.e., the rhizome apex ultimately tilts dorsally and develops into an erect leafy shoot accompanied by subsequent branching at the base of the aerial shoot and/or at intervals along the rhizome to continue the system. Stolons originate from the rhizomatous axes.

Kumar and Udar (1976) described the rhizomatous system of *H. dentatum* as being similar to those of *H. mnioides* and *H. indicum*, but added that the mode of shoot and stolon development is different in *H. dentatum*. They stated,

> "In this species a small spur like bud, at the base of [an] aerial shoot, develops into [a] creeping horizontal rhizome which is comparatively longer and gives rise to [a] new erect gametophore, but usually the tip of [the] rhizome does not get upturned." (Kumar and Udar 1976, p. 28).

The meaning of this statement is unclear as the plants illustrated (Kumar and Udar 1976, figs. 2, 3, 5, p. 25) clearly show erect leafy shoots which must have developed from upturned rhizome apices. Apparently the rhizome systems of each Indian species bear numerous slime cells since thick mucilage sheaths encompass the stolons and rhizomes. These slime layers have been reported by the original authors (Udar and S. Chandra 1965; Kumar and Udar 1976, 1977; Udar and Kumar 1982) and were present on many of the observed herbarium specimens (the mucilage rehydrated along with the dried plants when immersed in hot water).

Haplomitrium mnioides (Figures 45-47; 48-55). Of the strictly anisophyllous taxa, only *H. mnioides* is prostrate to obliquely ascending (Figs. 48, 49a, 49b). Complete descriptions for this species have been published by Lindberg (1874, 1875), Mitten (1891), Stephani (1900), Horikawa (1929), and Schuster (1967). The plants are deep green with pale green to yellowish rhizomatous systems. In general, *H. mnioides* is among the smallest of the strictly anisophyllous taxa, measuring 13.0-23.0 mm tall and 3.0-6.0 mm broad. The aerial shoots possess a distinct basal region characterized by much reduced basal leaves (Figs. 48, 50).

The lateral, succubously inserted leaves are slightly spreading and are positioned so that the leaves of each row partially overlap those of the row below (Figs. 49a, 49b). The lateral leaves are orbicular to subrotund and measure 2.5-3.4 mm long X 2.4-3.4 mm wide X 1.2-1.6 mm from the base to the broadest portion of the leaf (Figs. 45c, 45d, 46, 49a, 49b). Unicellular, slime cells frequently occur on the leaf margins and occasional sessile or single-stalked slime cells form on the laminae of the leaves (Fig. 51). Although the leaf is mostly unistratose, a median 2-3 stratose zone exists on the basal lamina (Figs. 47a, 47b). This field averages about 3-7 cells in length and 11-13 cells in width, but varies depending on the individual. The median leaf cells measure 48.5-95.8 µm long X 37.0-62.5 µm wide. The transversely inserted dorsal leaves are commonly ovate (Figs. 45a, 49a) and are typically 1.4-2.0 mm long X 1.0-1.9 mm wide X 0.5-0.6 mm from the base to the broadest portion of the leaf. Thus, the dorsal leaves are slightly longer than broad and are as broad to slightly broader than the stem (Fig. 49a).

Although the leafy shoots usually remain unbranched, very young branches lie dormant at the bases of the leafy shoots (Fig. 52). Although similar in position to the spurs of H. gibbsiae and H. chilensis, the young branches of H. mnioides remain more globose and do not elongate to the extent seen in H. gibbsiae.

The well-developed, intertwined rhizomatous system is robust and freely branched (Fig. 48). Additional aerial shoots are formed as the apex of the rhizome turns upward and becomes leafy (Figs. 48, 53). Minute "bumps" representing young, unelongated branches like those at the bases of the leafy shoots are generously formed at intervals along the rhizome (Fig. 54).

The stolons are robust, but rarely exceed 8.0 mm in length. Both the rhizome and stolons bear numerous clavate slime cells which secrete a thick layer of mucilage (Fig. 55).

Haplomitrium blumii (Figures 56-60; 61-62). Nees (1830, 1844), Goebel (1891, 1930), Mitten (1891), Schiffner (1893, 1900), Stephani (1900), D. H. Campbell (1920), Schuster (1967), Udar et al. (1968) and Piippo (1984) have all provided descriptions of H. blumii. This taxon closely resembles H. mnioides, but grows erect (Fig. 56). The plants are deep green and very fleshy with a pale green to yellowish rhizomatous system. Depending on environmental conditions, the plants may be very robust or slightly more spindly. Their sizes range from 14.0-50.0 mm tall and 3.9-5.9 mm broad. The leaves of the erect axes are spreading and the leaves of one row may slightly overlap those of the row below (Fig. 56).

The succubously inserted lateral leaves are mostly obovate, but some subrotund to orbicular outlines may also occur (Figs. 56, 61a, 61b). The leaves are 2.8-3.2 mm long by 1.8-2.2 mm wide by 2.0-2.2 mm from the base to the broadest region of the leaf. The polystratose basal fields of the H. blumii leaves are more extensive and often slightly thicker than those generally found in H. mnioides (Figs. 62a, 62b). The basal field in H. blumii is 2-4 stratose and often extends as much as 8-10 cells in length and 13-20 cells in width. The median cells of the leaf average 50.5-88.0 µm long X 45.0-80.0 µm wide. The slime papillae are like those of H. mnioides in structure and distribution. The transversely inserted dorsal leaves range from lingulate-triangular in the three most apical rows to more ovate lower on the shoot (Figs. 61c, 61d). The sizes of these leaves are highly variable. Although they are typically broader than the stem (1.2-2.4 mm long X 1.0-1.7 mm wide) (Fig. 58), some are as broad or slightly more narrow than the stem. In rare cases, leaves of many different sizes may be distributed at various loci along the dorsal row of a single shoot.

The leafy shoots are mostly unbranched, but, like H. mnioides, small, green, globose, unelongated branches are present at the bases of leafy shoots (Fig. 57) and at intervals along the rhizome. The stout rhizomatous system is freely branched and the descending stolons originate from the rhizome in direct association with the leafy shoots (Figs. 56, 59). Both the rhizome and stolons are generously covered with slime cells (Fig. 60).

Haplomitrium andinum (Figures 63-65; 66-67). Descriptions of H. andinum have been provided by Spruce (1885), Stephani (1900), Fulford (1962), and Schuster

(1967). Haplomitrium andinum is also relatively fleshy and large with an erect habit (Fig. 66). The plants measure 2.0-5.5 cm tall X 6.0-6.3 mm broad. The leafy axes of H. andinum are usually unbranched, but according to Spruce (1885), terminal branching occasionally occurs.

Like the other taxa of this complex, the leaves are spreading and distant, but may be slightly contiguous (Figs. 66, 67a). The lateral leaves are mostly orbicular, subrotund to broadly elliptical with dimensions of 2.3-3.2 mm long X 1.8-2.3 mm wide X 1.1-2.3 mm from the base to the broadest portion of the leaf (Figs. 63a, 64a, 64b, 66, 67a). The leaves of H. andinum are unistratose with a small bistratose zone near the point of insertion (Fig. 65a, 65b). This zone usually extends nearly the entire width of the leaf base and may include only the basal 2 or as many as the basal 8 rows of cells. The median leaf cells measure 50.5-100.5 µm long X 45.0-55.3 µm wide, thus are only slightly larger than those of H. mnioides and H. blumii. The 2 or 3 most apical dorsal leaves of most shoots are very reduced, measuring 0.8-1.3 mm long X 0.3-0.8 mm wide X 0.2-0.4 mm to the broadest portion of the leaf and are bluntly triangular in outline (Figs. 63b, 67a). The dorsal leaves located basal to the three apical rows are lingulate to elliptical (Figs. 63c, 67b) and measure 2.3-2.9 mm long X 1.0-2.4 mm wide X 2.2-2.4 mm from the base to the widest portion. Thus, they are different in shape and distinctly larger than the more apical dorsal leaves. The 2 or 3 most distal dorsal leaves are narrower than the stem while the dorsal leaves basal to them are as broad or broader than the stem (Fig. 67).

The size and shape differences among the 2 or 3 most apical dorsal leaves and those basal to them indicate an abrupt rather than gradual rate of dorsal leaf enlargement. In other taxa, leaf development is usually at a uniform pace so that a gradual increase in the size and shape of the younger and older dorsal leaves is displayed. However, in many H. andinum shoots, the development is abrupt. The youngest dorsal leaves remain in the early stages of development until at least two additional, younger leaves are produced, then they abruptly complete their development. So, either cell division is retarded or cell enlargement is slowed in the youngest dorsal leaves.

The rhizomatous system is freely branched as in H. mnioides and H. blumii. Likewise, stolons are produced from the rhizomatous axis in association with the leafy shoots and the entire system is covered with numerous slime cells.

Like H. indicum, H. dentatum, and H. grollei, H. giganteum (Stephani 1922, Grolle 1963; Schuster 1967) is known only from fertile plants. Similarly, Steereomitrium is also known only from fertile shoots (E. O. Campbell 1987). Therefore, the treatment of these taxa will be deferred until the section pertaining to the reproductive gametophyte.

Apical Organization

Both the leafy and leafless shoots of Haplomitrium grow through the activity of tetrahedral apical cells with three cutting faces (Figs. 68-75; 78-80). In typical Metzgerialian fashion, these generative cells lie slightly sunken in the shoot tips. In cross sections, the apical cell can appear as either an equilateral (Figs. 69a, 69b, 80) or isosceles triangle (Figs. 71, 72), depending on the stage in the division cycle and/or the species viewed. In longitudinal sections, the apical cell appears as an isosceles triangle with the height always greater than the width (Figs. 70, 73a, 79). The lateral faces of the apical cell are curved inward at the free surface so that the free surface diameter of the cell is less than its median diameter (Fig. 68). This is apparent in median longitudinal section (Fig. 73a), as well as in serial transections through the apex (Figs. 69a, 69b, 71a, 71b, 72a, 72b). For example, at the free surface level the apical cell of H. hookeri measures approximately 15.4-17.0 µm in width, while 10 µm basal to the free surface it is approximately 20.0 µm wide (Figs. 69a,

69b). The bulging midregion of the apical cell is even more apparent in H. mnioides (Fig. 73a), where the width at the free surface is approximately 19.8-26.4 µm and 10-20 µm below the free surface, is about 33.0 µm. The stolon apical cells display the same morphology as those of the leafy shoots although their sizes may differ slightly (Table II, Figs. 74, 75).

Pyramidal apical cells with three almost equal cutting faces have been reported in the isophyllous shoots of H. hookeri (Gottsche 1843; Leitgeb 1875) and H. gibbsiae (E. O. Campbell 1959), as well as in the anisophyllous species, H. blumii (Goebel 1891; D. H. Campbell 1920) and H. mnioides (Bartholomew 1987). Likewise, longitudinal sections through leafy apices, showing apical cells with reduced free surfaces and bulging midregions, have been illustrated in H. hookeri (Leitgeb 1875), H. blumii (D. H. Campbell 1920), and H. mnioides (Bartholomew 1987).

One aspect regarding the apical cell that has previously been overlooked is the slight dorsal tilt of the cell. The displaced angle of the cell as compared to the median longitudinal axis of the shoot is evident in median longitudinal sections which reveal that the dorsal height of the apical cell is not as great as its ventral height (Figs. 70, 73a). For example, the dorsal height of the H. hookeri apical cell shown in Figure 70 is 22.0 µm, while the ventral height is 24.2 µm. In Figure 73a, the dorsal height of the generative cell of H. mnioides is 48.4 µm, while the ventral height is 52.8 µm. The apical tilt also becomes apparent when sequential merophyte patterns are compared in serial transections. In the most distal cross section, one of the merophytes of a sequence is not visible, while the complete spiral of merophytes is visible in the next section, 10 µm basal to the first (Figs. 69a, 69b, merophyte X, 72a, 72b, merophyte VII). If no apical tilt existed, each transection through the apex would reveal the complete compliment of merophytes.

Derivatives are formed from the apical cell in three ranks in either a sinistrorse (Figs. 72a, 72b, 74, 78a) or dextrorse sequence (Figs. 69a, 69b, 71a, 71b).

TABLE II. APICAL CELL DIMENSIONS OF SELECT *HAPLOMITRIUM* SPECIES

	Leafy Shoot/Rhizome			Stolon		
	Width[a] (μm)	Depth[b] (μm)	Height[c] (μm)	Width (μm)	Depth (μm)	Height (μm)
H. hookeri	15.4-17.0	15.4-17.0	24.0-40.0	15.4	ND	19.8
H. monoicum	19.8-21.7	19.8-21.7	ND	24.2	22.0	ND
H. intermedium	19.8-22.0	19.8-28.2	44.0-60.4	20.8-24.2	21.0-24.2	23.0
H. gibbsiae/ *H. chilensis*	22.0-28.6	22.0-26.4	48.7-52.8	28.0-30.0	26.8-28.6	ND
H. mnioides	19.8-26.4	22.0-26.4	50.6-55.7	15.6-17.5	15.6-19.8	33.0
H. blumii	19.8-24.2	22.0-24.2	ND	19.8	ND	48.4
H. andinum	18.9	28.6	35.2	ND	ND	ND

[a]distance from one lateral face to another; [b]distance from ventral point of free surface to midpoint of dorsal segmenting face; [c]distance from point of tetrahedron to triangular free surface; ND, no data.

Although the apical cell is always tetrahedral, it is dynamic during segmentation, i.e., the sizes of the apical cell cutting faces do not remain constant during segmentation. When a derivative is formed, the wall which delimits it from the apical cell is slightly oblique so that the anodic portion of the segment is slightly broader than the kathodic region (Figs. 69a, 69b, 72a, 72b, 74, 78a, 78b). As evident from Figure 78b, the cell plate curves in at the kathodic end of the segment, so as the merophyte expands and the wall matures, the kathodic width is less than the anodic width. The obliquity of the wall forming the youngest derivative makes the cutting face preceding it temporarily longer than the two other segmenting faces (Figs. 71b, 74, 78a, 78b). This phenomenon, in which anodic portions of the segments are always broader than kathodic regions, occurs in all species of Haplomitrium. Leitgeb (1875) first recognized the existence of broader anodic portions in the merophytes of H. hookeri.

In some species, one cutting face, the dorsal face, is always shorter than the other two walls. This is probably a function of the enlargement potential of the dorsal merophytes. If the dorsal derivatives do not enlarge as rapidly as the lateral merophytes, physical limitations prohibit the dorsal face of the apical cell from expanding to the same extent as the lateral walls in preparation for the next sequence of divisions. The reduction of the dorsal face is most exaggerated in the anisophyllous species of H. mnioides, H. blumii, and H. andinum. Individuals bearing apical cells with one reduced cutting face were also reported in H. gibbsiae by E. O. Campbell (1959) and H. blumii by D. H. Campbell (1920). However, D. H. Campbell (1920) incorrectly assumed the shorter cutting face to be ventral in position as in anisophyllous Jungermannialian taxa instead of dorsal as is truely the case. None of these workers ventured an explanation for how one cutting face becomes reduced in length.

Basically, then, there are two major forces which influence apical cell dynamics. First, the oblique wall which delimits each derivative creates a broad anodic region in each segment with the consequence that the preceeding apical cell cutting face is temporarily longer than the other two. Thus, as the segmentation spiral continues, the sizes of the segmenting faces change alternately. Second, the rate of merophyte development places physical restrictions on the expansion of the apical cell, so that if one row of merophytes develops more slowly or to a lesser extent than the other two rows, the apical cell face adjacent to the reduced derivatives will be restricted during apical cell expansion, and, in cross section, the apical cell will appear as an isosceles triangle.

Each derivative cut from the apical cell is initially a pentahedron with triangular acroscopic and basiscopic surfaces and more or less rectangular anodic, kathodic, and free surfaces (Figs. 76, 77). Thus, the derivative morphology conforms with that described by Crandall-Stotler (1981) for hepatics. The merophytes develop like those of the Metzgeriales. The first division of the derivative is oblique anticlinal and extends from the free surface to the anodic wall (Figs. 73b, 76, 77, [walls 1-1]). Thus, the first wall forms in the kathodic portion of the segment and creates a larger anodic and smaller kathodic cell (Figs. 71a, 76, 77, 78a, [walls 1-1]). The second division of the merophyte occurs in the anodic cell (Figs. 71a, 76, 77, [walls 2-2]). Thus, a 3-celled merophyte is created such as that illustrated in Figure 76. The divisional sequence of the merophyte is accompanied by the expansion of the merophyte walls so that the free surface assumes a rectangular outline, i.e., the anodic and kathodic portions become more or less equal in width.

The exact sequence of future merophyte divisions is difficult to ascertain with certainty. However, each of the two rectangular-shaped, lateral cells of the 3-celled merophyte appear to divide twice periclinally to produce epidermal, cortex, and hydroid initials (Figs. 73a, 73b, 77, [walls 3-3, 3^1-3^1, 4-4, 4^1-4^1]). Continued divisions in each of these cells ultimately produces the corresponding tissue zone (Figs. 73a, 73b, 77, [walls 5-5, 5^1-5^1]).

The central wedge-shaped cell of the 3-celled merophyte enlarges and projects beyond the level of the apical cell (Fig. 73b). It is then cut by a sequence of anticlinal divisions which mirror the first two divisions of the derivative itself (Fig. 73b, walls 3a-3a, 4a-4a). The resulting basal cells contribute to the stem tissue directly below the leaf, while divisions in the distal portion of the central cell delimit the wedge-shaped apical cell of the leaf (Fig. 73b, wall 5a-5a). This leaf apical cell segments alternately in a single plane to produce the leaf lamina. In this manner, one leaf is produced per segment, resulting in the 3-ranked leaf arrangement of Haplomitrium. In the stolon, the central cell of each merophyte yields a slime papilla instead of a leaf (Fig. 75). Thus, in Haplomitrium, the slime papilla and the leaf are homologous structures and provide evidence supporting the theory that slime cells are archaic attempts at leaves (Schuster 1979, 1984b).

Leaf sizes are directly related to the sizes of the derivatives and corresponding segmenting faces of the apical cell, while leaf insertion is mainly a result of apical tilt. As first described in Jungermannialian representatives by Buch (1930, 1932), if the apical cell is not tilted, i.e., the median axis of the apex is directly in line with that of the shoot, the cutting faces are not inclined and all three rows of leaves will be transversely inserted. However, in species or individuals with a dorsally tilted apical cell, the lateral cutting faces are inclined antically. As the merophytes develop, this oblique insertion is enhanced and the leaves are succubously inserted. Therefore, the gradations in leaf insertions present in Haplomitrium, namely, from transverse to subtransverse to succubous, are largely dependent upon the degree of apical cell tilt.

On the other hand, the size of the leaf is largely determined by the size of the derivative in which it originated. Individuals with apical cells whose segmenting faces remain more or less equilateral during segmentation, i.e., all three rows of derivatives are mostly equivalent in size, possess three rows of leaves which are similar in size. In contrast, axes with apical cells which appear as isosceles triangles in cross section, i.e., the two lateral faces are equal in size while the dorsal cutting face is shorter, possess two rows of large lateral leaves and one row of reduced dorsal leaves.

The extent of the apical tilt and the amount of reduction in the dorsal cutting face of the apical cell are quite variable in members of Haplomitrium. Species such as H. intermedium, H. hookeri, and H. monoicum display great variability in the sizes of the dorsal leaves, even on a single shoot, and exhibit leaf insertions which range from transverse to subtransverse to slightly succubous. Haplomitrium gibbsiae is also variable in these characters, but to a lesser extent. On the other hand, these characters are more highly fixed in H. mnioides, H. blumii, and H. andinum. In these taxa the apical cell exhibits a definite dorsal tilt so that the lateral leaves are unquestionably succubous. Also, the dorsal derivatives and cutting face of the apical cell are distinctly reduced and the dorsal leaves are uniformly smaller than the lateral leaves.

Apical cells with this tetrahedral geometry are responsible for growth in all taxa of the Jungermanniales except Pleurozia Dum. (Crandall-Stotler 1976, 1981), while the only Metzgerialian taxa possessing tetrahedral apical cells are Treubia Goeb., Apotreubia Hatt. & Mizut., Noteroclada Tayl., Sewardiella Kash. (Renzaglia 1982), and the sporelings of Fossombronia Raddi (Lampa 1902; Renzaglia and Bartholomew 1985). Some individuals of Takakia reportedly possess a tetrahedral apical cell (Crandall-Stotler and Bozzola 1989; McFarland et al. 1989), while others grow via a meristem

(Crandall-Stotler and Bozzola 1988, 1989). However, Haplomitrium differs from both Takakia and Jungermannialian taxa in its pattern of merophyte development. The Monocleales, Marchantiales, and all members of the Anthocerotophyta except Dendroceros Nees possess wedge shaped adult apical cells (Proskauer 1951; Campbell 1954; Renzaglia 1978; Crandall-Stotler 1981), while mosses differ in possessing obovoidal to fusiform apical cells (Frey 1971; Crandall-Stotler 1986). In both apical cell geometry and pattern of merophyte ontogeny, Haplomitrium resembles only some members of the Metzgeriales.

The fine structure of the apical cell and young derivatives of Haplomitrium is exemplified here by H. hookeri and H. gibbsiae (Figs. 78-85). Both species display similar ultrastructural characteristics. The free wall of the apical cell is much thicker than are the cutting faces and is devoid of plasmodesmata (Figs. 79, 81a, 81b). This outer wall is approximately 0.21 µm in thickness while the cutting faces average from 0.7-0.16 µm depending on their age. The cutting faces are interrupted at numerous points by plasmodesmata as are the walls of the cells comprising the adjacent merophytes (Figs. 78-81a, 82-85). As many as 20-25 plasmodesmata were counted within a 20 µm stretch of an apical cell cutting face, seen in section in H. hookeri. The channels are 0.03-0.04 µm wide and are lined with plasmalemma and ER (Figs. 82-85). These desmotubules are formed when portions of the ER become trapped as vesicles converge and fuse with one another during cell wall formation (Fig. 82) (Robards 1976). This construction is like that found in the plasmodesmata of vascular plants (Ledbetter and Porter 1970; Robards 1976) and certain charalian algae, but differs from the simple plasma membrane-lined plasmodesmata found in most green algae (Pickett-Heaps 1975; Marchant 1976; Duckett 1986) and Takakia (Crandall-Stotler and Bozzola 1988).

A thin, invaginated plasma membrane closely follows the outlines of the cell walls. At numerous intervals along the apical cell free surface, groups of 10-28 saccate, vesicle-like structures, which measure 33.0-35.0 nm X 20.0-23.0 nm, occur within enlarged invaginations of the plasmalemma (Fig. 81b). These structures are likely responsible for adding materials to the thickened external wall or for the additional deposition of polysaccharides to the mucilage layer which encompasses the plant.

All organelles typically found in the meristematic cells of vascular plants such as nuclei, plastids, vacuoles, mitochondria, smooth and rough ER, free ribosomes, and dictyosomes (Ledbetter and Porter 1970; Esau 1977) are also present in the apical cells and merophytes of Haplomitrium. Interphase nuclei are located centrally within the cells but their positions vary in accordance with the future position of the next division wall (Figs. 78a, 79, 80). For example, the nucleus of merophyte II in Figure 78a is positioned near the anodic wall in preparation for the ensuing division which will cut the kathodic portion of the merophyte. Each nucleus is spherical with an average diameter of 5.0-7.0 µm and contains a central nucleolus of about 3.0 µm in diameter. Numerous large patches of heterochromatin are a prevalent feature of the nucleoplasm. A karyosome, i.e., a small osmiophilic spherule lying outside the nucleolus is visible in some sections (Fig. 86). These structures have also been reported in meristematic cells of angiosperms (Barlow 1983), Takakia, and apparently also occur in Pallavicinia lyellii (Hook.) Carruth., and the sporophytic meristem of Phaeoceros laevis (L.) Prosk. (Crandall-Stotler and Bozzola 1988). According to Barlow (1983), the karyosomes of angiosperms occur with greatest frequency in the more slowly dividing cells of the meristem's quiescent center in species with low DNA content.

Multiple discoidal to pleiomorphic plastids are scattered along the peripheries of the apical cells and the cells of the young merophytes (Figs. 78-80). A single transection through an apical cell of Haplomitrium reveals 4-8 of these organelles which are characterized by their weakly developed internal lamellar systems (Figs. 81a, 84, 85). The young plastids of the apical region typically measure 1.0-3.6 µm long X 0.5-2.0 µm at the broadest portion of the organelle. They are traversed by 2-4 almost evenly spaced groups of lamellae and often contain 1-3 minute (80-100 nm in diameter) oil droplets (Figs. 78-81a, 84, 85). No plastids containing starch grains were present in the apical cell or the youngest series of derivatives (Figs. 78-81a). However, Duckett and Renzaglia (1988b) reported chloroplasts with small grana and starch grains in immediate apical derivatives of *H. hookeri*. The number and maturity of the plastids increase with the increasing age of the merophytes. Stacks of thylakoids become recognizable, and starting in the second youngest series of merophytes, 1-2 starch grains with diameters of 0.2-0.3 µm begin to form within the plastids (Figs. 84-88). New plastids are produced through constriction (Figs. 78b, 85, 86). No budding or partition formation as described by Diers (1970) and Berrie and Webster (1982), respectively, were seen in *Haplomitrium*.

The basic fine structure of the chloroplasts of *Haplomitrium* is typical of that in other hepatics, mosses, green algae, and vascular plants (Ledbetter and Porter 1970; Pickett-Heaps 1975; Esau 1977; Duckett 1986). The ultrastructure of the chloroplasts of *Haplomitrium* appears to be most similar with that of the mature plastids of the Metzgerialian taxa of *Pallavicinia* and *Aneura*, although Duckett and Renzaglia (1988b) stated that the plastid ultrastructure of *H. hookeri* is similar to that of *Blasia*. In contrast, the chloroplasts of *Haplomitrium* are structurally different from the laminate to plate-like plastids of the anthocerotes and the similarly shaped monoplastids in the meristems of *Takakia* (Crandall-Stotler and Bozzola 1988).

The vacuolar system of the apical cell and first series of derivatives is very small (Figs. 78-80, 86). However, in older merophytes the vacuoles become more centrally located and occupy a larger volume of the cell (Fig. 88).

Mitochondria are dispersed throughout the cells of the apical region. These organelles may be spherical with diameters of 0.43-0.86 µm, or oval to cylindrical with dimensions of 0.4-2.1 µm X 0.3-0.8 µm and bear internal saccate to irregular cristae (Figs. 78-81a, 82, 84-87). As many as 13-20 mitochondria are visible in a single section through an apical cell or derivative. Like the plastids, the mitochondria divide by constriction (Fig. 85).

The dictyosomes of Haplomitrium are composed of 4-6 cisternae and their associated vesicles (Figs. 82, 83). In longitudinal section, each cisterna is approximately 0.81-0.93 µm long X 10.0-18.1 nm broad. They are most numerous near sites of cell wall formation or slime secretion (Figs. 82, 83) (Ledbetter and Porter 1970; Esau 1977).

Tubules of smooth and rough ER lie just internal to the plasmalemma of both the apical cell and the derivatives. These tubules are often arranged almost parallel to the plasmalemma (Figs. 81a, 81b). In addition, both ER and free ribosomes are prevalent thoughout the cytoplasm.

No distinct oil bodies were observed in the apical cell or its immediate derivatives, but, single membrane-bound structures with stromata reminiscent of the stromata of mature oil bodies were apparent and perhaps represent undifferentiated oil bodies or oil bodies which have been partially destroyed during the fixation procedure (Figs. 79, 81a). Lipid droplets are also extremely rare. Only one such droplet was seen in the apical cell of one specimen of H. gibbsiae (Fig. 80). Lipids are more frequent in the older cells of the apical region where they are distributed along the edges of the cytoplasm.

Mature vegetative cells commonly contain all the typical organelles and endomembrane systems found in meristematic cells (Ledbetter and Porter 1970). However, some, like the plastids and oil bodies, are more highly differentiated with maturity. The ultrastructural changes that occur in cells as they mature are exemplified in Figures 86-88. This series of micrographs is taken from a single longitudinal section through the apical portion of a leafy shoot of H. hookeri. The first micrograph (Fig. 86) is of a portion of the young merophyte located about 18.0 µm posterior to the most basal point of the apical cell. The next micrograph (Fig. 87) shows the adjacent cells approximately 32.0 µm posterior to the apical cell, and the third (Fig. 88) shows those about 43.0 µm from the generative cell. The maturation process is most evident by the changes in plastid ultrastructure, and in the increase in vacuolar volume and oil body abundance. The loose lamellar structure characteristic of the young plastids becomes organized into thylakoids with distinct stacks of grana, and starch grains become larger and more numerous as the plastids mature. The starch grains of the chloroplasts in the youngest merophytes (Fig. 86) number 3-4 per organelle and measure approximately 0.26-0.53 µm long X 0.15-0.25 µm broad, while those in the cells located 43.0 µm from the apical cell (Fig. 88) number about 5-8 per chloroplast and are approximately 0.59-1.24 µm long X 0.26-0.45 µm wide. Interestingly, the starch grains become more osmiophilic as they age. Plastid and mitochondrial replication also becomes less frequent in more mature cells. The vacuolar system which is originally highly dissected and occupies a relatively small volume of the cell becomes less dissected and increases in volume. Rough ER and free ribosomes become less apparent with decreasing meristematic activity. A typical mature vegetative cell of a H. blumii shoot is shown in Figure 89. Mature oil bodies and chloroplasts are present along with large vacuolated areas. Notice the distinct thylakoids, grana stacks, and the large starch reserves in Figure 90.

No cytoskeletal elements were observed in any of the material examined in this investigation. These structures were apparently inadvertently destroyed during the fixation procedure. However, from the ultrastructural evidence that was obtained, it appears that the fine structure of the apical cell, derivatives, and vegetative cells of Haplomitrium is like that reported for other hepatics and mosses (Duckett 1986). However, Haplomitrium can be distinguished from Takakia and the anthocerotes by plastid structure and from all other land plants, except other liverworts, in the possession of oil bodies (Duckett 1986; Crandall-Stotler and Bozzola 1988).

Stem Development and Anatomy

The shoot system of Haplomitrium consists of prostrate, leafless, rhizomatous axes, erect to obliquely ascending leafy shoots, and positively geotropic, hyaline, leafless stolons. As previously discussed, each of these shoots is formed through the activity of a single tetrahedral apical cell.

The rhizome is not truely one continuous axis, but actually is constructed by the sequential branching of the axis, often after leafy shoot production. After growing prostrate for a time, the apex of the rhizome becomes erect and develops into a leafy shoot (Figs. 35, 48, 53). During this transformation process, i.e., during the time in which the apical cell derivatives are just beginning to function in leafy shoot production, minute scales are formed by irregular divisions in the basal cells of the apical slime papillae (Figs. 10, 91). Consequently, a few scales are left on the prostrate portion of the axis as it turns upward (Fig. 92). Concomitant branching from the base of this erect shoot, then, extends the creeping rhizome. Sometimes, even when no leafy shoot is produced, the rhizome forms rhizomatous branches at all angles. Additional leafy shoots may occasionally develop from some of these branches which are vertically ascending (Figs. 93, 94) or dorsally tilted at their apices. Branches which descend vertically from the rhizome or those which are initially lateral, but soon grow downwards remain leafless and function as stolons (Figs. 9, 26, 27, 35, 48, 56, 59). Because of its role as a base from which new branches arise, Gottsche (1843) considered the rhizome to be a perennating structure.

The mode of rhizome formation and branching described here has been reported for *H. hookeri* (Gottsche 1843; Lilienfeld 1911; Worley 1969), *H. blumii* (Goebel 1891; D. H. Campbell 1920), *H. intermedium* (Berrie 1962; Grubb 1970), *H. gibbsiae* (E. O. Campbell 1959), *H. mnioides* (Schuster 1967; Grubb 1970), *H. indicum* (Udar and S. Chandra 1961), and *H. grollei* (Kumar and Udar 1977). Only *H. intermedium* exhibits an alternate mode of stolon production, in that, in some individuals, stolons originate from leafy shoots, not rhizomatous axes (Figs. 26, 32).

Although Grubb (1970) elaborated on the differences between the rhizome systems of *H. mnioides*, *H. gibbsiae*, *H. hookeri*, and *H. intermedium*, the differences in the appearances of the rhizomes of various taxa and individuals merely result from variations in the number, position, and extent of elongation of the newly formed shoots. For example, under different environmental conditions, the rhizomatous system of a single species may be very elongated or extremely stunted, respectively. In the latter case, the branches remain so abbreviated that their ultimate ontogenetic fate is undeterminable. Also, the point of origin of the new rhizome branch may vary, i.e., the rhizome branch may form at the very base of a leafy shoot or at some point higher on the shoot. In fact, E. O. Campbell (1959) reported that *H. gibbsiae* plants growing under crowded conditions produce only one strong lateral shoot from the base of each aerial axis which eventually turns upward and becomes leafy. If the surrounding plants threaten to overtop the *H. gibbsiae* plants, the lateral shoots may grow out from higher on the stem and develop immediately into leafy shoots. On the other hand, when the rhizome is stressed due to dessication, several lateral shoots are produced resulting in an interwoven tangle of branches (E. O. Campbell 1959).

The leafy and leafless shoots of Haplomitrium appear to be homologous structures as evidenced not only by their common apical patterns, developmental sequences and mature anatomies, but also by the fact that leafless shoots have the potential to become leafy. Lilienfeld (1911) demonstrated the continuity between the shoot systems by placing a typical leafy shoot of H. hookeri in strong shade. As the shoot grew, remote, rudimentary "leaves", or scales formed. Conversely, Lilienfeld (1911) showed that leafless rhizomes begin to form normal leaves when exposed to light. Even stolons, when exposed to light under controlled conditions have the potential to become chlorophyllous at their apices and produce branches and leaves (Fig. 95).

The leafy and leafless shoots of all species of Haplomitrium are anatomically identical (Figs. 96-104). Each is composed of a single epidermal row of cells, several rows of cortical cells and a central conducting strand of hydroids. The numbers of cells constituting the cortex as well as the conducting strand varies depending on the diameters of the shoots (Table III; Figs. 99, 100). The fundamental shoot anatomy described here concurs with that previously reported for H. hookeri (Gottsche 1843; Lilienfeld 1911; Grubb 1970), H. blumii (Goebel 1891; Campbell 1920), H. mnioides (Grubb 1970), H gibbsiae (Campbell 1959), H. indicum (Udar and S. Chandra 1965), H. dentatum (Kumar and Udar 1976), H. grollei (Kumar and Udar 1977), and H. kashyapii (Udar and Kumar 1982).

The leafy shoots of Haplomitrium are usually circular in cross section (Figs. 98, 101, 103), while the rhizomatous axes and stolons may appear either circular (Fig. 97) or broadly oval (Fig. 104). The flattening of the axis is likely due to its being slightly crushed by the physical pressures placed on its cells as it grows through the substrate. As presented in Table III, the diameters of the rhizomes and aerial shoots may be as small as 200 µm in

TABLE III. ANATOMICAL FEATURES OF THE SHOOTS OF *HAPLOMITRIUM* SPECIES

	Leafy Shoot and Rhizome Diamaters (µm)	Cortex Cell Row No.	Central Strand Cell Row No.
H. hookeri	280-510	3-4	6-8
H. ovalifolium[a]	220-360	ND	ND
H. monoicum	200-460	2-3	4-5
H. intermedium	220-400	1-2	2-5
H. gibbsiae/ *H. chilensis*	410-1015	3-7	6-16
H. indicum[b]	ND	5-6	6-7
H. dentatum[c]	660-1045	6	4-12
H. grollei[d]	858-946	6-7	10-11
H. kashyapii[e]	250-315	6-8	12-19
H. mnioides	800-1100	3-4	6-8
H. blumii	425-1000	3-4	6-12
H. andinum	450-1200	3-4	6-8

[a]from Schuster 1971; [b]from Udar and Chandra 1964; [c]from Kumar and Udar 1976; [d]from Kumar and Udar 1977; [e]from Udar and Kumar 1982; ND, no data.

H. moniocum, to as large as 1200 µm in H. andinum. The stolon dimensions are much more variable because most taper greatly toward the apex. Therefore, the diameters vary with the length of the stolon. In general, however, the stolons are more slender than the leafy shoots (compare Figures 103 and 104).

The epidermal cells of the leafy and leafless shoots are quadrangular in cross section and measure approximately 26 X 28-32 X 54 µm. In longitudinal section, these cells are elongate and measure approximately 51-120 µm in length, but are usually shorter than the adjacent cortical cells.

The fine structure of the epidermal cells of the leafy shoots (Fig. 105) and stolons (Fig. 106) is virtually identical except for the more highly vacuolated condition of the stolon epidermal cell. They each possess a centrally located, heterochromatin-rich nucleus with a diameter of about 5.5-6.0 µm, a large nucleolus and accompanying karyosome, complete endomembrane systems, mitochondia, oil bodies, and chloroplasts. Like the apical cell free surface, the external walls of the epidermal cells, which measure 0.23-0.39 µm thick are thicker than the walls of adjacent epidermal cells which measure only 0.09-0.13 µm. Numerous plasmodesmata like those described in the walls of the apical cell and derivatives traverse all walls of the epidermal cells except the external wall.

Slime cells are distributed on both leafy and leafless shoots in all species except the leafy shoots of H. intermedium and H. monoicum, where they occur only in apical regions. In all species, slime papillae are particularly numerous at the apices of growing shoots where they overarch and immerse the apical cell in protective slime. However, the slime papillae are probably most abundant over the stolons where they form in lieu of leaves. Lilienfeld (1911) noted the thick stand of slime cells covering the rhizome and stolon apices of H. hookeri and suggested that they almost formed a protective pseudoparenchymatous tissue which she considered functionally equivalent with the protective root caps of vascular plants.

The mucilage cells of the shoots are usually sessile, but may also be composed of one or sometimes two short stalk cells terminated by a clavate slime cell (Figs. 34b, 55, 60, 107, 108a). Schuster (1967) stated that only stalked slime papillae occur within the family, but sessile slime cells have been reported on the rhizomes of H. intermedium (Berrie 1962) and H. moniocum (Engel 1981), and were illustrated on the leaves of H. indicum by Udar and S. Chandra (1965).

The young slime cells of leafy shoots possess numerous organelles which are clustered in the terminal third of the cell (Fig. 107). Mitochondria, ER, ribosomes, and dictyosomes are prevalent near the tip of the cell. The rest of the cell contains numerous small vacuoles and a few oil bodies. As the mucilage cells mature, many of the organelles deteriorate and those that remain are clustered as the vacuolar system becomes less dissected and increases in volume (Figs. 108a, 108b). The nucleus and mitochondria are positioned near the base of the cell (Fig. 108a). The nucleus measures about 5.4 X 3.0 µm and possesses a heterochromatin-rich nucleoplasm. Thus, it is similar in size and construction to the nuclei of the vegetative cells of Haplomitrium. The mitochondria range in form from cylindrical (0.83-0.77 µm long X 0.33-0.45 µm broad) to very elongate (2.9-3.8 µm long X 0.19-0.32 µm broad). The only other organelles still recognizable at this stage are the large vacuoles and a compact system of peripheral ER cisternae, which are tightly packed and parallel to one another (Figs. 108a, 108b). Similar ER cisternae have been reported in the nectary cells of Euphorbia millii Ch. des Moulins (Crown-of-Thorns) (Ledbetter and Porter 1970), which suggests that such ER elements are designed to support the active production of secretory materials. Galatis and Apostolakos (1977), who described a similar developmental sequence for the slime papillae of Marchantia, also reported a similar ER membrane system and noted that it is a prominent feature of cells which secrete lipids, hydrophilic substances, and proteins. However, according to Galatis and Apostolakos (1977), the ER membranes are not directly involved in secretion, but play an intermediate role in the process. The slime cell is delimited by a wall which measures 0.22-0.28 µm in thickness. The tangential section which grazed the slime cell wall shown in Figure 108b reveals the homogeneous rather than fibrous constitution of the wall.

The cortex in all shoot systems is composed of polygonal cells which are usually slightly larger and more elongate than the epidermal cells. This region is composed of (1-) 3-4 (-8) rows of cells in transverse section, depending on shoot robustness (Table III). Typical cortical cells from the leafy shoots of H. hookeri and H. blumii are shown in Figures 109 and 110, respectively. The thin walls of the cortical cells are 0.13-0.23 µm thick and possess numerous plasmodesmata. Like the epidermal cells, a single large nucleus (5.7-7.4 µm in diameter), and numerous mitochondria, ER, ribosomes, oil bodies, and small vacuoles are present in each cell. Multiple plastids with numerous, large, osmiophilic starch grains are very prominent within the cells. The starch reserves of H. hookeri average 5-6 per plastid and measure 0.2-2.3 µm long X 0.10-0.2 µm wide, while those in H. blumii average only 2-3 per organelle, but measure 1.3-2.4 µm long X 0.5-0.9 µm wide. Lipids and oil bodies are also abundant.

The rich starch content of the cortex has been recognized by numerous workers using iodine staining techniques (Gottsche 1843; Goebel 1891; Lilienfeld 1911; D. H. Campbell 1920; E. O. Campbell 1959; Udar and S. Chandra 1965; Grubb 1970; Kumar and Udar 1976, 1977; Udar and Kumar 1982). Also, fungal hyphae have been reported within the epidermal and cortical cells of the rhizome and stolon of H. hookeri by Lilienfeld (1911). Based on this finding, she suggested the possibility of a mycorrhizal association, but also reported the presence of the parasitic endophytes

Pythium, Olpidium, and a Chlorochytrium-type organism. Although in this study, fungal hyphae along with bacteria, cyanobacteria, and unicellular green algae were often observed in the mucilage surrounding the rhizome and stolons of plants maintained on vermiculite, none were found within the cells of the shoots.

The central strand of the shoot is composed of thin-walled, elongate hydroids. The number of hydroids comprising the strand is dependant to a greater degree on the diameters of the shoots than on the species involved (Table III). For instance, reproductive shoots, especially female shoots are more robust than vegetative or male shoots and therefore contain a larger number of more well-developed hydroids (compare Figures 99 and 100). This was first noticed by Hébant (1977) in the shoots of H. gibbsiae. The development of the conducting strand is also dependent on environmental conditions, particularly on water availability (de Sloover 1959; Grubb 1970). The conducting strands normally present in both Moerckia hibernica (Hook.) Gott. (as Moerckia flotoviana [Nees] Schiffn.) and H. hookeri gametophytes reportedly disappeared when grown experimentally under conditions of strong humidity or in liquid culture (de Sloover 1959; Grubb 1970). Therefore, the number of hydroids per strand, the extent of their development, and even the presence or absence of a central strand is not always constant within a species.

The hydroids differentiate from cells that are derived from the innermost segments of the apical derivatives (Figs. 77, 86). This origin has also been demonstrated for the hydroids of Metzgerialian taxa (McCormick 1914; Campbell and Williams 1914; Smith 1966; Hébant 1977; Renzaglia 1982). The distance below the apex at which the conducting strand first becomes differentiated was found to be about 220-300 µm in H. hookeri, 180-410 µm in H. andinum, 170-540 µm in H. blumii, and 250-500 µm in H. mnioides. Grubb (1970), who found differentiation of the hydroids to begin ca. 200 µm basal to the apex in H. hookeri and ca. 300-500 µm in *H. mnioides*, reported that the distance varies not only with the species, but also with environmental conditions.

In this study, hydroid ontogeny was investigated ultrastructurally in *H. mnioides* and *H. gibbsiae*. The hydroids differentiate from living cells located in the center of the shoot. These cells are originally indistinguishable both morphologically and ultrastructurally from the cortical cells exterior to them (compare Figs. 86-88 and 110). As differentiation proceeds, these cells increase greatly in length but very little in width. The vacuolar systems increase in volume, first forcing the other organelles into a cluster (Figs. 111, 112), then isolating them within a small cytoplasmic zone between the tonoplast and the plasma membrane (Figs. 113, 114). No additional materials are deposited on the walls during differentiation so they do not increase in thickness (Figs. 111, 113). However, the lateral walls, which measure 0.22-0.25 µm in thickness are usually slightly thicker than the end walls which are 0.11-0.13 µm thick (Fig. 113). Most hydroid end walls remain transverse although some become slightly oblique (Figs. 111, 113). Plasmodesmata occur on all walls of the initial cells and are in turn present on both lateral and end walls of the immature hydroids. The wall material which surrounds each plasmodesma is removed and the channels are enlarged from their original diameters of 0.03-0.04 µm to about 0.54 µm. As seen in Figure 115, the early stages of this process are recognizable by the darkening of the wall material adjacent to each plasmodesma being hydrolyzed. A slight depression becomes evident as wall material is removed from around the plasmodesma (Figs. 116, 118), resulting in an open canal containing only an isolated desmotubule (Figs. 117, 118). Ultimately, this strand of ER is digested away leaving a true perforation in the wall at the point where the plasmodesma previously existed (Figs. 119, 120, 121). The pits now measure about 0.70-1.55 µm in diameter. Residual materials are trapped between the plasmalemma and cell wall as pit formation occurs (Fig. 116). In general, each pit is formed from a single plasmodesma, although if two plasmodesmata are in close proximity to one another, the removal of wall material to enlarge the channels could include both plasmodesmata, thus resulting in one large perforation. Pits form in both the end and lateral walls of adjacent hydroids but are less frequent in the latter (Figs. 122, 123). As seen in scanning micrographs of the hydroids of H. intermedium and H. blumii, the perforations in both the lateral and end walls are circular in surface view (Figs. 124-126).

During the pit-forming period, the protoplast of the cell gradually deteriorates and finally, in the last stages of hydroid formation, completely degenerates leaving only lightly staining protoplasmic remnants in the cells (Figs. 120, 121, 123). This final breakdown of protoplast has been linked with a peak of strong lysosomal acid phosphatase activity (Hébant 1973a, 1975, 1977; Burr et al. 1974). Often, the hydroid walls appear to be in various stages of collapse (Fig. 123). This is due to the difference in turgor pressure between the conducting cells and the adjacent parenchyma cells causing the parenchyma cells to protrude into the thin-walled hydroids (Hébant 1977).

So, mature Haplomitrium hydroids are elongate cells devoid of cellular contents, i.e., they are dead at maturity. Both the end walls and lateral walls are thin, with the lateral walls often just slightly thicker than the end walls. The terminal walls are transverse to slightly tapering and all walls are pitted with true perforations, each derived from one or rarely two plasmodesmata. These findings agree with the structure described at the light microscope level for the mature hydroids of H. gibbsiae by Campbell (1959) and Grubb (1970), and with the findings from the ultrastructural studies of the mature hydroids of H. mnioides and H. gibbsiae (Burr et al. 1974; Hébant 1977).

The developmental scheme described by Smith (1966) for the hydroids of the simple thalloid Symphyogyna is very much like the hydroid ontogeny of Haplomitrium. In both taxa, the terminal and lateral walls of the hydroids possess pits each of which originate from the enlargement and subsequent breakdown of one or two plasmodesmata. However, in Symphyogyna, the resulting pits are lanceolate or diamond shaped while those of Haplomitrium are circular. In addition, during hydroid elongation in Symphyogyna, the end walls become oblique and secondary wall materials are deposited (Smith 1966). In Haplomitrium, end walls remain transverse to only slightly oblique and no materials are deposited on the walls so they remain thin. This corroborates Hébant's (1977) findings that lignin is absent in the hydroid walls of Haplomitrium and Takakia. Thus, although the hydroids of Haplomitrium function in the movement of water and minerals, they are likely not homologous with the tracheary elements of higher land plants. The hydroids of Haplomitrium do exhibit some developmental similarities with the sieve elements of leptosporangiate ferns in that each pore of the sieve-area also forms from a single ER-lined plasmodesma by a process that involves the widening of the plasmodesmatal canal. However, the patterns differ in that leptosporangiate ferns develop barrel-shaped periplasmodesmatal regions bordered by electron dense bands and a callose or calloselike substance is often deposited around the prepore plasmodesma (Evert et al. 1989).

Although conducting tissue is relatively common in mosses, Haplomitrium is one of the few hepatics to possess such a differentiation of tissues (Grubb 1970; Hébant 1977; Crandall-Stotler 1981). Within the Hepatophyta, conducting strands occur only in one family of the Metzgeriales, that being the Pallaviciniaceae (including Hymenophyton Dum.) (Smith 1964, 1966; Hébant 1964; 1973b; 1977; Schuster 1966; Winkler 1969; Burr et al. 1974; Crandall-Stotler 1981; Renzaglia 1982). Takakia also bears a strand of conducting tissue (Winkler 1969; Grubb 1970; Hébant 1973b, 1977; Crandall-Stotler 1981).

The conducting strands of mosses are much more diverse than those of hepatics. These strands may traverse the stems, leaves, and even setae and the differentiation of conducting tissue ranges from the weakly organized pitted conducting parenchyma of Andreaea Hedw. (Hébant 1977; Murray 1988) to the extremely well-developed and differentiated strands of the Polytrichales which not only include hydroids, but also leptoids and stereids (Hébant 1973b; 1977). Except for the lack of lignified secondary sculpturing, moss hydroids have much in common with the tracheids of primitive vascular plants. Moss hydroids arise from quadrate cells in the center of the stem, which elongate rapidly and develop steeply inclined tracheid-like end walls. Polyphenolics are deposited on the lateral walls while the terminal walls degenerate leaving only a few thin fibers. Pits are absent. As in all hydroids, the protoplasm degenerates through lysosomal activity (Hébant 1977). It is, therefore, apparent that both the ontogeny and the mature morphology of the moss hydroids differ from that of Haplomitrium.

The hydroid structure of the simple thalloids and Takakia is similar to that of Haplomitrium, in that the cells are axially elongated, lack protoplasm at maturity, and, with the exception of Moerckia Gott., possess numerous pits in their walls. (The walls of Moerckia possess an unusual swollen appearance which according to Hébant [1977] is probably a consequence of a partial hydrolysis resulting in the removal of interfibrillar substances). Unlike the hydroids of Haplomitrium, those of the Pallaviciniaceae (except Moerckia) possess sharply oblique end walls, lanceolate pits, and develop thickenings on their lateral walls (Smith 1966; Hébant 1973b, 1977, 1978; Burr et al. 1974).

Grubb (1970) and Hébant (1977) considered the mature conducting strand of Takakia to be like that of Haplomitrium. However, unlike Haplomitrium, the lateral walls of *Takakia* hydroids are apparently almost twice as thick as their end walls and have few or no pits (Crandall-Stotler, personnal communication).

A comparison of the existing ontogenetic and structural information regarding hydroids suggests that the conducting strand of *Haplomitrium* is significantly different from that of the conducting strands present in some members of the Metzgeriales. Decisive statements regarding the relationships between the central strands of *Takakia* and *Haplomitrium* must be deferred until the hydroid ontogeny of *Takakia* has been determined.

Leaf Development and Anatomy

As the rhizome curves upward to produce the aerial shoot, irregular divisions occur in the basal cells of the 1- to 2-celled slime papillae that overarch and protect the generative cell of the shoot (Fig. 91). This sequence of divisions, which transforms the apical slime papillae into scales, precedes the onset of leaf production. Therefore, in all species of *Haplomitrium*, scales are apparent at the bases of the erect shoots, posterior to the zone of leaf formation.

Leaf ontogeny in *Haplomitrium* is as described for *H. mnioides* by Bartholomew (1987). The initial development of the leaf is through the activity of a single depressed-ovoid apical cell with one cutting face, which is derived from the central leaf initial of each merophyte. Longitudinal divisions which begin in the oldest segments produce a triangular leaf terminated by a single mucilage cell (Fig. 127). Pressures exerted by the actively dividing segments transform the leaf apical cell into one with two cutting faces as seen in older stages of leaf development (Fig. 128). Apical segmentation soon ceases and the subsequent growth of the leaf is due to a regular pattern of divisions in the basal cells of the lamina to produce the typical entire leaf (Fig. 128). When present, the reduced dorsal leaves develop by the same pattern except that apical segmentation is arrested at an earlier stage than in the lateral leaves (Fig. 129). The multistratose basal fields found in some taxa (Table I) are brought about by divisions in multiple planes in the cells of the basal lamina (Fig. 130). Lobed and incised leaves, in taxa which have been studied, e.g., *H. hookeri*, *H. gibbsiae*, and *H. dentatum*, or culture-induced bifid leaves (Kitagawa 1984; Bartholomew 1987) are produced through a modification of this ontogenetic pattern. The initial growth is still via a single apical cell, but, to form bifid leaves, the apical cell cuts itself in two (Fig. 131) and each daughter cell then segments independently to form a lobe of the bifid leaf (Fig. 132). The longer the segmentation sequence, the more bifid the leaf becomes. Final expansion of the leaf is by divisions in the basal cells of the lamina as previously described. Leaves with lateral lobes as seen in *H. hookeri* and *H. gibbsiae* as well as fertile bracts with lateral projections form when a cell on the margin of an already mature leaf begins to act as an apical cell. This newly activated apical cell segments a few times to form a projection often terminated by a slime cell (Figs. 133, 134). The longer the period of apical activity, the longer the lobe or lateral projection becomes. The cellular arrangements of the mature leaf reflect this ontogenetic sequence. The fields of meristematic activity, both of the apex and the lateral projections, are clearly distinct from the straight rows of cells which result from the regular pattern of divisions in the basal cells of the leaf. Similarly, the straight rows of basal growth are uninterrupted by the lateral projections, thus indicating that these outgrowths were added after the basic leaf morphology had been attained (Fig. 134).

Slime cells are present on the leaf margins of all species of *Haplomitrium* and are also present on the leaf laminae of some species. The size and abundance of the slime papillae range from the rare, tiny, unicellular mucilage cells which are restricted to the leaf margins in H. monoicum (Figs. 26a, 26b), to the numerous, larger, 1- to 2-celled slime papillae on the margins and over the leaf laminae of H. blumii and H. mnioides (Fig. 51). The abundance of slime papillae present on the leaves is contingent upon the age of the cells and environmental conditions, particularly the amount of moisture to which they are exposed (E. O. Campbell 1959; Schuster 1966).

Despite the shape of the leaves, the ontogenetic sequence in Haplomitrium is like that exhibited by the Metzgeriales. Although the dissected leaves of Haplomitrium are superficially suggestive of Jungermannialian leaf ontogeny in which two apical cells are involved in the earliest stages of leaf development, it appears that only one generative cell is initially responsible for the entire, lobed, incised, and bifid leaves of Haplomitrium.

Although many workers have alluded to the numerous protective apical slime papillae (Gottsche 1843; Leitgeb 1875; Goebel 1891; Lilienfeld 1911; D. H. Campbell 1920; E. O. Campbell 1959), no one has described the formation of scales. However, although she did not specifically discuss leaf ontogeny, Lilienfeld (1911) did report finding what she termed rudimentary leaves at the junction of the rhizome and the erect axes of H. hookeri. She described these "leaves" as being composed of rows of small united cells with each row terminated by a clavate papilla. Considering her description and illustration (Lilienfeld 1911, p. 328, fig. 7), these structures are probably equivalent to the scales described here.

While Gottsche (1843) attributed leaf origin in H. hookeri to the club-shaped papillae associated with them, Leitgeb (1875) determined that the young leaf acquires a triangular shape due to the segmentation of an apical cell with one cutting face followed by longitudinal divisions in the newly formed segments. This same leaf ontogeny was described by D. H. Campbell (1920) for H. blumii. Furthermore, it was emphasized that at least in H. hookeri (Leitgeb 1875) and H. blumii (Goebel 1891; D. H. Campbell 1920), the leaf is never bifid as in the leafy hepatics. Although mature culture-induced bifid leaves were reported for H. mnioides by Kitagawa (1984) and Bartholomew (1987) and, in this study were also found on H. gibbsiae plants growing on moistened vermiculite, it has been clearly shown that these leaves still develop according to the Metzgerialian pattern. The Metzgerialian leaf ontogeny of Haplomitrium supports the idea that Haplomitrium and Takakia are not as closely related phylogenetically or taxonomically as previously hypothesized. The leaves of Takakia lepidozioides are produced in an irregular spiral, not in ranks, and are cylindrical and paired, with each segment developing from a single surface initial (Crandall-Stotler 1986).

In contrast to leaf outline, which is modified by environment, leaf anatomy is rather stable. The lateral leaves of Haplomitrium range from being unistratose throughout, except at the extreme base where a bistratose zone exists, to bearing a large, polystratose basal field composed of as many as four cells in thickness (Table I). The rudimentary basal leaves of the aerial shoots as well as the reduced dorsal leaves of distinctly anisophyllous taxa are unistratose throughout even in species whose lateral leaves possess a polystratose basal field. The lack of a bistratose insertion zone may be correlated to their abbreviated development. Except for Campbell (1959), who reported the unistratose condition of the small leaves at the bases of the aerial shoots of H. gibbsiae, the unistratose composition of reduced leaves in taxa whose leaves normally possess multistratose basal fields has gone unmentioned.

Like the meristematic cells of the shoot, the fine structure of the young leaf cells, i.e., those just basal to the terminal slime cell, is characterized by plastids with loosely defined thylakoids, each containing 1 or 2 oil droplets, numerous cylindrical to elongate mitochondria, and relatively small, highly dissected vacuolar systems (Figs. 135, 136). A few oil bodies may also be present (Fig. 135). The other features of the cells are typical vegetative cell characteristics.

Plasmodesmata link the leaf cells, including the young terminal slime cell of the triangular-shaped leaf to create a single symplasm. The young slime cell in Figure 135 is characterized by dividing plastids, cylindrical mitochondria, dictyosomes, oil bodies, small vacuoles, ribosomes, and ER. The ER cisternae are positioned along the tip of the cell where they are loosely clustered in a parallel arrangement. The slime papilla in Figure 136 is slightly older than the one in Figure 135. This cell is highly vacuolated and most of the organelles have deteriorated beyond recognition. The elements of the ER are still present, however, and are now more numerous and compactly arranged. Thus, the differentiation of slime cells associated with the leaves closely parallels that of stem-associated slime cells.

As is typical of vegetative cells, the mature leaf cells (Fig. 137) possess larger, less dissected vacuoles, and numerous oil bodies. In addition, the lamellar system of the plastids is organized into thylakoids with grana and the starch reserves are more prominent (Figs. 137, 138). These mature leaf cells also possess numerous large, lipid droplets (0.9-1.5 um in diameter) which are located in the peripheral cytoplasm (Fig. 137).

Although the ultrastucture of the leaves of Haplomitrium is far from unique among hepatics, it does differ from that of Takakia. According to Crandall-Stotler (1986), the young leaf cells of the latter taxon are unlike those known in any other bryophyte group, in that each young leaf cell possesses only two large, starch-containing plastids which lie at opposite poles of the cell, in close proximity to the central nucleus. Further, Takakia lacks the oil bodies that are characteristic of Haplomitrium.

Oil Bodies

The oil bodies of *Haplomitrium* were first described in *H. hookeri* by Müller (1939). Since then, Schuster (1966, 1967) and Worley (1969), have supplied additional oil body information for this species. Oil body descriptions have also been published for *H. mnioides* (Hattori 1951, 1953; Schuster 1967; Yang 1968; Inoue 1974), *H. gibbsiae* (Schuster 1967), *H. intermedium* (Schuster 1967), *H. ovalifolium* (Schuster 1971), *H. dentatum* (Kumar & Udar 1976), and *H. kashyapii* (Udar & Kumar 1982), as well as for *Steereomitrium minutum* (Campbell 1987). This limited information, undoubtedly, stems from the fact that living material is required for oil body investigations. This study was similarly restricted, limiting oil body investigations to *H. blumii*, *H. gibbsiae*, *H. hookeri*, *H. intermedium*, *H. mnioides*, and *Steereomitrium minutum*.

Like most other hepatics, the oil bodies of *Haplomitrium* are prominent cell elements (Figs. 139-152). When viewed with the light microscope, the oil bodies appear as glistening, opaque to gray organelles dispersed throughout the midregion of the cell (Figs. 139-151). They are easily recognized at the ultrastructural level by their extremely electron dense outer membrane. In fact, this membrane is more dense than any other membrane in the cell (Fig. 152). Also, the clustering of the lipophilic globules and the difference in the granularity of the oil body matrices compared to that of the surrounding cytosol serve to distinguish these organelles from other cell elements. In addition, ER membranes and rough ER are frequently seen encircling the oil bodies (Figs. 153, 156-159) and vesicles line the stroma-side of the outer oil body membrane (Fig. 154). These unique organelles are most easily seen in the cells of leaves, although they occur in virtually every cell of the differentiated gametophyte and sporophyte including stems, stolons, slime cells, gametangial jacket cells, setae, and young capsule wall cells.

Oil bodies have also been reported in the egg cells, spores, elaters, and gemmae in other hepatics (Müller 1939; Schuster 1966; Stewart 1978).

As is evident at the ultrastructural level (Figs. 152-159), a mature oil body is a membrane-bound organelle composed of osmiophilic droplets dispersed in a granular matrix (Verdier 1965; Pihakaski 1966, 1968, 1972a, 1972b; Hölzl 1969; Suire 1970; Crandall-Stotler 1974; Galatis et al. 1978, 1978b). At first glance, a comparison of Figures 152 and 155-159 suggest a difference in the electron densities of the lipid globules among the oil bodies of some species. However, these apparent differences are probably due to the varying degrees of preservation during fixation rather than to species differences. The fact that oil body lipids are difficult to retain during traditional electron microscopy fixation techniques has been well documented (Galatis et al. 1978; Duckett and Renzaglia 1987; Duckett et al. 1988). The fact that within the same cell and sometimes within the same oil body, lipid spherules display different gradations in electron densities suggest that these lipids are exhibiting varying degrees of preservation (Figs. 152, 154, 155, 158, 159).

Oil bodies form during the early ontogeny of bryophyte organs. Precursory oil bodies are present even in the apical cell and youngest derivatives of the shoots of Haplomitrium (Figs. 79, 81a). The oil bodies have reached maturity by about the fourth or fifth series of merophytes. No Golgi apparati were observed in association with the oil bodies in this study, but vesicles were seen along the inner face of the oil body membrane (Fig. 154). Also, rough ER was seen in close association with the oil bodies, often encircling them (Figs. 153, 157-159). Therefore, it is likely that the oils are synthesized by the ER and packaged and deposited in the developing oil body by vesicles as postulated by Pihakaski (1966, 1968, 1972b) and Galatis et al. (1978).

Oil body morphology, which probably reflects chemical composition in part, can provide useful systematic information when used in conjunction with other characters. Oil body size, shape, and contents are all useful taxonomic criteria (Müller 1939; Pihakaski 1966; Schuster 1966). Of these, oil body size is the most stable species character (Pfeffer 1874; Müller 1939; Schuster 1966). The number of oil bodies per cell fluctuates greatly from one cell to the next depending on the cell's location in the plant (Schuster 1966; Worley 1969; Galatis et al. 1978a). In this study, the number of oil bodies observed in the cells of the shoot system was typically slightly less than that of the median leaf cells of the same species (Table IV). To minimize these fluctuations and allow comparative evaluations, oil body observations have traditionally been made from median leaf cells. In this study, median leaf cells were defined as those comprising the second apical series of leaves while the "old" leaves were those located in the basal half of the shoot. In addition, cells located midway between the apex and the base of both leafy shoots and stolons were examined.

The "texture" of an oil body is determined by the number and size of the inner oil droplets along with the thickness of the stroma (Pihakaski 1968; Stewart 1978). Those with a single, large inner droplet appear smooth and homogeneous, those with numerous globules appear granular, and those with globules which bulge out of the stroma appear botryoidal (Pihakaski 1968; Stewart 1978). These characters along with oil body shape can provide valuable taxonomic information if evaluated carefully, since they vary with age (Slavik 1950; Worley 1969). Although in this study, age appeared to have little effect on oil body outline, it did influence the contents, and in many cases, the size of the organelles (Table IV). The oil bodies of older cells tend to be slightly larger than those of younger cells and the oil bodies become more granular to botryoidal

TABLE IV. OIL BODY CHARACTERISTICS OF SELECT HAPLOMITRIUM AND STEEREOMITRIUM SPECIES

	No./ Cell	Spher[a]	Spindle[b]	Trian[c]	Cont
		(average measurements in µm)			
H. intermedium					
young leaves	3-10	5-11	11-12 X 5- 9	8-11	g
old leaves	3-10	6-10	12-24 X 5-10	9-10	g
leafy shoots	3-12	5- 8	6-10 X 5- 8	ND	tb
stolons	4- 9	4-11	12-14 X 5- 8	8-10	tb
oH. hookeri					
young leaves	12-22	2- 3	3- 6 X 1- 3	3- 4	h-fg
old leaves	12-24	2- 4	3- 6 X 1- 3	3- 5	h-fg
leafy shoots	10-26	2- 3	4- 7 X 1- 3	ND	h-fg
stolons	6-14	2- 3	3- 5 X 1- 2	3- 4	h-fg
S. minutum					
young leaves	18-40	2- 3	2- 5 X 1- 3	2- 3	h-fg
old leaves	ND	ND	ND	ND	ND
leafy shoots	12-24	2- 3	4- 5 X 1- 1	ND	h-fg
stolons	ND	ND	ND	ND	ND
H. gibbsiae					
young leaves	18-33	1- 3	5- 7 X 2- 3	2- 4	fg
old leaves	20-38	1- 3	6- 9 X 3- 3	2- 4	cg
leafy shoots	16-27	1- 2	4- 7 X 2- 2	ND	fg
stolons	21-30	2- 3	4- 6 X 1- 4	3- 4	fg
H. mnioides					
young leaves	28-50	3- 5	6- 7 X 3- 4	5- 7	fg
old leaves	28-50	4- 5	9-10 X 4- 5	6- 7	b
leafy shoots	14-17	4- 5	5- 7 X 3- 4	5- 7	fg
stolons	21-40	6- 9	4- 7 X 3- 4	ND	fg
H. blumii					
young leaves	58-74	2- 5	9-18 X 2- 4	4- 7	g
old leaves	50-62	2- 3	9-16 X 2- 4	4- 8	g-cg
leafy shoots	61-70	2- 4	9-11 X 2- 4	4- 6	g
stolons	14-38	3- 4	9-11 X 2- 4	4- 5	g

[a]includes both spherical and ellipsoidal outlines; [b]includes both spindle and naviculate outlines; [c]length on a side; b, botryoidal; cg, coarsely granular; Cont, Contents; fg, finely granular; g, granular; h, homogeneous; ND, no data; Spher, spherical; tb, tiny botryoidal; Trian, triangular.

as they senesce or are stressed in some way (Table IV). The botryoidal morphology is probably a result of the natural disintegration process of oil bodies.

The taxonomically useful oil body characteristics observed for the Haplomitrium and Steereomitrium species examined in this investigation are presented in Table IV. Comparison of the data demonstrates that within a single cell, each species possesses oil bodies that are spherical to ellipsoidal, spindle-shaped to naviculate, or triangular in outline (Table IV; Figs. 139-151). These basic morphologies are constant not only in the leaf cells, but also in the stems (Fig. 139), stolons (Fig. 140), slime cells (Fig. 141), and setae (Fig. 142) of the plants. Regardless of cell type, the triangular oil bodies are always very rare. The published descriptions of Haplomitrium oil bodies also suggest a great diversity of oil body shapes (Schuster 1966, 1967, 1971; Yang 1968; Worley 1969; Kumar & Udar 1976; Udar & Kumar 1982). However, some of the reported variability may be due, in part, to the angle at which the organelles are being viewed, and in part, to the choice of terminology employed by the authors.

The taxa observed in this investigation may be grouped based on the number, size range, and contents of the oil bodies. As is evident from Table IV, H. intermedium (Figs. 143, 144) is characterized by each cell possessing a small number of very large, opaque to gray, granular oil bodies. In fact, this species bears the largest and most polymorphic oil bodies of any reported Haplomitrium species. Only 3-10 oil bodies are present per leaf cell, but these organelles are extremely large, with some reaching 24 µm in length (Table IV). The morphologies are expressed in a multitude of outlines, including spindle-shaped, naviculate, ellipsoidal, ovate, citron-shaped, lunulate, triangular, and reniform. Occasionally, the oil bodies of older leaves are seen to contain a few large spherules (Fig. 144). This is a common phenomenon in oil bodies of older cells regardless of the species.

Sections through H. intermedium oil bodies revealed more than 150 spherical oil droplets ranging from 0.42-0.83 µm in diameter dispersed throughout the stroma (Fig. 155). The large number and size of these lipophilic globules correlates to the granular appearance of the oil bodies at the light microscope level.

The oil body characters described here for H. intermedium are almost identical to those outlined by Schuster (1967). He emphasized the large size of the organelles, reporting the dimensions as 5.0-9.0 X 6.0-9.0 to 8.0-10.0 X 10.0-12.0 (-20.0) µm. Likewise, he stressed their polymorphic shapes, listing them as ovoid to ellipsoidal to citron-shaped or fusiform, all of which are granular to botryoidal in texture (Schuster 1967).

Haplomitrium hookeri (Fig. 145, 156) can be recognized by its possession of 12-26 homogeneous to finely granular, minute oil bodies per cell (Table IV). As pointed out by Schuster (1966, 1967), the chloroplasts, which measure 4.0-4.5 µm in diameter, are larger than the oil bodies. Worley (1969) qualified this observation stating that the chloroplasts are larger than the spheroid oil bodies, but smaller than the other forms. On the average, the median leaf cells possess approximately 29-57 chloroplasts per cell. This falls within the (20-) 40 (-70) chloroplast per cell range reported by Worley (1969). Electron micrographs show that these finely granular oil bodies contain a small number of minute, spherical oil droplets (Fig. 156). About 24-86 globules with diameters of approximately 0.13-0.21 µm are present.

The oil body data obtained in this investigation for H. hookeri fall nicely into the descriptive range reported for this species by other workers. In his classical study on oil bodies, Müller (1939) created nine morphological categories, of which, H. hookeri was classified within the "Schrotkorntyp" (translated literally as the crushed grain type). This group is characterized by minute oil bodies, measuring only

2.0-4.0 µm, with mostly homogeneous contents. More specifically, Müller (1939) described the oil bodies of H. hookeri as being circular to egg-shaped with diameters of 2.0 µm, numbering up to 20 per cell, and appearing homogeneous. Similarly, Schuster (1966, 1967) reported the oil bodies of this species as being spherical with diameters of 2.0-2.4 µm and ovoid or ellipsoidal, measuring 2.0 X 3.0-3.5 µm, numbering from 16-20 up to 25-42 per cell, and being homogeneous. Worley (1969) counted (12-) 23 (-40) oil bodies per cell in H. hookeri, and in addition to observing spheroid oil bodies, noted three other morphologies, namely, triangulate, fusiform, and lunulate. The sizes of these forms were cited as: spheroid, 3.0-3.5 µm; triangulate, 4.0-6.0 µm on a side; fusiform, 4.0-9.0 X 1.5-2.5 µm; lunulate, 4.0-9.0 X 1.5-2.5 µm. Worley (1969) also noted that, although the oil bodies were mostly homogeneous, the triangulate ones sometimes contained a clear droplet, and the larger fusiform and lunulate bodies were comprised of small spherules imparting a "bubbly" appearance to the organelles. Although these characters were not seen in the H. hookeri oil bodies observed in this study, they were seen in the older cells of the other Haplomitrium species examined. Therefore, it is highly probable that this trait also exists in senescent or highly stressed H. hookeri cells.

The oil bodies of Steereomitrium minutum (Fig. 146) are very similar in number, shape, size, and texture to those of H. hookeri (Table IV). Approximately (18-) 25-40 homogeneous to finely granular oil bodies exist in each cell. E. O. Campbell (1987) reported that S. minutum possesses finely granular oil bodies which number (0-) 11-21 (-40) per cell and are ellipsoidal with diameters of 2.6-4.0 µm. However, all gametophytic cells examined in this investigation contained numerous oil bodies and each morphology typical for Haplomitrium oil bodies, i.e., spherical to ellipsoidal, spindle-shaped to naviculate, and triangular outlines, were present.

Haplomitrium gibbsiae, H. mnioides, and H. blumii can be grouped together based on the similarities of their oil body characteristics (Table IV). These species usually possess a larger number of oil bodies per cell (20-74) than do H. intermedium, H. hookeri, and Steereomitrium minutum. The oil bodies are finely to coarsely granular, and are of intermediate size (5.4-17.8 X 2.4-4.5 µm) between those of H. intermedium and H. hookeri.

The oil bodies of H. gibbsiae are less numerous and smaller than those of H. mnioides and H. blumii (Table IV; Figs. 147-148, 157). The leaves of H. gibbsiae possess an almost equal mixture of spherical and spindle-shaped to naviculate oil bodies, while the triangular form is much less abundant. As is true for the other species examined, the spherical organelles are smaller than the other forms (Table IV). Each cell contains approximately (18-) 20-38 finely granular oil bodies. The chloroplasts are larger (5.5-7.5 X 2.2-4.4 µm) than the spherical oil bodies, and approximately equal in size to the spindle to naviculate forms (Table IV; Fig. 147).

The aging process of oil bodies was nicely exemplified in this species. The oil bodies progressed in contents from finely granular organelles in young leaves (Fig. 147), to coarsely granular in older leaves (Fig. 148), and finally, to botryoidal in a severely stressed, chlorotic, dying leaf (Fig. 149). Although the chloroplasts of the senescent leaf had degenerated, there were approximately 20 oil bodies present in each median leaf cell. Most of these were large and spindle-shaped or naviculate in outline with dimensions of 5.4-7.2 X 2.7-4.5 µm. The spherical forms were less abundant but also large, with diameters of about 4.5 µm. Electron micrographs of thin sections through H. gibbsiae oil bodies (Figs. 152, 157) show more than 25-50 droplets measuring 0.17-0.32 µm in diameter.

Schuster (1967) has provided the only other description of H. gibbsiae oil bodies. The findings presented here generally corroborate his observations.

Schuster (1967) stated that the oil bodies are finely granular with subspherical (1.8-2.0 X 2.0-3.0 µm), ellipsoidal (ca. 1.5-2.0 X 6.0-8.0 µm to 1.8-2.0-5.0 µm), slenderly fusiform, paramecium-like, or euglenoid (ca. 1.5-2.0 X 6.0-8.0 µm) outlines. The greatest discrepancy involves the number of oil bodies present per cell. However, this difference is not critical since oil body number is known to be highly variable (Schuster 1966; Worley 1969; Galatis et al. 1978a). Schuster (1967) counted 32-50 per cell while only about (18-) 20-38 per cell were observed in this investigation. Also, Schuster stated that the chloroplasts which measured 4.0-4.2 X 5.0-5.4 µm are larger than the oil bodies. This, however, seems to be variable (e.g. compare Figs. 147 and 149).

Recently, Hässel de Menéndez (1975) has synonymized _H. chilensis_ with _H. gibbsiae_. The oil body characters given by Schuster (1971) for _H. chilensis_ are virtually identical to the oil body characters of _H. gibbsiae_ and can be construed as evidence supporting the synonymy of these two species. According to Schuster (1971), the oil bodies of _H. chilensis_ number 20-28 per cell, appear botryoidal, and are globose to ellipsoidal with occasional subtriangular forms present. These oil bodies reportedly measure 4.0-6.0 X 4.0-5.0 µm (Schuster 1971). In comparison, the _H. gibbsiae_ oil bodies observed in this study number (18-) 20-38 per cell, grade from finely granular to botryoidal depending on age, and are spherical to ellipsoidal, spindle to naviculate, or triangular in outline, with the elongate forms measuring 4.6-6.5 X 1.8-3.0 µm and the spherical forms measuring 2.0-3.6 µm in diameter.

The oil bodies of _H. mnioides_ (Figs. 150, 158) are smaller and slightly less abundant than those of _H. blumii_, but are larger and more numerous than those of _H. gibbsiae_ (Table IV). A typical leaf cell of _H. mnioides_ contains an almost equal mixture of spherical, ellipsoidal, and spindle-shaped to naviculate forms, while the triangular form is rare. As in the other species, the spindle-shaped to naviculate oil bodies are almost twice as large as the spherical organelles, and, in general, the oil bodies of the older leaves are slightly larger than those of younger leaves (Table IV). The oil bodies of young leaves are minutely granular, but become botyroidal with age. The chloroplasts, which measure 8.8-12.1 X 7.7-8.8 μm, are larger than all of the spherical oil bodies and most of the spindle-shaped to naviculate and triangular forms. At the ultrastructural level, a thin section through an oil body of H. mnioides (Fig. 158) reveals more than 40-50 inner droplets with diameters of 0.12-0.24 μm.

Differences in the number, shapes, and contents of H. mnioides oil bodies are reflected in the literature. Hattori (1951) and Inoue (1974) illustrated only spindle-shaped oil bodies, some of which are granular and some botryoidal. In contrast, Schuster (1967) described these organelles as fusiform, ellipsoidal or ovate, and more or less homogeneous. Likewise, Yang (1968) reported two types of oil bodies in the cells of aberrant gametophytes, the more numerous spindle-shaped along with some spherical forms. The variability in oil body number is reflected in the literature with reports ranging from 31-40 (Inoue 1974), to 50-90 (Schuster 1967), to 70-77 per cell (illustrated by Hattori 1951, 1953).

The oil bodies of H. blumii are described here for the first time (Table IV; Figs. 151, 159). In general, these organelles are more granular in appearance, larger in size, and more numerous than those of H. mnioides or H. gibbsiae (Table IV). Each cell possesses approximately 60-74 oil bodies which are opaque to gray in color and granular in appearance. Of the oil body forms found in H. blumii, the spindle to naviculate outlines reach the greatest size (9.2-17.8 X 2.2-4.4 μm) and are usually the most abundant form. The spherical and ellipsoidal oil bodies average 2.2-4.6 μm in size and are smaller and less numerous than the spindle to naviculate forms (Fig. 151). Triangular and lunulate outlines also occur, although these forms are rare. Most of the oil bodies of older cells possess a single, central spherule, but some contain 2-3 spherules as observed in older H. intermedium oil bodies. In both cases, the spherical, ellipsoidal, and triangular oil bodies are smaller than the chloroplasts which measure 4.4-8.8 X 5.6-10.6 µm. In contrast, the spindle-shaped to naviculate forms are subequal to slightly larger than the chloroplasts. Electron micrographs reveal more than 35-57 lipid globules with diameters of 0.16-0.18 µm in the sectioned oil bodies of H. blumii (Fig. 159).

The oil bodies of three other Haplomitrium species have been described at the light microscope level, but were not examined in this investigation. Kumar and Udar (1976) described the oil bodies of H. dentatum as being homogeneous and bearing a central spherical or granular projection. They described the oil body shapes as fusiform and triangular, ranging in size from 4.7-12.15 X 1.35-3.54 µm (Kumar and Udar 1976).

The oil bodies of H. kashyapii were described as numbering 40-70 per cell, being fusiform, elliptic, lunulate, and triangular in outline, and ranging in size from 8.0-9.0 (-13.0) X 2.0-5.0 (-7.0) µm (Udar and Kumar 1982). Based on these descriptions, the Indian species of H. dentatum (Kumar and Udar 1976) and H. kashyapii (Udar and Kumar 1982) possess oil bodies which are most similar to those of the H. gibbsiae-H. mnioides-H. blumii complex.

The oil bodies of H. ovalifolium were described by Schuster (1971) as numbering 25-35 per cell and being slightly to clearly botryoidal. He reported the oil body morphologies as subspherical to ovoid with dimensions of 2.5-2.8 (-3.5) µm for the former and 2.2 X 3.5 to 3.0-3.2 X 4-5 (-6.5) µm for the latter (Schuster 1971). The oil body characteristics of H. ovalifolium are one of two traits used by Schuster to distinguish this species from H. hookeri. He emphasized the fact that these organelles are homogeneous or nearly so in H. hookeri while they are clearly botryoidal in H. ovalifolium (Schuster 1971). However, as seen in this investigation, the botryoidal appearance may simply indicate aging or stress and is not necessarily species specific. Therefore, this character is questionable. Although Schuster (1971) reported the oil bodies of *H. hookeri* as being slightly smaller than those of *H. ovalifolium*, they fall into the size range observed in *H. hookeri* in this study (Table IV). A reinvestigation of the oil bodies of *H. ovalifolium* is warranted before any definitive taxonomic descisions based on oil body characters can be reached.

At both the light microscope and ultrastructural levels, species distinctions among oil body characters are ambiguous. When comparing *Haplomitrium* species, only *H. intermedium* is noticeably different. However, it is unknown if this reflects a significant biochemical divergence between *H. intermedium* and the other *Haplomitrium* species.

The mere fact that *Haplomitrium* possesses oil bodies ensures its position within the Hepatophyta since these organelles are known to occur exclusively in hepatics (Schuster 1966; Watson 1971; Galatis et al. 1978b; Crandall-Stotler 1984; Duckett and Renzaglia 1987, 1988a). The oil body characters of *Haplomitrium* are most similar to those of the Metzgeriales. In contrast, oil body data suggests systematic differences between *Haplomitrium* and the Monocleales, Marchantiales, and Treubiales. Members of each of these orders exhibit cellular dimorphism in possessing specialized oil cells which contain very large oil bodies that nearly fill the entire cell lumen (Schuster 1984b). Similarly, the systematic position of *Takakia* is clarified by oil body characters. No membrane-bound oil bodies are present in *Takakia*, therefore supporting its removal from the Haplomitriales (Crandall-Stotler 1986). In addition, Asakawa, Tokunaga et al. (1979) compared the biochemistry of *H. mnioides* and *Takakia* and found more similarities between *Takakia* and mosses than between *Haplomitrium* and *Takakia*. This implies a rather large systematic gap between *Haplomitrium* and *Takakia* (Asakawa, Hattori et al. 1979).

Branching

As discussed earlier, branching in *Haplomitrium* is most frequent from the rhizomatous axes (Figs. 54, 160) and near the bases of the leafy shoots (Figs. 52, 57, 161). However, occasionally, branches do arise at various heights along the aerial shoots (Figs. 162-166). The branches potentially arise from all three merophyte rows (Figs. 54, 163, 164a). Those of the rhizomes and lower portions of the leafy shoots diverge at 90° angles to the parent axes (Figs. 52, 57, 160, 161, 163). Since the rhizome is leafless, there is no branch-leaf association (Fig. 54). Leafless branches may grow laterally to continue the rhizome, become positively geotropic and function as stolons, or remain as small dormant buds. Other rhizomatous branches may ultimately become erect and leafy. The ontogenetic fates of the branches formed at the extreme bases of the aerial shoots are similar to those of the rhizome branches, except that no stolons are formed except in *H. intermedium*. The stolons themselves are unbranched except in *H. intermedium*, where they branch at 45° angles to the parent stolons (Fig. 34a).

The branches of the leafy shoots may be axillary (Figs. 164-166), supra-axillary (Figs. 52, 162), or positioned midway between two leaves (Figs. 163, 170, 171). They diverge from the main axis at either 90° (Figs. 162, 163) or occasionally 45° angles (Fig. 166). These branches may be leafless and stunted (Figs. 54, 161), or elongate and leafy (Figs. 165, 166), and in some cases, impart an almost dichotomous branching appearance to the plant (Figs. 165, 166).

Serial sections through *H. hookeri*, *H. gibbsiae*, and *H. mnioides* shoots (Figs. 167-173) reveal that all of these branch orientations result from a single ontogenetic pattern. In all cases, branching is exogenous with the branch initial originating a short distance (less than 275 μm or in the third or fourth merophyte) from the apical cell in the kathodic portion of either a lateral or dorsal merophyte (Figs.

167, 168). Each initial, which can be distinguished from neighboring cells by its slightly larger size and distinctive shape (Fig. 167), arises from an epidermal cell in the second row of stem cells basiscopic to the young leaf (Fig. 167). Since the leaf apical cell has already been formed by this stage, mature leaf morphology is unmodified by branch formation. Three oblique divisions soon traverse the branch initial to form a tetrahedral branch apical cell (Figs. 167, 168).

The epidermal cells distal to the branch apical cell divide to displace the cell to its final location (Figs. 169, 170). Extensive divisions in these cells position the branch apical cell in the axil of the next oldest leaf, while less extensive divisions may position it slightly above the leaf axil (supra-axillary), or midway between the leaves giving the impression that there is no direct association of the branch with the leaf (Fig. 170).

Although the branch apical cell is formed near the apex of the shoot, it does not begin to segment immediately. Instead, it remains dormant for a time while the main shoot continues to grow. Thus, the branch apical cell is "pushed" to a more basal position on the shoot as new merophytes are formed by the main shoot apical cell (Fig. 171). This phenomenon in which branch apical cell segmentation is delayed to older derivative stages is not unique to Haplomitrium. Other taxa in which this occurs include Treubia, Apotreubia Hatt. et Mizut., Noteroclada, and Phyllothallia Hodgs. (Crandall-Stotler 1981; Renzaglia 1982).

Like the main shoot apical cell, the branch apical cell is slightly sunken in the thallus. It is tetrahedral in shape (Figs. 167-171) and measures about 14.0-24.2 µm in width, 28.0-38.2 µm in height, and 26.4-30.4 µm in depth. Derivatives are produced in a spiral sequence (Fig. 171) and numerous slime cells which overarch and ensheath the generative cell in protective mucilage form from these segments (Fig. 172). The first derivative forms toward the basiscopic side of the branch, the second toward the acroscopic side, and the third on the third side of the branch. At least in H. mnioides, the direction of the branch apical cell spiral was found to be the same as that of the main shoot apical cell; thus the plant is homodromous (Fig. 168).

The merophytes of the branch apical cell develop as they do in the main shoots, either resulting in elongate, leafless shoots with slime papillae (Fig. 160), or elongate shoots with basal scales and distal leafy portions. Sometimes the branch apical cell cuts off a small, limited number of derivatives then ceases its activity, causing the branch to remain as a small bump or short spur. Whether or not the apical cell activity resumes at some point to continue the elongation of the branch is unknown. In any case, due to the delayed segmentation of the branch apical cell, the tissue of the newly formed branch does not merge with that of the parent axis, but instead, abruptly meets the already well-developed tissues of the main shoot (Figs. 172-173).

The angle of branch divergence depends on the height of the branch apical cell on the shoot when branch growth begins. If the branch apical cell begins to segment relatively soon after its formation, while it is still positioned near the shoot apex where it is oriented at a 45° angle, the divergence of the branch will also be 45° (Fig. 167). If the elongation of such a branch keeps pace with the growth of the main axis, the shoot appears dichotomous (Figs. 164a, 164b, 166). However, if the main shoot grows and the branch apical cell remains dormant, its angle of orientation is changed to 90° (Figs. 169, 170), as is the angle of mature branch divergence (Figs. 161-163).

This exogenous pattern of branch ontogeny coincides with the Fontinalis-type of branching described by Crandall (1969). She reported this pattern for H. mnioides, H. blumii, and Takakia lepidozioides.

Branching in Haplomitrium has also been discussed, at least in part, in H. hookeri by Leitgeb

(1875) and Lilienfeld (1911), *H. gibbsiae* by E. O. Campbell (1959), and was generally reviewed by Schuster (1966, 1967, 1982, 1984a). All workers agreed that the branches are exogenous and arise through the segmentation of an apical cell with three cutting faces. However, the precise origin of the branch initial was uncertain. Lilienfeld (1911) described the branch generative cell as arising through three divisions from a single superficial cell located indiscriminately at almost any distance from the vegetative apex, independent of the leaves on both aerial and subterranean shoots. She believed that terminal branching and axillary branching characteristic of other leafy hepatics was absent in *H. hookeri*.

As a result of her work with *H. gibbsiae*, E. O. Campbell (1959) reported that the branch apical cell formed "... on the side of a stem at a point midway between two leaves of a vertical row." She maintained that the number of apical cells, their time of formation, and their height on the aerial shoot varies from individual to individual.

Schuster (1966, 1967) proposed that a relatively mature cortical cell situated midway between two leaves of a vertical row dedifferentiated and was transformed by three divisions into a branch apical cell. He called this branching pattern the *Haplomitrium*-type. Later, he reported that two types of branching possibly occur, one type in basal regions of the plant where branches arise at right angles from the axis, and one distally, where the stem is furcate and the branches diverge at acute angles from the main axis (Schuster 1982). He suggested that the former resulted from the *Haplomitrium*-type of branching, i.e., from a dedifferentiated cortical cell, while the latter resulted from terminal-furcate ("pseudodichotomous") branching (Schuster 1982).

However, as first shown by Crandall (1969) and confirmed in this investigation, the branch apical cell is not derived from a dedifferentiated cortical cell, but is formed close to the main shoot apical cell in a branch initial which originates in the kathodic portions of the epidermal initials of derivatives in all three merophyte rows. Following its creation, the branch apical cell is displaced to its ultimate position by divisions in the adjacent epidermal cells and is positioned at various "heights" on the shoot by the development of later formed apical merophytes. This latter phenomenon combined with the delayed segmentation of the branch apical cells accounts for the differences in the angles of branch divergence in distal and basal portions of the plant. Therefore, a single ontogenetic pattern can explain the branching of Haplomitrium. Such a pattern, in which branching occurs from all three merophyte rows of a radially symmetrical, erect axis resulting in a three dimensional, almost bushlike plant, is exemplary of the presumed ancestral mode of branching (Schuster 1966, 1984a).

Two distinct basic types of branching occur in the Jungermanniales, i.e., axillary athecal branches which are often associated with stem leaf modifications, and gyrothecal branches (Leitgeb 1895; Evans 1912; Crandall 1969; Schuster 1966, 1984a). Neither of these patterns occurs in Haplomitrium. Branching modalities in the Metzgeriales are more varied, ranging from those of ancestral taxa such as Treubia and Phyllothallia with lateral or latero-ventral branches exhibiting branch apical cell dormancy like Haplomitrium, to the terminal bifurcate type of branching common in the majority of Metzgerialian taxa, or the collared, ventral, endogenous branches found in members of the derived Metzgeriaceae (Renzaglia 1982). In addition, the branching pattern displayed by Haplomitrium, i.e., Fontinalis-type branching is also the most common type of branching in the mosses (Leitgeb 1868; Crandall 1969). Therefore, as aptly pointed out by Schuster (1967), based on branching modalities, Haplomitrium is more distinctly isolated from the Jungermanniales than from the simple thalloid hepatics and/or mosses.

The Reproductive Gametophyte

Gross Morphology

The reproductive shoots of *Haplomitrium* are erect in stature and, with the exception of *H. monoicum*, all species are dioicous. Sexual dimorphism is common within the genus, having been reported in *H. intermedium* (Berrie 1962; Hewson 1973; Schuster 1967), *H. hookeri* (Gottsche 1843; Macvicar 1926; Frye and Clark 1943; Schuster 1966, 1967), *H. gibbsiae* (Campbell 1959; Allison and Child 1975), *H. chilensis* (Schuster 1971), and *H. dentatum* (Kumar and Udar 1976). In these taxa, the female plants are larger, possess thicker stems, and bear larger, more variable leaves than do the males.

Traditionally, species of *Haplomitrium* have been placed in subgeneric categories, based on gametangial positions, with the archegonia and antheridia of each species being produced at similar loci on their respective shoots. Even in the monoicous *H. monoicum*, antheridia and archegonia never occur on the same shoot. Taxa of the subgenera *Archibryum* Schust. and *Haplomitrium*, which include *H. intermedium*, and *H. hookeri*, *H. ovalifolium* (Schuster 1971), and *H. monoicum*, respectively, typically produce only a small number of gametangia and display great variability in gametangial positions. Usually, their gametangia are axillary, or supra-axillary, and may be either dispersed within the distal portion of the shoot in unmodified stem leaves, loosely arranged in terminal groups, or organized at swollen nodes along the stem in specialized bracts. Occasionally, gametangia are scattered along the stem with no apparent association with a bract. Such lack of fixity in gametangial positions is often regarded as plesiomorphic. In contrast, the location of the gametangia is more highly fixed in species of the subgenus *Calobryum* (*H. gibbsiae*, *H. indicum*, *H. dentatum*, *H. grollei*, *H. kashyapii*, *H. mnioides*, *H. blumii*, *H. andinum*, and *H.*

giganteum), which, in nature, typically produce well-defined, terminal gametangial discs, some with the potential for apical proliferation. Often, the fertile shoots of this group possess only a few (3-6) cycles of leaves which, on small shoots, may be clustered distally on the stem. Larger shoots bear distant leaves with relatively long internodes except at the gametangial disc, where they become clustered. Fertile shoots may be isophyllous, slightly anisophyllous, or express varying degrees of anisophylly depending on the species. When expressed, anisophylly is more exaggerated on the female shoots than on the males. The apex of the gametophore is dilated and bears 20 to as many as 125 gametangia depending on the species. One or 2 series of enlarged bracts subtend the terminal disc and 1 or 2 reduced bracts are within the rosette. Rudimentary bractlets and slime hairs are often interspersed among the gametangia and slime papillae are always associated with the gametangia. The bracts overarch and protect the young gametangia, but become erect-spreading and eventually expose the gametangia at maturity.

Female gametophores are known from all *Haplomitrium* species, while males are known from each species except *H. giganteum* (Stephani 1922; Grolle 1963, 1964; Schuster 1967) and *H. grollei* (Kumar and Udar 1977). A summary of the distinguishing male gametophytic characters of all known *Haplomitrium* species is provided in Table V, while the female characters are summarized in Table VI.

Haplomitrium hookeri (Figures 174-178; 179-180). The reproductive gametophytes of *H. hookeri* are heteromorphic. Schuster (1966, 1967) reported that male and female plants grow in separate patches, but Worley (1969) found both unisexual turfs, and turfs with the sexes intermixed. The smaller male gametophores are about 15-20 mm tall X 2.0-3.0 mm broad, while the larger female shoots are 16.5-26.0 mm long X 2.0-3.0 mm wide. The leaves of the fertile shoots are imbricate, but, the general habits,

TABLE V. MALE GAMETOPHYTIC CHARACTERS OF HAPLOMITRIUM AND STEEREOMITRIUM SPECIES[a]

	Shoot (LXW; mm)	Antheridial Position	Bracts	An/ Bract	Assoc Struct	Apical Prolif
H. hookeri	15.0-20.2 X 2.0-3.0	ax or sax; loose ap cluster; scattered distally; naked	like stem leaves or enlarged	2-4 to 6(15)	P, H, B	+
H. ovalifolium[b]	ND	like HH	variable	1-2	P, H, B	+
H. monoicum	4.8- 7.2 X 1.2-1.5	ax or sax; scattered distally	like stem leaves	1-2	P, H, B	+
H. intermedium	9.0-17.0 X 1.5-2.0	ax or sax; ap discs or nodes; scattered; naked[c]	like stem leaves or enlarged	2-4 to 6-8	P, H, B	+/-
H. gibbsiae	6.0-12.0 X 2.5-3.0[d]	dilated ap disc	ob ovate, rhom; ib ling-rhom	30-35 or more	P, H, B	+
H. chilensis	like HG	like HG	like HG	50-60	P, H, B	+/-
H. indicum[e]	6.0-15.0 X 0.4-0.5	dilated ap disc	enlarged	125	H	+
H. dentatum[f]	6.0-12.0 X 2.0-4.5[d]	dilated ap disc	ob dentate	15-80	B	+

TABLE V. CONTINUED

	Shoot (LXW; mm)	Antheridial Position	Bracts	An/ Bract	Assoc Struct	Apical Prolif
H. kashyapii[g]	9.0-30.0 X 2.0-5.5[d]	dilated ap disc	ob rhom; ib ling-lanc	many	B	+
H. mnioides	13.0-20.0 X 4.2-5.0[d]	dilated ap disc	ob subro; ib ling	20-35 or more	P, H, B	+
H. blumii	12.0-35.0 X 3.5-5.0[d]	dilated ap disc	ob subro, obov; ib ling	80-105	P	-
H. andinum	16.0-25.0 X 2.0-5.2[d]	dilated ap disc	ob subro, ellip; ib ling	25-40 or more	P, H, B	-
S. minutum	10.0-12.0 X 0.9-2.5	ax or sax; loose ap clusters; scattered	like stem leaves or enlarged	3 to 25 or more	P	+/-

[a]males are unknown for H. grollei and H. giganteum; [b]from Schuster 1971; [c]Schuster 1967; [d]width of apical disc; [e]from Udar & Chandra 1961, 1965; [f]from Kumar & Udar 1976; [g]from Udar & Kumar 1982; An, Antheridia; Assoc Struct, Associated Structures; ap, apical; ax, axillary; B, bracteoles; ellip, ellipsoidal; H, slime hairs; HH, H. hookeri; HG, H. gibbsiae; ib, inner bract; LXW, length X width; lanc, lanceolate; ling, lingulate; ND, no data; ob, outer bract; obov, obovate; P, slime papillae; Prolif, Proliferation; rhom, rhomboidal; sax, supra-axillary.

TABLE VI. FEMALE GAMETOPHYTIC CHARACTERS OF HAPLOMITRIUM AND STEERIOMITRIUM SPECIES

	Shoot (LXW; mm)	Archegonial Position	Bracts	Ar/ Bract	Assoc Struct	Apical Prolif
H. hookeri	16.5-26.0 X 2.0-3.3	ax or sax; loose ap cluster; scattered distally; naked	like stem leaves or smaller	1-3 to 10	P, B	+/-
H. ovalifolium[a]	ND	ax; ap; scattered	ovate, ellip	1; 6-8	ND	+
H. monoicum	4.6- 7.2 X 1.2-1.5	ax or sax; scattered distally	like stem leaves	1-2	P, B	+
H. intermedium	19.0-21.0 X 2.0-2.3	ax or sax; ap discs or nodes; scattered	like stem leaves or enlarged	1-3	P, H, B	+/-
H. gibbsiae	14.0-20.0 X 4.7-7.0[b]	dilated ap disc	ob ling-ovate, rhom; ib ling-trian	15-20	P, H, B	+
H. chilensis	like HG	like HG	like HG	like HG	P	+/-
H. indicum	6.1- 8.0 X 4.0-5.0[b]	dilated ap disc	like HG	14-35	P	ND

TABLE VI. CONTINUED

	Shoot (LXW; mm)	Archegonial Position	Bracts	Ar/ Bract	Assoc Struct	Apical Prolif
H. dentatum	8.0-12.0 X 3.0-4.5[b]	dilated ap disc	like HG	20-30	ND	ND
H. grollei	10.0-22.0 X 3.0-4.5[b]	dilated ap disc	like HG	18-20	ND	+/-
H. kashyapii	12.0-25.0 X 2.0-4.0[b]	dilated ap disc ax or sax; scattered	like HG	ND	P	+/-
H. mnioides	17.0-23.0 X 7.1-7.3[b]	dilated ap disc	ob subro; ib ling-trian	20-35	P, H,	+
H. blumii	17.0-37.0 X 6.5-7.4[b]	dilated ap disc	ob subro; ib ling-trian	80-100	P, H, B	-
H. andinum	13.0-20.0 X 3.0-4.2[b]	dilated ap disc	ob obov, subro; ib ling-trian	20 or more	P, H, B	-

TABLE VI. CONTINUED

	Shoot (LXW; mm)	Archegonial Position	Bracts	Ar/ Bract	Assoc Struct	Apical Prolif
H. giganteum	26.0-60.0 X 7.0-9.0[b]	dilated ap disc	ob obov-ling; ib ling-trian	25-40	P	-
S. minutum[c]	6.0- 8.0 X 1.5-2.0	loose ap cluster	like stem leaves	ND	P, B	+

[a]Schuster 1971; [b]width of apical disc; [c]Campbell 1987; ap, apical; Ar, Archegonia; Assoc Struct, Associated Structures; ax, axillary; B, bractlets; ellip, ellipsoidal; H, slime hairs; HG, H. gibbsiae; ib, inner bract; LXW, length X width; ling, lingulate; ND, no data; ob, outer bract; obov, obovate; P, slime papillae; Prolif, proliferation; rhom, rhomboidal; sax, supra-axillary; trian, triangular.

symmetries, and leaf characters are as in the vegetative gametophytes (Figs. 174a, 174b, 176a, 177).

The gametangia of H. hookeri are axillary or supra-axillary and are usually located in the terminal quarter of the shoot. The antheridia are sometimes loosely grouped near the apex of the shoot (Figs. 174a, 175), but more often they are scattered along the stem in the axils of bracts (Figs. 174b, 176a, 176b). The leaves associated with the antheridia may be like the stem leaves in size and shape, or slightly enlarged and spreading, often becoming recurved and exposing the antheridia at maturity. The enlarged bracts measure approximately 1.2-1.3 mm long X 1.4-1.5 mm wide X 0.8 mm to the broadest portion of the leaf. The antheridia are arranged in groups in or near the axils of the bracts (Fig. 174b, 176a, 176b). As first noted by Gottsche (1843), 3-4-celled slime hairs are associated with the antheridia. In addition, papilla-like slime cells are also present (Figs. 179). The number of antheridia per bract varies from 2-4 (-6) (Figs. 174b, 175, 176a, 176b). Gottsche (1843), Macvicar (1926), and Frye and Clark (1943) reported 2-3 antheridia per bract, while Lilienfeld (1911) and Inoue (1971) counted 2-5 in each axil. Cooke (1894) and Carrington (1874) each found 5 or 6 antheridia per apical bract, while Worley (1969) reported 4-15. Worley (1969) also noted that the antheridia comprising such a "clump" matured alternately, i.e., a group might consist of 3-4 bright orange, ripe antheridia, 1-2 pale yellow, nearly ripe antheridia, while the rest remained green and immature. He also described the ripest antheridia as most often being arranged in a row transverse to the axis of the plant (Worley 1969).

Gottsche (1843), Evans (1917) and Schuster (1967) reported antheridia with no specific relation to bracts scattered along the stem. Worley (1969) specifically stated that no such isolated antheridia occurred in the western North American material which he examined. In this investigation, only one plant which had been collected in Austria (Loitlesberger

1348), was found to possess a single, old, free-standing antheridium. The production of non-axillary antheridia is probably rare, and the few that are formed are doubtless easily detached with specimen handling.

Archegonial positioning in *H. hookeri* is similar to that of the antheridia, except that the archegonia are most often axillary, in loose groups at the summit of the shoot (Figs. 177, 178, 180) as opposed to being scattered along the shoot in the axils of stem leaves. As pointed out by Gottsche (1843) and Lilienfeld (1911), the scattered position seems to result from continued growth and gametangial production when fertilization does not occur. As reported by Gottsche (1843), Lilienfeld (1911), and Frye and Clark (1943) free-standing, unprotected archegonia are also rarely formed. Unlike the male bracts of *H. hookeri* and the perichaetial bracts of most of the other species, the apical female bracts of *H. hookeri* are slightly smaller than the vegetative leaves. These bracts are about 1.0-1.2 mm long X 0.7-1.4 mm wide X 0.4-0.5 mm to the broadest portion of the leaf and vary in shape from rhomboidal to oval to rarely lingulate (Figs. 31, 33, 34, 40). Thus, they may be either broader than long as reported by Worley (1969) or longer than broad as described by Pearson (1902) and Schuster (1967). Generally, 2-3 archegonia are produced per bract when scattered (Fig. 178). Gottsche (1843) also counted 2-3 per bract, while Lilienfeld (1911) reported 2-5. Loose apical groups are often composed of as many as 10 archegonia (Gottsche 1843; Evans 1917; Lindberg 1885). Single-stalked mucilage papillae and minute bractlets are interspersed with the archegonia of the gynoecia (Figs. 178, 180).

Haplomitrium ovalifolium. According to Schuster (1971), the fertile shoots of *H. ovalifolium* are similar to those of *H. hookeri*. However, no sexual dimorphism exists in *H. ovalifolium*. The two taxa are similar in gametangial positions, but male shoots differ in having the perigonial bracts arranged in an imperfect spiral, i.e., the 3-ranked arrangement of bracts and leaves of the androecial region is destroyed (Schuster 1971). Also, fewer antheridia are produced per bract in H. ovalifolium than in H. hookeri (Table V). In general, the female plants each bear 6-8 archegonia, and the perichaetial bracts are longer than broad, ovate to elliptical to occasionally obovate, and are about 900-1100 µm long X 740-925 µm wide (Schuster 1971).

Haplomitrium monoicum (Figures 181; 182-184). Unlike other Haplomitrium species, H. monoicum is monoicous. Male and female gametangia, however, are never intermixed, but rather occur on separate shoots of the same plant. Due to the difficulties involved in separating rehydrated rhizomatous axes from the substrate without tearing the shoots, male and female gametophores were not found in connection in this study. However, male and female shoots branching from a shared rhizomatous axis were clearly illustrated by Engel (1981, Fig. 8, p. 671), leaving no reason to doubt the monoicous status of this species.

As observed by Engel (1981), this taxon does not exhibit sexual dimorphism. The dimensions of the fertile shoots (Tables V, VI), as well as the habits, symmetries, and leaf characters of the plants fall within the range typical of vegetative axes (compare Figs. 21, 22 and Fig. 182). The gametangia are produced singly or in pairs, in or near the axils of unmodified bracts in the distal quarters or halves of the shoots (Figs. 183a, 183b, 184). Engel (1981) reported that occasionally, female shoots bear archegonia exclusively at the apex of the shoot and the shoot appears to be acrogynous. However, in this study, all shoots seen with apical archegonia also possessed archegonia at one or more loci basal to the terminal cluster (Fig. 184). Sessile and single-stalked slime cells, as well as uniseriate hairs and rudimentary bractlets are associated with the antheridia (Fig. 181), while only slime cells were seen with the archegonia. However, Engel (1981) also reported minute accessory bractlets in association with unfertilized gynoecia.

Haplomitrium intermedium (Figures 185-188; 189-191). Of all the taxa, H. intermedium displays the greatest variability in gametangial positioning. The habit, shoot symmetry, and leaf characters of the fertile plants are as in the vegetative gametophytes (compare Figs. 26, 28 and 185). As reported by Berrie (1962), marked sexual dimorphism occurs in this species with the male shoots being shorter and slightly more slender than the female shoots (Tables V, VI). Hewson (1973) showed the size differences to be statistically significant and associated this dimorphism with a dimorphic sex chromosome mechanism in which females have an n=8+X genotype and males, an n=8+Y genotype.

As discussed by Berrie (1962), the gametangia of H. intermedium are most often axillary or occasionally supra-axillary (Figs. 186-188). Schuster (1967) additionally reported antheridia which were scattered along the stem and remote from leaves. No such "naked" antheridia were found on the plants examined in this study. However, only a few collections were available and these did not include those collected by Schuster.

Gametangia are generally positioned at or near the apex of the shoot, in the axils of enlarged, spreading bracts (Figs. 186a, 186b). The perigonial bracts are longer and broader than the vegetative leaves, measuring 1.4-2.0 mm long X 1.2-1.5 mm broad X 0.9-1.0 mm to the broadest portion of the leaf. These bracts are usually entire, but may rarely be lobed (Fig. 187a). As few as 2-4 to as many as 6-8 antheridia are associated with each bract. The production of these clusters is associated with a dilation of the stem tissue below (Figs. 186a, 188c). This is not the case in H. hookeri, H. ovalifolium, or H. monoicum in which the stem region below the gametangial cluster remains unmodified. As noted by both Berrie (1962) and Schuster (1967), the crowding of the enlarged bracts along with the slight dilation of the stem apex creates an ill-defined apical rosette (Figs. 186a, 186b).

Archegonia are also produced on swollen discs in ill-defined apical rosettes. However, each perichaetial bract usually bears only 1-3 archegonia (Figs. 188a, 188b) as opposed to the 2-4 to 6-8 antheridia produced in perigonial bracts. As first noted by Berrie (1962), the perichaetial bracts differ from the male bracts in being more or less imbricate and erect, as well as being larger and more variable in morphology. The female bracts are generally ovate to somewhat lingulate, or asymmetrical and measure 1.5-3.0 mm long X 0.7-1.2 mm wide X 0.8-1.3 mm to the broadest portion of the leaf. The margins may be entire, undulate, or shallowly lobed (Fig. 188b).

Uniseriate hairs and small, rudimentary bractlets are associated with the gametangia of both the male and female apical inflorescences (Figs. 189, 191). This confirms the findings of Schuster (1967), but is contradictory with those of Berrie (1962) who reported no slime hairs in association with the archegonia.

Repeated apical proliferation after gametangial production, resulting in a series of fertile nodes separated by internodal regions composed of 2-3 cycles of leaves may or may not occur (Figs. 185, 188a-c). When proliferation does occur, each node is characterized by a slight dilation of the stem and enlarged bracts similar to those subtending the apical rosettes.

Sometimes, apical rosettes are totally lacking and instead, 2 or 3 gametangia, either antheridia or archegonia, are produced in each axil of a few unmodified bracts along the shoot (Figs. 187a, 187b, 190). The gametangia that are dispersed in such a manner are associated with slime cells, but no bractlets (Fig. 190).

Haplomitrium gibbsiae/H. chilensis (Figures 192-194; 195-207). Male and female H. gibbsiae plants may grow intermixed with one another or in separate clumps (E. O. Campbell 1959). In either case, male plants can usually be distinguished by their smaller stature (Tables V, VI). This size differential is particularly evident in the diameter of the terminal gametangial discs and the dimensions of the subtending bracts. The fertile shoots of H. gibbsiae and H. chilensis are virtually indistinguishable (Tables V, VI), and the variability in habits, symmetries, insertions, and sizes and shapes of the stem leaves is as in the vegetative shoots except that they are distant (Fig. 195). Often, on small plants, only 2-3 cycles of leaves exist. In these cases, all the leaves are clustered distally, embracing the gametangia (Figs. 196, 201, 203). Female shoots which are leafy throughout bear leaves that are less distant and often more evenly spaced than in male shoots (Figs. 200, 202). Furthermore, on female shoots, the leaves of two rows often show a more marked tendency toward succubous insertion.

One or two cycles of enlarged, broadly ovate to rhomboidal, perigonial bracts delimit the male gametangial receptacle. These leaves are lobed and often have dentate margins (Figs. 197, 199). They are typically isophyllous and measure about 1.2-1.5 mm long X 1.3-0.6 mm wide, but occasionally, they may be slightly anisophyllous. At least one inner reduced bract which is lingulate-rhomboidal, often with 1 or 2 lateral projections tipped with slime cells, is internal to the enlarged peripheral bracts (Fig. 198). The gynoecium is similarly organized (Figs. 204-207), but the outer bracts which measure about 3.5-4.7 µm long X 3.0-3.2 µm wide X 1.0-1.5 µm to the broadest portion of the leaf, are much larger than those of the androecium. Occasionally, these bracts are ovate-lingulate in outline and may be lobed or have dentate margins. Although usually isophyllous, occasionally, an outer dorsal bracteole can be distinguished. Similarly, Schuster (1971) reported that sometimes the gynoecia of H. chilensis are anisophyllous. In this case, the bracteole is smaller than the bracts and is broadly lanceolate to ovate-lanceolate in outline.

In _H. gibbsiae_, 30-35 or more antheridia are axillary to the bracts, while 50-60 or more were commonly found on _H. chilensis_. Unicellular slime cells, 2-5-celled mucilage hairs, as well as rudimentary, irregularly shaped bractlets are interspersed among the antheridia (Figs. 193) . Schuster (1971) noted that in _H. chilensis_, only 1 or 2 lingulate to obovate-obcuneate reduced bractlets exist among the antheridia. However, unless sectioned androecia are observed, the tiny rudimentary bractlets, slime hairs, and mucilage papillae often escape detection.

Female _H. gibbsiae_ and _H. chilensis_ shoots bear 15-20 archegonia along with intermixed slime papillae, 2-3-celled mucilage hairs, and rudimentary bractlets (Fig. 192). Only 1- and 2-celled slime papillae were found among the archegonia of _H. chilensis_ (Fig. 194), but only a small number of plants were available for sectioning. Similarly, Schuster (1971) reported no accessory bractlets within the gynoecia of _H. chilensis_.

Campbell (1959) reported a knob-like proliferation bearing three rows of leaves emerging from the center of both antheridial and archegonial discs of _H. gibbsiae_. Numerous male _H. gibbsiae_ shoots with apical proliferations were also observed in this investigation (Fig. 199). However, Schuster (1971) maintained that no apical proliferation occurs in male or female _H. chilensis_ plants. Instead, he observed new growth from one or rarely two axillary branches which originate below the gametangial discs. From this, Schuster (1971) suggested that, unlike _H. gibbsiae_, further apical growth in _H. chilensis_ is precluded by gametangial production and instead, growth is continued through lateral branching. Of the fertile _H. chilensis_ plants observed in this study, none were observed with apical innovations or with branches originating from below the gametangial rosettes. However, serial transections reveal apical cells having widths and depths of approximately 23 µm which are positioned in the terminal receptacles of both mature androecia and gynoecia (Figs. 193, 194). Since these dimensions fall within the range of vegetative apical cells (Table III), there is no evidence to suggest that the apical cell is being destroyed by gametangial production. This implies that the potential for apical proliferation exists in at least some individuals of H. chilensis.

Haplomitrium indicum (Figure 208). Udar and S. Chandra (1961, 1965) and Mehra and Kumar (1990) have described both male and female H. indicum plants, but only female plants were made available for this study. Therefore, the following descriptions of males have been taken from the works of Udar and S. Chandra (1961, 1965) and Mehra and Kumar (1990). According to these investigators, H. indicum is dioicous and heteromorphic, with the males being erect, more slender and possessing more distantly spaced leaves than female shoots and measuring 6-8 (-15) mm tall X 0.4-0.5 mm broad. The female shoots are equivalent in height, but are more robust than the male shoots (Tables V, VI). Regardless of the sex, the leaves are clustered distally and the shoots are isophyllous or slightly anisophyllous (Fig. 208). Insertion is transverse at the apex to succubous at the base. The leaves are usually subrotund, orbicular, ovate, or rhomboidal and are about 1.0 mm long X 1.0 mm broad at the widest portion of the leaf in male shoots (Udar and S. Chandra 1965) and, are about 1.7-2.5 mm long X 1.8-2.5 mm wide X 0.5-1.3 mm to the broadest portion of the leaf on female shoots. As in some H. gibbsiae plants, the leaves become more crowded toward the apex and the gametangia are arranged in distinct terminal discs, axillary to the enlarged bracts. No other information concerning the perigonial bracts, except that they are 2- to 3-stratose at the base, is given by Udar and S. Chandra (1961, 1965) and Mehra and Kumar (1990).

As many as 125 antheridia reportedly comprise the androecial rosette and 2-3 "crops" of these are successively formed. Although not interpreted as such by Udar and S. Chandra (1961, 1965), these successive "crops" of androecia are the result of apical proliferation. Thus, as in H. gibbsiae, antheridial production does not preclude further growth of the shoot. In addition, Udar and S. Chandra (1965) reported young lateral shoots often arising from the bases of the male gametophore and occasionally from near the apex of the stem. Perhaps these branches are analogous to those described by Schuster (1971) in H. chilensis.

Only about 25 archegonia are produced in the terminal rosettes of female shoots. Due to the paucity of specimens, no sectioning was done through the gynoecia. Therefore, the presence or absence of intermixed slime hairs could not be determines. However, Mehra and Kumar (1990) reported slime papillae interspersed among the archegonia. The perichaetial bracts are similar to the vegetative leaves but are larger, more elongate, and often dentate. Occasionally, two of the bracts may be slightly more lingulate than the third. These measure about 2.9-3.0 mm long X 1.6-2.2 mm wide X 1.5-2.0 mm to the broadest portion of the leaf. Two reduced inner bracts are typically present. Both of these are lingulate-triangular in outline with the larger one being about 2.0 mm long X 0.7 mm wide X 1.0 mm to the widest portion of the leaf, while the smaller one is only approximately 0.5 mm long X 0.2 mm wide X 0.2 mm to the broadest region of the leaf.

In any case, from the descriptions, illustrations, and photographs presented by Udar and S. Chandra (1961, 1965) and Mehra and Kumar (1990), as well as my own observations, the fertile plants of H. indicum appear to be identical with those of H. gibbsiae.

Haplomitrium dentatum (Figure 209). Both male and female H. dentatum plants have been described by

Kumar and Udar (1976), but again only female plants were made available for study. Therefore, information regarding male shoots is based on the descriptions of Kumar and Udar (1976). Male and female plants of this taxon grow in separate but juxtaposed patches (Kumar and Udar 1976). The males are typically small and slender, while the females are larger and more robust (Tables V, VI). Although Kumar and Udar (1976) described the males as having the leaves crowded distally, and the females being leafy throughout, the females of the type material possess only 3-6 cycles of leaves which are crowded terminally (Fig. 209). The gametophores are slightly anisophyllous with the dorsal leaves being slightly smaller than the suborbicular to rhomboidal-shaped lateral leaves. The dorsal leaves of male shoots are approximately 1.6 mm long X 2.8 mm wide and the lateral leaves are about 2.0-2.2 mm long X 3.0 mm wide (Udar and Kumar 1977). When distinguishable, the dorsal leaves of the female shoots are about 1.5-1.8 mm long X 1.4-2.0 mm wide X 0.6-1.0 mm to the broadest portion of the leaf, while the lateral leaves are approximately 2.0-2.3 mm long X 1.5-2.2 mm wide X 1.1-1.2 mm to the widest portion of the leaf. The margins of the stem leaves are entire to occasionally dentate.

According to Kumar and Udar (1976), 50-80 antheridia are arranged in a terminal rosette or 15-30 antheridia may be borne in the axils of 1-2 cycles of bracts. However, the illustrations provided by Kumar and Udar (1976, Figs. 1-5, p. 25) suggest that, as in H. indicum and H. gibbsiae, the antheridia in the second cycle of bracts were at one time organized into a distinct terminal rosette and have been displaced to their present location by the apical proliferation of the shoot. Also illustrated are tiny intra-androecial bracts (probably bractlets). One or two cycles of enlarged perigonial bracts encircle the inflorescence. They are spreading to patent, possess dentate margins, and measure 1.8-2.5 mm long x 2.2-3.2 mm wide (Kumar and Udar 1976).

Approximately 20-30 archegonia are produced on the dilated terminal receptacle and are subtended by 1-2 cycles of enlarged outer bracts which are erect, and generally rhomboidal in outline with distinctly dentate margins. They are similar to each other and are about 2.6-4.0 mm long X 1.8-3.1 mm wide X 1.7-1.8 mm to the broadest portion of the leaf. This description concurs with that of Kumar and Udar (1976). The reduced inner bract is lingulate, dentate, and about 0.8 mm long X 0.3 mm wide X 0.3 mm to the broadest region of the leaf.

Kumar and Udar (1976) considered the dentate bract margins and the antheridial positions as important species characters. However, as already discussed, the antheridial positions are likely the result of apical proliferations as in *H. gibbsiae*. In addition, the dentate margins are not uniformly expressed and are also displayed by *H. gibbsiae*, *H. mnioides*, and some individuals of *H. blumii* and *H. andinum*.

Haplomitrium grollei (Figures 210-211). This taxon is known only from female plants. The shoots are erect, 10-22 mm tall, 1.3-1.6 mm broad in vegetative areas, and 3.0-4.5 mm broad at the gynoecia including the bracts. The aerial axes possess 4-6 cycles of leaves which are clustered distally. According to Kumar and Udar (1977), the leaves are transversely inserted, but the shoot symmetry is anisophyllous. The plants examined in this investigation generally appear to be isophyllous or only slightly anisophyllous. The leaves are elliptical to ovate to subrotund with undulate to shallowly lobed margins. They are about 1.2-2.6 mm long X 1.0-1.7 mm wide X 0.5-1.0 mm to the broadest portion of the leaf. According to Kumar and Udar (1977), the leaves are unistratose with 3-4-stratose basal fields. A lack of adequate specimens precluded sectioning of the leaves, so this report could not be tested.

Eighteen to 20 archegonia are grouped into a well-defined terminal inflorescence which is subtended by 1-2 cycles of large bracts (Fig. 210). Although all three bracts of a series are longer than broad, one is often ovate to rhomboidal, while the other two are ovate-lingulate. The former is about 4.0-4.4 mm long X 3.0-3.3 mm wide X 1.2-1.5 mm to the broadest portion of the leaf. The latter are about 4.2-5.0 mm long X 1.7-2.2 mm wide X 3.2-3.5 mm to the widest portion of the leaf. The margins are shallowly lobed, angular, and often dentate. Kumar and Udar (1977) reported a dorsal bracteole which is exceptionally small (2.3-3.0 mm long X 1.2-1.9 mm wide to 3.5-3.8 mm long X 2.4-3.3 mm wide) and with a variable shape described as lingulate, rhomboidal, and sometimes irregularly lobed. The only bract that approaches this description is the inner reduced bract just described. This leaf is lingulate-triangular and measures about 2.8 mm long X 1.0 mm wide X 2.1 mm to the broadest region of the leaf.

Although the shoots just described are at least seemingly acrogynous, other shoots exist in which apical proliferation has occurred resulting in swollen gynoecial nodes which alternate with internodes composed of 2-3 cycles of leaves (Fig. 211). Kumar and Udar (1977) pointed out that these shoots are reminiscent of *H. intermedium*, but that *H. grollei* differs in its larger size, in not being leafy throughout, in possessing larger leaves with 3-4-stratose bases, and in lacking terminal branching. In addition, the fertile nodes of *H. grollei* typically possess 2 or 3 cycles of enlarged, imbricate, dentate bracts. Only a single series of bracts delimits the nodal inflorescences of *H. intermedium*. Based on female gametophytic features, then, *H. grollei* is indistinguishable from *H. gibbsiae* (Table VI).

Haplomitrium kashyapii (Figure 212). Udar and Kumar (1982) have described both male and female *H. kashyapii* plants, but, only a few female shoots were made available for this study. Therefore, the following description of the male gametophyte has been derived from that of Udar and Kumar (1982). These workers stated that H. kashyapii is dioicous, but not "distinctly" heteromorphic. The male gametophores are (9-) 11-24 (-30) mm tall and 2.0-4.0 (-5.5) mm wide at the androecia (Udar and Kumar 1982), while the female shoots are about 13-25 mm tall and 2.0-4.0 mm wide at the gynoecia. The larger shoots are leafy throughout, i.e., there are more than five cycles of more or less evenly distributed leaves present (Fig. 212), while smaller plants possess only 2-3 cycles of leaves which are crowded distally. Symmetry is isophyllous to slightly anisophyllous and the leaves are transversely inserted and erect to erect-spreading. Although leaf shape and size are variable, the leaves are generally suborbicular to orbicular or rhomboidal and measure 2.0 X 2.0-2.7 mm.

Udar and Kumar (1982) stated that several antheridial patterns exist. Large plants possess either a single terminal androecial rosette, or two rosettes as a result of apical proliferation. In the smaller plants, the antheridia are clustered within very large bracts at the apex of the shoot. There are no legitimate reasons for interpreting these "antheridial patterns" as distinct from one another. The antheridia of H. kashyapii are simply organized into well-defined, dilated terminal discs with apical proliferation continuing the growth of the shoot. The perigonial bracts are enlarged, erect spreading, broader than long, of indefinite shape to broadly rhomboidal, and measure about 3.2-3.6 mm long X 3.9-4.3 mm wide. Tiny, lingulate to narrowly lanceolate inner bracts occur within the rosette.

Although only two poorly preserved female shoots were available for study, they did demonstrate a range of archegonial positioning which is similar to that described by Udar and Kumar (1982). One shoot bore archegonia on a dilated, terminal disc, while the second had archegonia scattered over the stem, either remote from or in association with a bract. According to Udar and Kumar (1982), the scattered pattern is the most common, with archegonia only rarely aggregated terminally. The perichaetial bracts are enlarged (ca. 3.0-5.7 µm long X 1.8-4.5 µm wide), broadly ovate to rhomboidal, and have undulate or dentate margins (Udar and Kumar 1982). Due to the depauperate condition of the plants, these measurements could not be verified, nor could inner reduced bracts be observed. In addition, the number of archegonia located on the terminal disc could not be accurately determined.

The bracts and the gametangial positioning are again reminiscent of H. gibbsiae. Although no field grown H. gibbsiae plants are known to possess scattered gametangia, when male plants were grown under artificial environmental conditions, the antheridia were produced in a scattered fashion similar to that seen in field grown H. kashyapii plants (see experimental section).

A comparison of the characters listed in Tables V and VI reveal that the gametophores of H. kashyapii, H. dentatum, H. indicum, and H. grollei share many features with H. gibbsiae. In fact, the males of H. kashyapii, H. dentatum, and H. indicum are virtually indistinguishable from those of H. gibbsiae. The males of each are similar in shoot size, symmetry, and leaf characters. They have numerous antheridia (30-50 to 125) arranged in well-defined terminal rosettes with perigonial bracts that are enlarged, spreading and typically broadly ovate-rhomboidal and lobed, often with dentate margins. Intra-androecial bracts are present and each taxon possesses the potential for apical proliferation which results in additional antheridial rosettes.

Unlike the condition in H. gibbsiae males, the larger male shoots of H. kashyapii are leafy throughout. However, the existence of smaller shoots with only 2-3 cycles of distally clustered leaves suggests that this feature is likely modified by environmental conditions. In fact, as will be discussed later, H. gibbsiae plants also possess the potential to become leafy throughout when placed on moistened vermiculite under controlled conditions.

Therefore, at least in this instance, this does not appear to be a valid character for distinguishing between the males of these taxa.

The females are also quite similar. With the possible exception of *H. kashyapii*, the other Indian endemics and *H. gibbsiae* all possess an average of 20 archegonia aggregated into well-defined terminal rosettes. Each taxon displays similar shoot habits, symmetries, and leaf characters. However, there are some differences in the sizes of the shoots, as *H. indicum*, *H. dentatum*, and *H. grollei* tend to be smaller than *H. kashyapii* and *H. gibbsiae*. Known gametangial characters suggest that these taxa are extremely closely related and make their status as separate species doubtful.

Haplomitrium mnioides (Figures 213-218; 219-220). Unlike the prostrate to obliquely ascending vegetative shoots of *H. mnioides*, fertile shoots are erect and possess distinct terminal gametangial rosettes (Figs. 213, 217). The absence of any remnant of an anisophyllous prostrate portion on reproductive shoots indicates that these axes were morphogenetically determined to be fertile at the time of their origin.

Superficially, the male and female shoots of *H. mnioides* are very similar. Both are erect and bear leaves which are often distantly spaced except near the apex where often the upper 1 or 2 cycles of leaves become enlarged and imbricate (Figs. 213, 217). The shapes and insertions of the leaves are as on vegetative shoots. However, the anisophylly of the vegetative shoot is often less distinct in the erect, fertile axes in that the dorsal leaves are typically less reduced. In fact, the sizes of the dorsal leaves are quite variable. In some cases, these leaves are only slightly broader than the stem, while in other cases they may be almost as large as the lateral leaves.

As is apparent from a comparison of the habit information in Tables V and VI, the female gametophore tends to be just slightly larger and more robust than the male. However, the gynoecia are distinctly larger than the androecia. The dilated terminal disc bears about 20-35 or more gametangia with interspersed unicellular slime cells, 2-3-celled slime hairs, and minute bractlets (Figs. 214, 215, 218a, 218b, 219). One, or occasionally 2 rings of enlarged, broadly subrotund, bracts subtend the gametangial cluster and overarch the young sex organs (Fig. 214). Eventually, these bracts become erect-spreading and expose the mature gametangia. The perigonial bracts are usually similar to one another in shape and size, measure approximately 2.5-3.2 mm long X 3.0-3.4 mm wide X 0.9-1.5 mm to the broadest portion of the leaf, and the margins are often undulate or dentate. The inner reduced bracts are more lingulate than the outer bracts and possess lateral projections with terminal slime cells (Fig. 215).

The rosette of perichaetial bracts is composed of 1 or sometimes 2 cycles of large peripheral bracts and 1 or 2 reduced inner bracts (Figs. 218a, 218b). Usually, the outer bracts are isophyllous, but occasionally a slightly smaller dorsal bracteole is distinguishable. Like the male bracts, the outer bracts are subrotund to nearly rhomboidal with blunt angles (Fig. 218a), but the perichaetial bracts are larger and measure approximately 3.2-3.9 mm long X 3.3-3.6 mm wide X 1.7-2.1 mm to the broadest portion of the leaf. When distinguishable, the dorsal bracteoles are 2.9-3.4 mm long X 2.7-3.0 mm wide X 1.3-1.5 mm to the widest region of the leaf. Schuster (1967) has also discussed the anisophylly of the gynoecium and has given similar measurements for the outer bracts and dorsal bracteole. The inner bracts are similar to those described in H. gibbsiae females. The larger inner bract is narrower than the outer bracts and is about 1.0-2.1 mm long X 0.6-0.8 mm wide X 0.7-1.0 mm to the broadest portion of the leaf (Fig. 218a), while the smaller one is triangular and measures about 0.3 mm long X 0.2 mm wide X 0.09 mm to the broadest region of the leaf (Fig. 218b). The lateral projections on all of the bracts are tipped with slime cells.

Apical innovations arise from the center of the old antheridial rosettes (Fig. 216) and transections through mature gametangial discs reveal apical cells positioned centrally within the receptacles (Figs. 220a, 220b). The apical cells are about 20 µm in width, 30 µm in depth, and 44 µm in height. These dimensions are well within the range of the vegetative generative cell. This implies that this cell is not destroyed by gametangial production so that the potential for renewed apical growth after gametangial production exists. A similar potential for proliferation exists in female shoots.

Haplomitrium blumii (Figures 221-222; 223-227). As reported by Udar et al. (1968) sexual dimorphism occurs in this taxon. The female plants are more robust than the males and possess larger terminal discs (Tables V, VI). The erect fertile shoots are similar to their vegetative counterparts, except they are terminated by well-developed, enlarged gametangial rosettes, the stem leaves are widely spaced, and the anisophyllous symmetry is often not as obvious as in sterile axes. Like H. mnioides, H. blumii displays a wide array of dorsal leaf sizes. The dorsal leaves range from being barely as broad as the stem to being almost as large as the lateral leaves. In addition, sometimes a mixture of different sized dorsal leaves exist on the same shoot. For example, one or two extremely reduced dorsal leaves may occur on a shoot whose dorsal row is composed mostly of leaves which are broader than the stem.

Approximately 80-105 antheridia comprise the terminal androecium (Figs. 223, 225). The enlarged, outer perigonial bracts are often isophyllous, but anisophyllous rosettes exist in some individuals. The outer bracts are similar to, but larger than the vegetative leaves, usually with undulate margins, or are rarely shallowly lobed (Fig. 225). One or 2 reduced, narrowly rhomboidal inner bracts occur within the confines of the outer bracts. However, as first pointed out by Goebel (1891) and later by D. H. Campbell (1920), no additional rudimentary bractlets are interspersed among the antheridia. However, 1-2-celled slime papillae are intermixed with the antheridia (Fig. 224).

The terminal gynoecium is about 6.5-7.4 mm in diameter including the bracts and contains approximately 80-100 archegonia (Figs. 227, 228). Fulford et al. (1958) and Schuster (1967) reported as few as 30 archegonia per disc. Uniseriate slime hairs composed of 3-12 cells, unicellular and single-stalked slime cells, and rudimentary bractlets occur amidst the female gametangia (Figs. 221, 222, 228). In contrast, Goebel (1891) reported only slime papillae within the gynoecium.

Like the male bracts, those of the female rosette may be isophyllous or anisophyllous. However, Udar et al. (1968) noted that anisophylly is usually more common in gynoecia than androecia. The large, outer perichaetial bracts of the gynoecium are subrotund with entire or undulate margins and are about 4.0-4.8 mm long X 3.9-4.1 mm wide X 2.3-3.0 mm to the broadest portion of the leaf. When present, the smaller dorsal bracteole is narrower and approximately 3.7-4.0 mm long X 1.9-2.8 mm wide X 1.0-1.3 mm to the widest portion of the leaf. A reduced inner lateral bract exists which is lingulate-triangular and measures about 1.5-2.0 mm long X 0.5-1.0 mm wide.

No apical proliferations were observed arising from the gametangial discs, nor were any apical cells found in the mature receptacles. Therefore, as suggested by Goebel (1891, 1930), D. H. Campbell (1920), Smith (1955), and Schuster (1967), the apical cell may be eliminated by the prolific production of gametangia.

<u>Haplomitrium andinum (Figures 228-230)</u>. Sexual dimorphism appears to be absent in this species as the shoots of both sexes display approximately the same dimensions (Tables V, VI). Like the sterile shoots,

the reproductive axes are erect and anisophyllous. Leaf insertions, morphologies, and dimensions are also similar to those of vegetative shoots (Fig. 228). As in the other species of this complex, the stem leaves are widely spaced except for the bracts which are crowded around the terminal gametangial disc (Fig. 228). As in H. mnioides and H. blumii, dorsal leaf size is variable. The leaves of this row may be narrower than the stem, as broad as the stem, or almost as large as the lateral leaves. In some individuals, the most acroscopic dorsal vegetative leaf is obviously more reduced than the other dorsal leaves (Fig. 228), but this does not appear to be constant on all fertile shoots.

Twenty-five to 40 or more gametangia are produced on the terminal receptacles (Figs. 229a, 229b, 230) and are subtended by 1 to 2 cycles of enlarged bracts. The outer bracts of both the androecia and gynoecia may be isophyllous or anisophyllous. The published descriptions by Fulford (1962) and Schuster (1967) illustrate this variability as the former author noted bracts and a dorsal bracteole, while the latter described the lateral and dorsal bracts as being identical. The dorsal bracteole of anisophyllous rosettes (which measures about 1.8-2.1 mm long X 1.0-1.8 mm wide X 1.0-1.2 mm to the broadest part of the leaf) can be distinguished from the lateral bracts (which are about 2.5-3.0 mm long X 1.9-3.2 mm wide X 1.4-1.9 mm to the widest portion of the leaf) by its reduced size. The outer perigonial bracts are usually similar to the vegetative leaves but larger, while the outer perichaetial bracts may be obovate to elliptical with margins that are entire, undulate, or dentate. The reduced inner bracts are lingulate with undulate to shallowly lobed margins, and measure about 1.0-1.6 mm long X 0.4-0.6 mm wide X 0.2-0.4 mm to the broadest portion of the leaf. Rudimentary bractlets, slime papillae, and 2-3-celled slime hairs are dispersed among the gametangia. No apical proliferation was observed.

Haplomitrium giganteum (Figure 231). This taxon is known only from three female shoots which were originally described by Spruce (1885) under Tylimanthus Mitten and later discussed by Grolle (1963) and Schuster (1967). The shoots are erect and measure 26.0-60.0 mm tall X 6.1-6.3 mm broad at the midregion of the shoot, expanding to 7.0-9.0 mm wide at the gynoecium including the bracts. The anisophylly of the shoot is unquestionable. The dorsal leaves are more reduced than in any other taxon (Fig. 231). The lateral leaves are succubously inserted, slightly decurrent on the stem and vary in outline from obovate to obcuneate (Figs. 107a, 107b, 108). They are approximately 2.6-3.5 mm long X 2.7-3.2 mm wide X 1.8-2.0 mm to the broadest portion of the leaf. The laminae are flattened and appear to be unistratose to the base, although sectioning was prohibited by the scarcity of plant material. Grolle (1963) and Schuster (1967) also deemed the leaves to be entirely unistratose. The dorsal leaves are always transversely inserted, narrower than the stem, bluntly triangular to lingulate in outline, and measure approximately 855-2070 µm long X 630-1282 µm wide X 405-1035 µm to the widest region of the leaf (Fig. 231). The minuteness of the dorsal leaves and their consequential adherance to the stem confer an overall flattened appearance to the shoot.

Twenty-five to 40 or more archegonia are produced on the terminal disc. The subtending rosette of outer bracts is clearly anisophyllous. The perichaetial bracts are narrowly obovate to lingulate with entire to undulate margins. The outer lateral bracts are larger than the stem leaves and measure about 5.0-6.4 mm long X 2.6-3.4 mm wide X 2.5-3.5 mm to the broadest portion of the leaf, while the dorsal bracteole is more reduced and measures about 3.4-4.9 mm long X 1.1-1.5 mm wide X 2.2-2.5 mm to the widest portion of the leaf. Reduced inner bracts are also present. The larger of these is lingulate and about 1.2-2.9 mm long X 0.4-1.0 mm wide, while the smaller is lingulate-triangular and only about 1.1 mm long X 0.4

mm wide X 0.9 mm to the widest region of the leaf. Single-stalked slime papillae are associated with the archegonia, but whether or not slime hairs or bractlets occur is unknown. No apical proliferation from the gynoecium is evident.

Steereomitrium minutum (Figures 232-234). The only description of S. minutum is that of E. O. Campbell (1987). This taxon is only known from fertile shoots which, according to Campbell (1987), are usually dioicous, but occasionally are monoicous with antheridial shoots arising as basal branches from archegonial shoots. Only dioicous individuals were seen in this investigation. Extreme morphological variability characterizes this taxon. The shoots are erect and range from 3-20 mm tall depending on the environmental conditions (Campbell 1987). Campbell (1987) correlated the taller sizes of some plants with their constant exposure to high humidity. The male shoots examined in this study measured about 10.0-12.0 mm long X 0.9-2.0 mm broad. No unfertilized female shoots were seen, but those with calyptrae were found to be about 6.0-8.0 mm long, excluding the length of the calyptrae. Campbell (1987) stated that male and female shoots are initially indistinguishable, but ultimately, the terminal part of the female stem broadens and the leaves widen. The shoots bear approximately 7-10 cycles of leaves which may be distributed subequally along the axis (Fig. 232a), or may be crowded toward the apex of the shoot (Fig. 233). The symmetry of the shoot ranges from isophyllous with all three rows of leaves transversely inserted, to slightly anisophyllous, with the dorsal leaves being transversely inserted, slightly smaller and more lingulate than the slightly succubous lateral leaves. As pointed out by Campbell (1987), it is difficult to determine the morphologies of the leaves since they are usually concave and squarrose and may be folded medianly. Generally, however, the leaves are obovate, ovate, elliptical, or subrotund, with bluntly acute, acuminate, or cuspidate apices and lobed or undulate margins (Fig. 234). They are about 2.0-2.6 mm long X 1.2-1.4 mm wide X 1.3-1.6 mm to the broadest portion of the leaf. The dorsal leaves, as well as the smaller leaves toward the base of the shoot are usually ovate to subrotund and entire. The unistratose laminae become bistratose near the base and tristratose at the point of insertion (Campbell 1987).

The antheridia are usually axillary or supra-axillary and are either aggregated into a loose apical cluster (Fig. 233), or scattered over the stem (Figs. 232a, 232b). The terminal clusters consist of 25 or more antheridia which are protected by two series of enlarged bracts (Fig. 232b). When dispersed along the stem, the antheridia originate in groups of three or more in the axils of stem leaves (Fig. 232b). The groups of the antheridia are associated with 1-celled slime papillae, but it is unknown whether or not bractlets and slime hairs are interspersed with the terminal antheridia since material was too scarce to section.

According to Campbell (1987), the archegonia are produced successively between leaves in a terminal cluster and are subtended by a rosette of closely crowded immature leaves. She also noted that slime hairs are associated with the archegonia. Apical proliferation occurs in the absence of fertilization.

From the available information, it is impossible to determine which morphologies and gametangial positions are "natural" for this taxon and which have been induced by the effects of artificial culture conditions. However, the gametangial positioning appears to be most like that of _H. hookeri_ and _H. ovalifolium_.

According to Schuster (1966, 1984a), the most primitive gametangial positioning within the Hepaticopsida is that in which "naked" gametangia are produced in acropetal sequence on the axis. The slightly more derived state is characterized by the production of gametangia in the axils of unspecialized bracts, and finally, through the aggregation of gametangia into specialized, protected androecia or gynoecia. This entire sequence is represented within the genus Haplomitrium and sequences of the evolutionary progression are even demonstrated within a species. Although rare, certain individuals of H. intermedium and H. hookeri do possess free-standing gametangia. In addition, some individuals of both these taxa exhibit the more advanced feature of having a small number of gametangia (2-6 antheridia, 1-3 archegonia) protected in the axils of bracts which are otherwise indistinguishable from vegetative leaves. This latter stage is also represented in H. monoicum which produces gametangia singly or in pairs in, or just above the axils of unmodified leaves. Still other H. intermedium individuals display the more advanced condition of having axillary gametangia produced in ill-defined rosettes complete with swollen stem tissue and enlarged bracts. In this case, internodal reduction occurs alternately with apical proliferation to produce the nodal inflorescences. Thus, the gametangial positions exhibited by H. intermedium form the bridge between species bearing scattered or loosely grouped gametangia and those with defined terminal rosettes. Extreme internodal reduction occurs in the derived members of the subgenus Calobryum. Here, axillary antheridia are produced in a cluster on a swollen terminal receptacle, all subtended by a ring of enlarged bracts. Within this complex, Schuster (1966) regarded those species displaying apical proliferation (H. gibbsiae and H. mnioides) as the least derived since they possess the potential for subsequent internodal elongation, and those in which apical proliferation appears to be absent (*H. blumii*, *H. andinum*, and *H. giganteum*) as the most derived.

In erect bryophyte taxa, there seems to be strong selection toward the inhibition of apical shoot growth subsequent to gametangial production (Schuster 1984b). This is especially apparent in female shoots and results in functional acrogyny. In addition to the metabolic demands of gametangial production, additional nutritional demands are placed on the female shoot after fertilization by the dependency of its sporophyte. These are best met if each shoot is limited to the production of only one sporophyte. The chances of meeting this condition are improved if only one series of archegonia is available for fertilization. As explained by Schuster (1984b), when successive gynoecia are produced, each is positioned higher than the next, thus reducing the chances for fertilization, and, if fertilization does occur, increases the problems of dessication and nutritional maintenance of the sporophyte. Therefore, acrogyny is selected for in erect taxa and simple acropetal development of archegonia is selected against (Schuster 1984b). The same selection pressures are not as strong in male shoots since the antheridia degenerate after releasing spermatozoids (Schuster 1984b). This further reinforces the presumption that within the genus, *H. hookeri*, *H. monoicum*, *H. ovalifolium*, and *H. intermedium* are the most ancestral species, while *H. blumii*, *H. andinum*, and *H. giganteum* are the most derived, since they are functionally acrogynous. These selection pressures are also apparent within some of the more plesiomorphic species of *Haplomitrium*. For example, in the male shoots of *H. hookeri*, antheridial positioning is highly variable, but, in female shoots, most individuals have archegonia clustered in loose terminal groups as opposed to being scattered acropetally. In *H. intermedium*, the entire realm of potentials is expressed, from simple acropetal archegonial development to ill-defined terminal inflorescences with repeated apical proliferation, to functional acrogyny.

Although the production of dilated terminal gametangial discs is considered to be the most derived state within the genus Haplomitrium, it is still a relatively primitive condition. This is evidenced by the fact that similar terminal bowl- or disc-shaped gametangial receptacles were produced by Lower Devonian gametophytes such as Sciadophyton Steinm. and Lyonophyton Remy & Remy (Remy and Remy 1980a, 1980b; Remy et al. 1980; Schweitzer 1981; Remy 1982). According to these workers, these plants also displayed other characters which are reminiscent of fertile Haplomitrium plants. Both Sciadophyton and Lyonophyton were bushy plants with a large number of erect gametangiophores radiating from a central point. The axes of Lyonophyton even had scalelike enations covering the erect axes, possessed a central strand of hydroids, and a few upturned lobes, which possibly represent involucral bracts, formed the margins of the disc-like receptacles. The antheridia were large, globose or clavate, with massive, multicellular stalks. Sometimes multicellular, branched paraphysis-like sterile tissue was intermixed with the antheridia on the disc. The archegoniophores were constructed similarly. In contrast, these plants differed from Haplomitrium in possessing a cuticle, stomata, biseriate antheridial jackets, having occasional sunken gametangia, and, in Sciadophyton, possessing a central strand of genuine tracheids (Remy et al. 1980; Stewart 1983). Based on these traits, Remy (1982) hypothesized that although these Devonian plants might not be the direct ancestors of extant bryophytes, they might be remnants of a plant group which preceded both bryophytes and pteridophytes. In any case, the fact that Haplomitrium and these ancient gametophytes bear numerous similar characters is evidence supporting the concept that Haplomitrium is a plesiomorphic taxon.

The gametangial positions displayed by Haplomitrium are reminiscent of those of the

Bryopsida. Here, the gametangia are either axillary as in Sphagnum L. and Polytrichum Hedw., or are produced in terminal rosettes (Crandall-Stotler 1984; Schofield and Hébant 1984). Like Haplomitrium, some moss species produce up to 100 antheridia or 60 archegonia within these terminal rosettes (Schofield and Hébant 1984). Another similarity with mosses is the common occurence of paraphyses or slime hairs which are intermixed among the gametangia (Schuster 1966, 1984a; Saito 1975; Crandall-Stotler 1984; Schofield and Hébant 1984).

In the Jungermanniales, gametangial positioning varies with the sex of the plant. In male plants, the antheridia are axillary, while female plants produce clusters of archegonia at the apices of shoots, i.e., the female shoots are acrogynous (Schuster 1966, 1984a; Watson 1971; Crandall-Stotler 1981, 1984). In contrast, a variety of antheridial positions are exhibited within the Metzgeriales (Schuster 1966; Crandall-Stotler 1981; Renzaglia 1982). An evolutionary progression similar to that found in Haplomitrium also occurs within the order (Schuster 1966). However, the thalloid nature and prostrate habit exhibited by most Metzgerialian taxa presented different evolutionary "problems" and demanded different "solutions". Therefore, instead of deriving protective structures from leaves, they were derived from the thallus itself and the only semblence of functional acrogyny occurs on the specialized sexual branches of the most derived taxa (Renzaglia 1982).

The gametangia of Takakia are produced singly or rarely in groups of 2-3. They may be axillary or irregularly dispersed with no apparent relationship with a leaf. No slime cells or bractlets are associated with the gametangia (Schuster 1967, 1984b; McFarland et al. 1989).

Antheridial Ontogeny

The antheridia of Haplomitrium are generated close to the apical cell (Figs. 235-239) in all three merophyte rows. They originate from cells derived from the central, wedge-shaped portion of the derivative, which in the vegetative aerial shoot, is exclusively responsible for leaf production (Figs. 235-236). Therefore, a leaf which is formed from the same merophyte as the antheridium may or may not be modified depending on when the antheridial initial is formed. For example, if the antheridial initial is delimited after some divisions have already occurred in the wedge-shaped cell of the derivative, i.e., after the leaf initial has already been formed, then the leaf is unmodified by antheridial production. In contrast, if the antheridial initial is delimited before the leaf initial is created, then the antheridial initial may include a portion of the derivative which would otherwise be involved in leaf production, thus resulting in a reduced leaf, papillae, or in extreme cases, no leaf at all depending on the number of antheridia produced from that merophyte (Figs. 237-238). One, 2, or 3 antheridia may form from a single segment according to one of the following patterns: leaf-antheridium-leaf (Figs. 237b), antheridium-antheridium-antheridium (Fig. 239), leaf-antheridium (Fig. 238b), antheridium (Fig. 238a), or antheridium-antheridium. In the case of leaf-antheridium-leaf, the anodic and kathodic portions of the segment produce leaf tissue, while the central portion forms an antheridium. When three antheridia are produced, the entire segment is used for gametangial construction, and when only a single leaf segment and antheridium are formed, the gametangium usually forms from the kathodic portion of the segment (Figs. 238a-c). When only a leaf and one antheridium or only two antheridia are formed, the portion of the segment not used for leaf or gametangial production yields bractlets, slime hairs, and/or slime papillae depending on the size of the unused portion. This explains the mixture of antheridia and slime hairs, bractlets, and slime papillae within the androecia.

These patterns of gametangial formation hold for all species regardless of androecial morphology. The positioning of the gametangia on the plant is controlled by which derivatives produce gametangia, the rate of apical segmentation, and the rate and extensiveness of merophyte development. If internodal growth is suppressed while the apical cell is still segmenting, and gametangia are produced from every newly formed merophyte, a terminal, swollen receptacle of stem tissue, upon which occur antheridia of various ages, will develop. Although most gametangia mature at approximately the same time, some younger antheridia are also present both among the older ones and at the periphery of the disc (Fig. 215). Basically, antheridial production is synchronous, i.e., antheridial differentiation is delayed until all of the antheridial initials of the disc have been formed and segmentation is complete. However, as antheridial differentiation progresses, the superficial cells of the receptacle continue to divide and some, particularly those located near the periphery of the disc, give rise to secondary antheridial initials. The antheridia which arise from these secondary initials are younger than those produced from the first set of initials and account for the presence of antheridia of various ages on the disc.

If internodal growth accompanies segmentation and gametangia are not produced in every cycle, they will be distributed along the stem, in or near the axils of leaves. As previously mentioned, the leaves which are associated with these antheridia are not modified because the leaf initials were delimited prior to the antheridial initials. If 1, 2, or 3 antheridia are produced in all three merophytes of a single cycle, the gametangia will appear to form a cluster of 3-9 antheridia around the stem at a single level as sometimes occurs in _H. hookeri_. If segmentation increases for a time while gametangial production occurs, and the internodal length is reduced, stem tissue will accumulate at the node as in *H. intermedium*. A return to "normal" internodal growth rates produces the elongated internodal region between androecia. If internodal lengths remain constant, and gametangia are not formed from all three merophytes of a given cycle, leaves will be produced from intervening series and the antheridia will be scattered among them as in *H. monoicum* and some individuals of *H. hookeri* and *H. intermedium*. On the other hand, if antheridia are produced from a few consecutive cycles, then loosely clustered, ill-defined gametangial inflorescences result such as in *H. intermedium*.

The association of the gametangia with modified leaves and/or slime papillae as a result of the replacement of a portion of the leaf by gametangial production, and the potential for gametangial construction in all three merophyte rows was first described in *H. hookeri* by Leitgeb (1875) and later by Lilienfeld (1911). In addition, Goebel (1891) realized that in the rosettes of *H. blumii*, the development of the small leaves inside the three large outer peripheral bracts was stopped by the formation of the gametangia. However, Goebel (1891) found only slime cells between the gametangia of *H. blumii* as opposed to the rudimentary bractlets and slime hairs reported in *H. hookeri* (Leitgeb 1875). He considered this to be a trait which separated the genus *Haplomitrium* from *Calobryum* (Goebel 1891). Actually, however, since terminal disc formation is coupled with the production of the maximum number of gametangia in each apical segment, virtually the entire initial is involved in gametangial construction and no portion remains available for slime hair or bractlet formation. Thus, the lack of these structures does not appear to be of great consequence.

Only the origin of the free-standing antheridia such as those reported in *H. hookeri* (Gottsche 1843, Evans 1917; Schuster 1967) and *H. intermedium*

(Schuster 1967) remains uncertain. Schuster (1967) suggested that this free-standing antheridium arises from a cortical cell of the stem. It is more likely that their distant positioning in relation to leaves is also controlled by levels of internodal elongation above and below their point of origin near the apex. However, due to the scarcity of plants bearing free-standing antheridia this claim could not be substantiated.

Serial sections through the androecia of H. monoicum, H. intermedium, H. gibbsiae, and H. mnioides indicate that antheridial construction is uniform among the species and is unique among known bryophytes (Figs. 240-259). The formation of the antheridium itself begins with the swelling of a superficial cell derived from the central wedge-shaped cell of a merophyte. A transverse wall soon delimits this cell from the axis (Fig. 240a), after which a second transverse division partitions it equally into an outer primary antheridial initial and an inner stalk initial (Fig. 240b). The former is destined to become the body of the mature antheridium, while the latter will become the stalk.

Usually, meristematic activity first commences in the antheridial initial. This cell is traversed by a curved vertical wall to form two cells (Figs. 241, 242). At this point, one of two developmental sequences may be followed to produce a single androgonial initial surrounded by three jacket initials. In some cases, two vertical walls which are parallel to the first, but oppositely curved, traverse the antheridial initial to yield a single, pyramidal central cell and three jacket cells (Figs. 244, 245). The vertical walls are then connected by a transverse division to delimit a fourth outer cell, i.e., the cover cell (Figs. 246, 247). In other cases, two curved vertical walls intersect the first wall to produce a central pyramidal cell surrounded by three jacket cells (Fig. 248). One of the three jacket cells is then partitioned to yield four jacket cells (Figs. 245, 248). From this stage on, a single pattern occurs. A series of transverse and vertical divisions build a unistratose jacket and allow further expansion of the central cell (Fig. 249). The enlarged androgonial cell is cut in two by a curved oblique-transverse wall (Fig. 250) to form upper and lower cells. Each of these is then divided by a wall perpendicular to the first wall to generate two tiers each composed of four quadrants (Figs. 251-255). Repeated synchronous vertical and transverse divisions cut the internal quadrants to produce the spermatogenous region of the mature antheridium (Figs. 256, 257). The surrounding uniseriate jacket cells keep pace with the enlarging central region through planar divisions in each cell (Fig. 257).

The stalk initial is first partitioned into two cells by a slightly curved vertical wall (Fig. 243). This division may occur just after the first division of the antheridial initial, or it may be suppressed until after the central cell has been formed (Figs. 246-248). In either case, each of the newly formed stalk cells is cut transversely to produce four cells in cross section (Figs. 258, 259). Repeated transverse divisions in each of the four rows of cells increases the length of the stalk by generating multiple tiers, each of which is composed of four cells (Figs. 251, 256). The number of tiers comprising the stalk is usually 3-4, but varies from 2 to as many as 10 depending on the species. Stalks which are relatively long often become twisted as the number of tiers increases.

Leitgeb (1875) reported that the development of the antheridia of H. hookeri followed the same pattern as that found in the Jungermanniales. Although E. O. Campbell (1959) and Mehra and Kumar (1990) recognized only one sequence in which a pyramidal central cell and three jacket cells are produced from two curved vertical walls which meet the first wall above, D. H. Campbell (1920) recognized the variability in the pattern and described the pyramidal androgonial cell and three jacket initials as being formed either by three eccentric intersecting divisions, or by three vertical walls followed by a fourth division which delimits the androgonial cell and cuts off a "cap cell".

Antheridial ontogeny in Haplomitrium is similar to that of all other hepatics in that no apical cell is involved in gametangial construction, while differing in this respect from the Bryopsida. However, unlike all other hepatics no filament forms at the onset of antheridial development in Haplomitrium. Haplomitrium differs further in possessing the potential for creating the androgonial initial by two different sequences of divisions. The first pattern in which three almost parallel vertical walls cut the antheridial initial is similar to the pattern of archegonial development in hepatics. However, the sequence in which the antheridial initial is traversed by three intersecting divisions is unique among the bryophytes and has been termed the Haplomitrium-type by Schuster (1966, 1984a).

In the Jungermanniales and Metzgeriales, the antheridial initial is first cut by a vertical wall followed by two oblique divisions which intersect distally to produce two androgonial initials and four jacket initials (Schuster 1966, 1984a; Crandall-Stotler 1981). In contrast, the antheridial initial of the Marchantiidae is cut into a quadrant of four cells, then divides further to yield four spermatogenous initials and four jacket initials (Schuster 1966, 1984a). The pattern of antheridial ontogeny in the Anthocerotophyta is similar to that of the Marchantiidae, but, the antheridia have a hypodermal origin and are subsequently submerged within the thallus tissue (Renzaglia 1978; Schuster 1966, 1984a). Of these patterns, Schuster (1984a) considered that of Haplomitrium to be the most primitive.

Nothing is known about antheridial ontogeny in Takakia. However, a similarity does exist in that each antheridium presumably originates near the apical cell from one of the cells usually involved in leaf formation (McFarland et al. 1989, paper presentation).

Structure of the Antheridium

The typical mature antheridium of Haplomitrium and Steereomitrium consists of a body that is ovoid or ellipsoid with a central spermatogenous region enclosed within a uniseriate jacket. The body is supported by a tiered stalk composed of four cells in cross section (Figs. 260-270). Slime papillae are produced at the base of the stalk in all species.

As illustrated by longitudinal sections through the antheridia of H. gibbsiae and H. mnioides, the fine structure of the jacket cells and the young spermatogenous tissue is generally similar to that of the vegetative cells of Haplomitrium except that no oil bodies occur in the spermatogenous cells (Figs. 273-275). The jacket is clearly unistratose and composed of cells which appear quadrate to rectangular in surface view (Figs. 260-266). Plasmodesmata connect adjacent jacket cells to create a symplasm, but, do not join the spermatogenous cells with the jacket cells. The plastids of young jacket cells house 1-2 starch grains which measure 0.4-1.0 µm long X 0.2-0.3 µm wide (Fig. 273b), while 1-3 larger grains which measure about 0.8-1.7 µm long X 0.5-1.7 µm wide occur in mature jacket cells. Numerous, irregularly shaped, small vacuoles (2.0-2.4 X 1.3-2.0 µm) are dispersed throughout the cells. Large lipid droplets (1.6-2.2 µm long X 1.3-2.3 µm wide) are scattered along the peripheries of the jacket cells and oil bodies are also occasionally present (Fig. 274).

The outer walls of the jacket cells measure about 0.2 µm thick in young stages (Fig. 273b) and increase to about 0.5 µm with age (Fig. 274). The adjoining walls of the jacket cells are thinner and measure about 0.07-0.18 µm depending on their ages. The inner walls which separate the spermatogenous zone from the jacket are just slightly thicker (about 0.09-0.20 µm depending on age) than the walls separating adjacent jacket cells.

The cells of the central portion of the young antheridium shown in Figure 273a are ultrastructurally similar to the peripheral jacket cells except they lack oil bodies. However, as discussed by Duckett et al. (1983), in later stages, these cells appear more like meristematic cells. The results of the synchronous divisions in the spermatogenous region are clearly evident in Figure 275. Each cell is virtually identical in size and shape, i.e., each is quadrate with opposite sides which measure about 7.3-7.9 µm and 9.0-9.9 µm, respectively, and is delimited by thin walls with plasmodesmata. The nuclei are typical for the genus. They are 6.0-7.0 µm in diameter, each with a central nucleolus of about 3.0 µm in diameter, often with an accompanying karyosome (Figs. 273a, 273b), and possess a heterochromatin-rich nucleoplasm. Mitochondria, plastids, and ER are peripheral to the nucleus. Pleiomorphic plastids are numerous (5-7 per cell per section) and measure approximately 1.5-1.7 µm long X 0.8-1.2 µm wide. They contain up to 5 minute osmiophilic droplets and occasionally a single starch grain (0.24-0.33 µm X 0.19-0.22 µm). As the spermatogenous cells mature, the rate of plastid divisions slows and the number of plastids decreases until only two exist in each spermatid mother cell. Ultimately this cell divides to produce two spermatozoids, each with a single plastid containing one starch grain (Duckett et al. 1982, 1983).

Elongate mitochondria, which measure about 0.67-3.0 µm X 0.32-0.44 µm, as well as dictyosomes, ER, and ribosomes are dispersed throughout each young spermatogenous cell (Fig. 275). A very small amount of the cell volume is occupied by vacuoles. In fact, a single section through a cell reveals about 6-8 small vacuoles which are only about 0.5-1.0 µm in diameter. A few peripheral lipids are present, but oil bodies are absent.

Most species of Haplomitrium possess antheridia which are ovoid or ellipsoid and have a 3-4-tiered stalk. However, some variation does exist among the species (Table VII). Haplomitrium monoicum possesses

TABLE VII. GAMETANGIAL CHARACTERS OF *HAPLOMITRIUM* AND *STEEREOMITRIUM* SPECIES

	Antheridia					Archegonia	
	Shape	Body Size (LXW; µm)	Stalk Size (LXW; µm)	# Tiers in Stalk	Color	Venter (LXW; µm)	Neck (LXW; µm)
H. hookeri	ovoid	242-340 X 196-242	69-100 X 38- 55	3- 4	orange	118-136 X 88-120	215-261 X 35- 45
H. ovalifolium[a]	ND	ND	ND	ND	yellow	ND	ND
H. monoicum	ovoid to ellip	104-155 X 80- 86	19- 24 X 26- 34	2- 3	white[b]	55- 99 X 63- 67	200-270 X 36- 45
H. intermedium	ovoid	112-155 X 99-199	46- 72 X 26- 41	3- 4	orange	95-110 X 90-101	161-219 X 52- 61
H. gibbsiae	ovoid to ellip	226-246 X 171-220	125-233 X 33- 44	3-10	white to cinnamon[c]	91-113 X 81-102	300-515 X 50- 57

TABLE VII. CONTINUED

	Antheridia					Archegonia	
	Shape	Body Size (LXW; µm)	Stalk Size (LXW; µm)	# Tiers in Stalk	Color	Venter (LXW; µm)	Neck (LXW; µm)
H. chilensis	ellip	133-387 X 93- 99	81- 99 X 46- 54	4- 5	white or yellow[a]	83- 90 X 76- 88	398-446 X 53- 56
H. indicum	ovoid[d]	ND	ND	ND	green to brown[d]	161-200 X 101-126	306-430 X 41- 60
H. dentatum	spher[e]	ND	ND	ND	orange[e]	180-211 X 131-151	414-630 X 63- 72
H. grollei	ND	ND	ND	ND	ND	180-216 X 157-162	450-459 X 54- 76
H. kashyapii	ovoid[f]	ND	ND	ND	yellow[f]	110-168 X 90-137	431-630 X 45- 56

TABLE VII. CONTINUED

	Antheridia					Archegonia	
	Shape	Body Size (LXW; µm)	Stalk Size (LXW; µm)	# Tiers in Stalk	Color	Venter (LXW; µm)	Neck (LXW; µm)
H. mnioides	ovoid	175-285 X 114-271	197-248 X 43- 66	3- 4	orange	72- 80 X 65- 72	284-296 X 56- 58
H. blumii	ovoid to ellip	175-333 X 157-207	117-252 X 27- 63	4- 5 (10)	ND	59-128 X 71- 85	358-634 X 36- 78
H. andinum	ellip	276-331 X 94-243	92-161 X 32- 37	3- 4 (7)	orange	160-180 X 126-134	361-634 X 45- 54
H. giganteum	ND	ND	ND	ND	ND	126 X 82	301 X 72
S. minutum	ovoid	260-270 X 171-252	117-201 X 36- 38	6-7	gold[g]	260-370[g]	

[a]Schuster 1971; [b]Engel 1981; [c]Campbell 1959; [d]Udar & Chandra 1961, 1965; [e]Kumar & Udar 1976; [f]Udar & Kumar 1982; [g]Campbell 1987; ellip, ellipsoid; ND, no data; spher, spherical.

the smallest antheridia of the genus. When mature, the body is ovoid to ellipsoid and is subtended by a very short stalk which is usually only 2 tiers high and measures only 19.2-24.2 µm long (Figs. 260, 261). As noted by Engel (1981), this is by far the shortest antheridial stalk in the genus. Engel (1981) also emphasized the fact that the stalk gradually widens and grades into the body of the antheridium and noted that similar dilations occur in H. gibbsiae antheridia. As seen in Figures 263-265 and 268-270, the antheridial stalks of all Haplomitrium and Steereomitrium species dilate as they join the antheridial body. However, the reduced stature of H. monoicum's stalk emphasizes this feature.

As evident from Table VII, H. hookeri (Fig. 262), H. intermedium (Fig. 263), and H. mnioides (Fig. 264) possess similar ovoid antheridia with 3-4-tiered stalks. Of these, H. hookeri possesses the largest antheridia, while of these taxa those of H. intermedium are the smallest (Table VII).

The antheridial morphology of H. gibbsiae is somewhat variable. Often, the antheridial stalks of this species are very long (Fig. 265). They typically consist of 6-10 tiers of cells and measure 125.4-233.3 µm long. However, 3-5-tiered stalks like those described by Campbell (1959) also occur, but less frequently. Similarly, the antheridial bodies of H. gibbsiae are variable, including both ovoid and ellipsoidal forms. The jacket cells are distinctly long rectangular and regularly tiered (Fig. 265). The only other Haplomitrium species with tiered, rectangular jacket cells is H. chilensis (Fig. 266). All other Haplomitrium taxa possess quadrate and irregularly arranged jacket cells (Figs. 261-264). The ellipsoid bodies and 4-5-tiered stalks which characterize H. chilensis antheridia also fall within the range of morphologies exhibited by H. gibbsiae (Table VII; Figs. 265, 266). In addition, the antheridia of H. gibbsiae are reportedly white when immature (Schuster 1967) and turn cinnamon to brownish with age (E. O. Campbell 1959; Schuster 1967; Allison and Child 1975). According to Schuster (1971), those of *H. chilensis* are ivory-white or yellowish. However, Schuster (1971) did not mention whether or not these antheridia had reached maturity. These taxa, along with *H. monoicum* are the only *Haplomitrium* species to possess white antheridia. In other species the antheridia are green when immature and yellowish to orange when mature (Gottsche 1843; Carrington 1874; Lindberg 1874; Evans 1917; Frye and Clark 1943; Udar and S. Chandra 1965; Schuster 1966, 1967, 1984a; Worley 1969; Kumar and Udar 1976; Udar and Kumar 1982).

The range of antheridial morphologies found in *H. blumii* and *H. andinum* also overlap. Both taxa possess ellipsoid antheridial bodies of similar dimensions (Figs. 268, 269), although some *H. blumii* individuals may also produce ovoid antheridia (Fig. 267). The antheridial stalks of both taxa are typically 3-5-tiered, but may be as many as 7 or 10 cells high in *H. andinum* and *H. blumii*, respectively (Table VII).

Little information exists regarding the antheridial morphologies of the Indian endemics. Apparently, the antheridial bodies of each species are globose to ovoid (Udar and S. Chandra 1965; Kumar and Udar 1976; Udar and Kumar 1982; Mehra and Kumar 1990) and at least the antheridia of *H. dentatum*, *H. indicum* and *H. kashyapii* are said to be long-stalked (Kumar and Udar 1976; Udar and Kumar 1982; Mehra and Kumar 1990).

The morphology and anatomy of the antheridia of *Steereomitrium* are as in *Haplomitrium* (Table VII; Fig. 270). The bodies are ovoid and are supported by long stalks composed of 6-7 tiers of 4 rows of cells. According to Campbell (1987), the jacket is uniseriate, gold-colored, and dehiscence is as in *Haplomitrium*.

Although *Haplomitrium* and the Bryopsida share similar antheridial positioning, their antheridial morphologies are very different. Since apical cells are involved in the formation of moss antheridia, these gametangia usually have long, narrow, bodies subtended by short stalks (Schuster 1966; Watson 1971). In contrast, apical cell activity is lacking in the construction of liverwort antheridia, so their bodies are globose to ovoid. Subtle differences do exist among the various groups of the Hepaticae, however. For example, the antheridia of the Metzgeriales are generally composed of an almost globose body and a short stalk, while those of the Jungermanniales usually possess long pedicels. In contrast, the sunken antheridia of the Marchantiales are ovoid with very short, broad stalks (Schuster 1966, 1984a; Watson 1971). The Metzgeriales and Jungermanniales commonly bear 2- or 1-seriate antheridial stalks (Müller 1948; Schuster 1966). However, although rare, some Jungermannialian taxa such as *Schistochila*, do possess 4-seriate stalks (Müller 1948; Schuster 1966, 1984a). In contrast, all *Haplomitrium* and *Steereomitrium* taxa uniformly produce 4-seriate antheridial stalks.

At the light microscope level, the spermatozoid of *Haplomitrium* is indistinguishable from that of other bryophytes, i.e., it is coiled and biflagellate (Fig. 271). However, its fine structure is highly distinctive (Carothers and Duckett 1978, 1979; Duckett et al. 1982, 1983; Carothers and Rushing 1988). To date, spermatozoid ultrastructure has been investigated in two species of the genus--*H. hookeri* (Carothers and Duckett 1978, 1979; Duckett et al. 1982, 1983; Carothers and Rushing 1988) and *H. gibbsiae* (Carothers and Rushing 1988). The recent study of the blepharoplast of *H. gibbsiae* by Carothers and Rushing (1988) and their re-examination of this structure in *H. hookeri* revealed that the blepharoplast morphologies of the two species are extremely similar. In fact, only quantitative characters such as spline width, basal body length, transition zone (TZ) lengths, and stellate pattern (SP) positions separate the species. According to Carothers and Rushing (1988), the spline of *H. gibbsiae* is usually 84 microtubules wide, the anterior TZ is ca. 0.75 µm long, the posterior TZ is ca. 0.55

µm long, the SP's proximal end is about 0.15 µm distal to that of the TZ, and the spline aperture is 2 microtubules wide. In contrast, the spline width in H. hookeri is usually 55 microtubules, the anterior and posterior TZ are both ca. 0.2 µm long, the SP and TZ are at about the same level, and the spline aperture is 1 microtubule wide (Carothers and Rushing 1988).

Numerous ultrasructural aspects which characterize the Haplomitrium spermatozoid are unique to the genus. For example, in all known hepatics except Haplomitrium, the spermatid chromatin consists of solid flexuous rods of uniform diameters which are oriented parallel to the axis (Duckett et al. 1983). In Haplomitrium, the chromatin rods are of different sizes and are not oriented in a distinct pattern. Furthermore, the shaping of the nucleus during spermatogenesis follows a pattern similar to that found in mosses as opposed to hepatics, i.e., the reduction in the diameter of the nucleus is uniform (Duckett et al. 1983).

According to Duckett et al. (1983), like other bryophytes, the spermatozoid of Haplomitrium is monoplastidic. However, in most known mosses and liverworts, numerous starch grains are contained within the plastid. Haplomitrium is one of only three known bryophytes (Phaeoceros and Sphagnum being the other two) whose plastid contains only one starch grain (Duckett et al. 1983).

The spline of Haplomitrium is the second widest known among bryophytes. Only the spline of Treubia, which consists of 104 parallel microtubules at its maximum width is broader (Carothers and Rushing 1989).

While the Haplomitrium spermatozoid exhibits the major hepatic blepharoplast characters of bilateral asymmetry, a bluntly rounded multilayered structure (MLS)-anterior, a relatively long-tapered lamellar strip, and a pair of staggered and subapical dimorphic basal bodies, it also posseses numerous unique features which distinguish it from those of other known hepatics, and in some cases, from all known bryophytes (Duckett et al. 1982, 1983; Carothers and Rushing 1988). According to Carothers and Rushing (1988), the spline is obliquely asymmetrical in shape, i.e., it widens in two stages. This shape is known only for _Haplomitrium_. In addition, _Haplomitrium_ is the only genus of hepatics with an open-type spline aperture and the only known bryophyte with the aperture positioned left of center (Carothers and Rushing 1988). Furthermore, the posterior and anterior basal bodies (BB) are also left of the spline's midline. This location is unique for the anterior BB, but usual for the posterior BB of all bryophytes (Carothers and Rushing 1988).

The _Haplomitrium_-type blepharoplast does share a few structural characteristics with at least some members of the Metzgeriales. For example, like _Pallavicinia lyellii_, the trio of proximally extended microtubular triplets in the anterior BB is ventrally positioned, while in other hepatics these triplets occupy a dorsal position (Carothers and Rushing (1988).

Duckett et al. (1982, 1983) described an accessory band of tubules which lies over the osmiophilic crest of the spline anterior in the mature blepharoplast of _H. hookeri_, and noted that this feature is known only in _Haplomitrium_ and in homosporous ferns. Along with the other attributes of the spermatozoid, this character led Duckett et al. (1982, 1983) to suggest that _Haplomitrium_ spermatozoids have little affinity with those of other hepatics. They suggested that spline geometry and the possession of an accessory band of microtubules may indicate a totally different line of evolution from that of other bryophytes and therefore negates the possibility of _Haplomitrium_ being a possible starting point for hepatic evolution. Instead, they conjectured that _Haplomitrium_ may be "... the closest living survivor to a missing link between bryophytes and pteridophytes." (Duckett et al. 1983).

Carothers and Rushing (1988) also greatly emphasized the uniqueness of the spermatozoid of

Haplomitrium. However, they warned against drawing phylogenetic conclusions based on current spermatozoid ultrastructural knowledge since only a small number of taxa have been investigated from the different bryophyte groups.

Dehiscence in Haplomitrium, as described in H. gibbsiae by E. O. Campbell (1959) occurs as follows. When water is present, the terminal jacket cells separate and fold outwards and the spermatozoids are extruded. The dehisced antheridia are usually brown and cup-shaped (Fig. 272).

In bryophytes, the transfer of sperm from antheridia to archegonia is dependent on a sufficient water supply. In dioicous species in which male and female gametangia are widely separated, raindrop splashes and microscopic animals are probably important dispersal agents (Harvey-Gibson and Miller-Brown 1927; Watson 1971; Clayton-Greene et al. 1977; Reynolds 1980). Perhaps Haplomitrium uses both of these methods to enhance the wider dissemination of sperm. The apical male rosettes of the members of the subgenus Calobryum could function as splash-cups (Schuster 1984a). Although these structures are common in mosses (Schofield and Hébant 1984) and splash-platforms exist in most Marchantiales, they are unknown in other hepatics (Schuster 1984a).

The intra-androecial slime hairs and mucilage papillae undoubtedly secrete mucilage which is important in protecting the gametangia from dessication and injury. An additional interesting function was discussed by Harvey-Gibson and Miller-Brown (1927). They reported that the materials secreted by the paraphyses of Dawsonia superba L. attract some insects and mites which may consequently play a significant role in sperm dispersal. This potential would be highly beneficial to an organism such as Haplomitrium which often grows as scattered individuals or in small unisexual turfs.

Archegonial Ontogeny

As discussed by D. H. Campbell (1920), E. O. Campbell (1959) and Mehra and Kumar (1990), the early stages of archegonial development in Haplomitrium are identical to those of the antheridium. In fact, if only young stages are observed, it is impossible to tell whether the gametangium will become an archegonium or an antheridium. In addition, both gametangia have a common origin, i.e., from cells derived from the central wedge-shaped portion of the apical derivative. So, like the antheridia, the archegonia are generated close to the apical cell, in any or each of the three merophyte rows, and may replace either a portion of the leaf or the entire leaf (Figs. 276-278). As in antheridial production, the unused portion of the segment is used for slime papilla, mucilage hair, and bractlet construction, thus explaining their intimate association with the archegonia. As already discussed, within a species, archegonia and antheridia are positioned at similar loci on their respective shoots. Therefore, the same apical segmentation patterns responsible for antheridial positioning are also responsible for archegonial positioning. Unlike as occurs in male rosettes, the development of the archegonia of a rosette is not synchronous. Instead, archegonial maturation begins immediately after each archegonial initial has been delimited (Figs. 276a-c).

A uniform pattern of archegonial construction occurs in H. hookeri, H. gibbsiae, H. mnioides, H. blumii, (Figs. 280-294) and presumably all other Haplomitrium species. Archegonial construction begins with the swelling of a superficial cell which is soon delimited from the thallus by a transverse wall (Fig. 280). Next, a second transverse division cuts this cell equally to yield an outer primary archegonial initial and an inner stalk initial (Fig. 281). The archegonial initial is then traversed by a curved vertical wall to form two cells (Figs. 282-284). As in antheridial ontogeny, a single pyramidal central cell and three peripheral jacket cells are delimited in the archegonial initial (Fig. 287) by one of two divisional sequences. In one pattern, the initial may be traversed sequentially by two vertical walls which are parallel to the first, but oppositely curved (Figs. 285-287). A transverse division then connects the vertical walls to delimit a cover cell (Fig. 288). In the alternate pattern, the archegonial initial is traversed by two curved walls which intersect distally with the first wall (Fig. 289). Therefore, as in the antheridial pattern, regardless of which sequence of divisions occurs, a central cell is produced that is surrounded by three jacket cells (Fig. 287). In both patterns, an additional radial division occurs in one of the three outer cells to produce four cells (Fig. 289) and the central cell expands and is cut transversely to produce a primary ventral canal cell and a primary neck canal cell (Figs. 290, 291). The primary neck canal cell is repeatedly cut by transverse divisions to form numerous neck canal cells (Fig. 279). The primary ventral canal cell eventually divides transversely to produce a venter canal cell and an egg.

After the four outer cells have been formed, transverse divisions occur in the neck region to form numerous tiers of neck cells with each tier being composed of four neck cells (Figs. 279, 292a). As observed by D. H. Campbell (1920) and E. O. Campbell (1959), occasionally, older neck cells in the distal region of the archegonium may divide vertically to form a few tiers of neck cells composed of 5 or 6 cell rows (Fig. 292b). However, no divisions of the terminal neck canal cells to form a polystratose canal region, such as described in _H. blumii_ by D. H. Campbell (1920), were observed in this study. The cells comprising the venter divide transversely and vertically so at maturity the venter is bistratose (Fig. 293).

The stalk initial is first traversed by a curved vertical wall (Figs. 284, 286, 291). Each resulting daughter cell is cut transversely to generate four cells in cross section (Fig. 294). One additional transverse division in each of the quadrants creates a second tier of cells and completes the construction of the stalk.

The fundamental patterns of archegonial development found for H. hookeri, H. gibbsiae, H. mnioides, and H. blumii are like those described for H. blumii (D. H. Campbell 1920), H. gibbsiae (E. O. Campbell 1959), and H. indicum (Mehra and Kumar 1990). The uniformity among these species implies that, like antheridial ontogeny, this pattern is constant for the genus.

The similarities of the early stages of antheridial and archegonial ontogeny in Haplomitrium not only support the hypothesis that the two types of gametangia are homologous in archegoniates (Goebel 1902), but also attests to the ancestral nature of Haplomitrium. The fact that the ontogenetic sequences are variable in Haplomitrium, i.e., two patterns of divisions occur which result in the same cellular arrangement, reflects the malleable genetic control inherent in the genus.

Within the hepatics, only Haplomitrium and Monoclea Hook. produce a central pyramidal axial cell and three outer cells through three curved, intersecting divisions during archegonial ontogeny (Campbell 1954; Schuster 1984a). The archegonia of all other hepatics investigated produce a single central cell surrounded by three jacket cells via three vertical, eccentric divisions followed by a transverse division to cut off a terminal cover cell (Schuster 1966, 1984a; Crandall-Stotler 1981). The fact that Haplomitrium also displays this latter pattern serves to link this taxon with the other hepatics.

The fine structure of a young H. hookeri archegonium is shown in Figures 295a-c. Although this longitudinal section is not median, young neck cells and the terminal cover cell are evident along with a portion of the axial row of cells. The neck cells contain the organelles typical of vegetative cells.

Nuclei, which are about 6.1-7.0 µm in diameter are centrally positioned within the cells. They are rich in heterochromatin and have nucleoli (ca. 2.1-3.0 µm in diameter) and occasionally karyosomes (ca. 0.7 µm in diameter). The plastids are about 2.0-3.0 µm long X 0.9-1.9 µm wide, have well-defined thylakoids, and contain (1-) 2-3 starch grains which measure approximately 0.23-0.46 µm long X 0.20-0.37 µm wide. A large, dissected vacuolar system is present along with elongate mitochondria, ER, ribosomes, and a few lipids and oil bodies. A thin, invaginated plasmalemma delimits the cytoplasm of each cell. Numerous vesicles occur between this plasmalemma and the cell wall which separates the neck cells from the axial cells. These vesicles are particularly abundant in the neck region of the archegonium (Figs. 296, 297).

A ventral canal cell marks the base of the axial row of cells (Figs. 295a, 295c). This cell appears ellipsoidal in shape, but little as far as organellar constitution is revealed by this section since most organelles are located centrally and only the edge of the cell was grazed when sectioned.

The most conspicuous organelles within the neck canal cells are plastids and large, autophagic vacuoles (Figs. 295b, 296, 297). The pleiomorphic plastids are about 1.8-2.0 µm long X 0.8-1.0 µm broad, contain 1-3 minute oil droplets and occasionally, 1 or 2 starch grains (Fig. 297). The autophagic vacuoles measure approximately 1.04-1.73 X 0.63-0.98 µm and are usually more or less spherical or ellipsoidal. However, numerous contorted forms occur when the tonoplast invaginates and the vacuole engulfs cytoplasmic constituents (Fig. 296). When the tonoplast reforms, the consumed vesicle remains suspended within the vacuole until it is digested. According to Esau (1977), these vacuoles play a role in metabolic processes such as senescence, differentiation, and the mobilization of reserves. In this case, it is difficult to say which process is underway. However, it is likely that the cells are breaking down as evidenced by the presence of spherical bodies composed of tightly packed concentric membranes.

At this young stage of development, all cells of the archegonium constitute a single symplasm. Plasmodesmata traverse the walls of adjacent neck cells, the walls between the neck cells and the axial cells, and the walls of adjacent neck canal cells.

The H. blumii archegonia depicted in Figures 298 and 299 appear to be mature, but are not yet ready for fertilization since the ventral canal cell has not yet divided, and the cover cells and neck canal cells are still intact. The neck and venter cells of this more mature H. blumii archegonium are more highly vacuolated and the plastids bear more well-developed thylakoids and contain more starch than their young counterparts (Figs. 298-299). A cross-section through the venter reveals its bistratose composition and ventral canal cell (Figs. 299a, 299b). Only a few organelles such as mitochondria, vacuoles, and starch-containing plastids are visible due to the grazing angle at which the cell has been cut. A large zone filled with an opaque, granular to fibrous material is visible between the ventral canal cell and the venter--a distance of about 1.6-2.6 µm (Figs. 299a, 299b). This region is also conspicuous in light micrographs (Figs. 300, 301). A scarcity of material precluded doing histochemical tests to determine its composition.

The fine structure of the axial row of neck canal cells is similar to that of the neck cells and venter cells, except that the vacuolar system is not as prominent and oil bodies are absent (Figs. 298a-c). Adjacent neck canal cells are connected by plasmodesmata, but such cytoplasmic channels no longer join the axial row of cells with the outer cells (Figs. 298b, 298c).

Structure of the Archegonium

The mature archegonia of *Haplomitrium* vary little among the species. The long, twisted necks are composed of numerous tiers of four neck cells and a central row of neck canal cells (Fig. 300). The lengths of the necks range from about 162-634 µm depending on the species. The venter is just slightly broader than the neck and is bistratose (Fig. 300-310). The entire structure is supported by a short stalk which is usually only two tiers tall. Each tier is composed of a quadrant of four cells (Fig. 300). A unicellular slime cell is associated with the stalk of each species and 1-3-celled slime papillae are in close proximity to the archegonium.

As is apparent from Figures 300-310 and Table VII, the most variable archegonial character is neck length. *Haplomitrium hookeri* (Fig. 306), *H. monoicum* (Fig. 301), *H. intermedium* (Fig. 307), and *H. mnioides* (Fig. 310) all produce necks which are about 162-296 µm long X 35-61 µm wide. *Haplomitrium andinum* and *H. giganteum* have slightly longer archegonial necks that measure approximately 301-370 µm long X 54-72 µm wide. The remaining species (*H. gibbsiae* [Figs. 300, 308], *H. chilensis* [Fig. 309], *H. indicum* [Fig. 302], *H. dentatum* [Fig. 303], *H. grollei* [Fig. 304], *H. kashyapii*, and *H. blumii* [Fig. 305]) possess necks which are about 310-635 µm long X 37-78 µm wide.

Usually there are 12-16 neck canal cells, but this number varies in direct association with the lengths of the archegonial necks. D. H. Campbell (1920) counted 16-20 neck canal cells in *H. blumii*, while E. O. Campbell (1959) found as many as 40 in *H. gibbsiae*.

The archegonia of *S. minutum* are apparently like those of *Haplomitrium*. According to Campbell (1987), the mature archegonium of this taxon is about 400-500 µm tall with a short stalk, swollen venter, and a long, twisted neck which is 260-370 µm long. The archegonia of *Steereomitrium* are further allied to those of *Haplomitrium* in that each tier of neck cells is composed of four rows of cells which may become secondarily 6-rowed near the apex.

The fundamental flask-shaped structure of the archegonium is similar among the archegoniates (Bold et al. 1980). However, as exemplified by the Anthocerotophyta, Bryophyta, and Hepatophyta, the ontogenies of the archegonia effect the general stature of the organs. Moss archegonia develop via apical cells and, therefore, are very elongate and slender with long stalks (Watson 1971). In contrast, those of hornworts are embedded in the thallus so their axial cells are in direct connection with the thallus tissue (Renzaglia 1978; Crandall-Stotler 1981). The archegonia of the hepatics are generally stouter with short stalks, as compared to moss archegonia (Crandall-Stotler 1981). The basic appearance of the Haplomitrium archegonia are, therefore, typical of hepatic archegonia. However, Haplomitrium is the only hepatic with four vertical rows of neck cells. Five rows of neck cells typically characterize the Metzgeriales and Jungermanniales, while 6 rows have been reported for the Monocleales, Marchantiales, and Sphaerocarpales (Schuster 1984a). Although Hattori and Mizutani (1958) reported 4 rows of neck cells for Takakia lepidozioides, Schuster (1967; 1984a) reported 6 rows and Crandall-Stotler (personal communication) noted as many as 7 rows of cells.

In preparation for fertilization, the ventral canal cell divides transversely to produce the egg and the venter canal cell. The latter cell, along with the neck canal cells will degenerate and the cover cell is lost.

Insufficient material precluded the determination of oogenesis and embryogenesis in Haplomitrium. No descriptions of oogenesis exist for the genus and only limited information is available pertaining to embryogenesis. The very early stages of embryogenesis were observed in H. gibbsiae by E. O. Campbell (1959). She described the first division of the zygote as being transverse and unequal to form a larger epibasal and a smaller hypobasal cell. Each of these cells is then cut by a vertical wall. Further divisions were described as rapid, but no details were presented (Campbell 1959). This pattern of development was also observed in H. indicum by Mehra and Kumar (1990). Schuster (1984a) reported irregular divisions after the formation of the epibasal and hypobasal cells with the former cell presumably yielding the capsule, seta, and foot, while the latter gives rise to a haustorium. The formation of the epibasal cell and hypobasal cell in Haplomitrium is as occurs in all hepatics, except perhaps Monoclea, which supposedly undergoes a free-nuclear stage (Campbell 1954). In the anthocerotes, the first wall is vertical not transverse (Renzaglia 1978; Crandall-Stotler 1981), and an apical cell is involved in the embryogenesis of mosses (Watson 1971). Therefore, embryogenesis in Haplomitrium is most similar to that of the remaining hepatics.

Sporophyte Protective Structures

In bryophytes, fertilization of the egg triggers not only sporophyte development, but also the production of gametophytically-derived protective structures around it. In Haplomitrium, these structures consist solely of archegonial and/or shoot derived tissues. Perianths are absent in all species.

The characteristics of the sporophyte protective structures of Haplomitrium and Steereomitrium are summarized in Table VIII. These structures are chlorophyllous, cylindrical to linear-oblong, and are exerted well above the bracts at maturity. Depending on the species, the lengths may range from 2.0-10.0 mm at maturity and the wall thicknesses are from 2- to 10-stratose. Unfertilized archegonia, and the structures intermixed with the archegonia in the gynoecium may be elevated and scattered over the entire surface of the protective structure, or restricted to its basal portions only, or remain

TABLE VIII. SPOROPHYTE PROTECTIVE STRUCTURE CHARACTERISTICS OF *HAPLOMITRIUM* AND *STEEREOMITRIUM*[a]

	Type	Wall Thickness[b]	Length (mm)	Base Width (mm)	Mouth Width (mm)	Elevated Structures
H. hookeri	calyptra	2-4 cells	2.1- 4.9	0.5-1.5	0.9-1.7	none
S. minutum	calyptra	ND	2.0	0.5	0.5	none
H. ovalifolium[c]	calyptra	ND	ND	ND	ND	none
H. monoicum	calyptra	ND	2.4- 2.8	0.3-0.5	0.7-0.9	none
H. intermedium	shoot-calyptra	3-4 cells	4.0- 5.0	0.7-1.0	0.7-1.0	I, B, A, P in basal 1/3; 2°
H. andinum	shoot-calyptra	3-4 cells	3.2-6.8	0.6-1.0	0.5-1.0	I, B, A, P in basal 1/3; 2°; or B, P over surf
H. blumii	shoot-calyptra	6-9 cells	1.0- 9.0	1.6-1.9	2.1-2.2	I, B, A, P in basal 1/3, AF or A, P over surf, 2°

TABLE VIII. CONTINUED

	Type	Wall Thickness[b]	Length (mm)	Base Width (mm)	Mouth Width (mm)	Elevated Structures
H. mnioides	shoot-calyptra	5-6 cells	4.3- 6.0	0.8-2.1	1.0-2.9	B, A, P, H in basal 1/3 or to apex
H. gibbsiae	perig	6-10 cells	5.0-10.0	1.4-1.6	1.2-1.8	B, A, P, H over surf, 2°
H. chilensis	perig	ND	4.5- 6.0	0.8-1.2	1.2-1.5	P, A over surf, rare B, 2°
H. indicum	perig	7-10 cells[d]	4.2	0.9	1.1	like *H. chilensis*
H. grollei	perig	ND	1.0	ND	ND	B, A in basal 1/3
H. kashyapii[e]	ND	ND	ND	ND	ND	B, A over surf

[a]sporophyte protective structures unknown for *H. dentatum* and *H. giganteum*; [b]approximate thickness of mid to basal portion of mature protective structure; [c]from Schuster 1971; [d]from Mehra and Kumar 1990; [e]from Udar and Kumar 1982; 2°, secondary elaboration; A, unfertilized archegonia; AF, apical foliar structures; B, bractlets; H, slime hairs; I, inner bracts; ND, no data; P, slime papillae; perig, perigynium; surf, surface.

unelevated, all depending on ontogeny. As the sporophytes mature and elongate, the mouths of the calyptrae are torn irregularly.

The protective structures of most species have traditionally been described either as smooth calyptrae or as calyptrae on which archegonia and/or foliar structures are elevated to various heights. However, except for that of H. blumii (Fulford et al. 1958), the ontogenetic patterns responsible for these morphologies have remained undescribed.

When associated with hepatics, the term "calyptra" has been interpreted in different ways. According to Evans (1939), Müller (1951), Parihar (1957) and Smith (1955) it is a structure derived from the growth of the venter of the archegonium after fertilization of the egg. In contrast, Herzog (1925), Macvicar (1926), Goebel (1930), and Campbell (1940) applied the term to any structure which protects the young sporophyte, whether it is derived from only the archegonial venter, or from the venter and stem tissue. In this study, the term calyptra is used in the first sense, i.e., to mean a sporophyte protective structure derived exclusively through divisions of the archegonium. The terms shoot-calyptra and perigynium are used as defined by Knapp (1930); i.e., a shoot-calyptra is a somewhat fleshy sporophyte protective structure derived from the growth of both archegonial and axial tissue peripheral to the archegonium, while a perigynium is derived solely from stem tissue. Each of these three sporophyte protective structures is represented within the genus Haplomitrium (Figs. 311a-e; Table VIII).

Haplomitrium hookeri (Figs. 312-315; 319), Steereomitrium minutum (Fig. 316), H. ovalifolium, and H. monoicum (Figs. 317-318) each produce true calyptrae, i.e., the meristematic activity which builds the protective structure is restricted to the distal portion of the swollen archegonial venter (Fig. 311a). Not only is this zone visible in serial sections of young sporophytes (Fig. 319), but it is also evidenced by the presence of small quadrate cells near the apex of the venter just basal to the old archegonial neck in more mature calyptrae. This zone of continued growth appears in sharp contrast to the elongate cells comprising the base of the structure (Figs. 312-314, 318). The resulting mature calyptrae are cylindrical, and due to their exclusive derivation from the fertilized archegonia, are smooth with walls composed of only about 2-4 layers of cells (Figs. 312, 314, 315, 317, 318).

The calyptra of H. hookeri has been described by numerous workers, most of whom simply stated that it is large, cylindrical or oblong, and completely smooth (Gottsche 1843; Carrington 1874; Pearson 1902; Evans 1917; Frye and Clark 1943; Worley 1969). Only Schuster (1969) postulated that this structure likely represents a calyptra in the strict sense, being derived largely, if not entirely from the archegonial wall.

Only a single, postmature calyptra of H. ovalifolium was seen by Schuster (1971). He described it as being small, cylindrical, and smooth. When this description is considered along with all the other similarities shared by H. ovalifolium and H. hookeri, it becomes highly likely that H. ovalifolium also develops a true calyptra.

The calyptra of S. minutum was briefly described by E. O. Campbell (1987). She stated that only the venter enlarges to produce a tubular calyptra which is terminated by the old archegonial neck. The conclusion drawn in this investigation concurs with Campbell's observations.

The only description of the calyptra of H. monoicum is that of Engel (1981). Although he did not call it a true calyptra, he did recognize that the walls are thin and smooth and bear no elevated unfertilized archegonia or bractlets.

Shoot-calyptrae are produced by H. intermedium (Figs. 324-325), H. andinum (Figs. 320; 326-330), H. blumii (Figs. 321-333; 331-336), and H. mnioides (Figs. 337-340; 341-343). These protective structures are created through the combined meristematic activity of the archegonial venter and the stem tissue directly beneath it (Fig. 311b). A small ring of stem tissue which is generally located just below the fertilized archegonium and its closely juxtaposed bractlets, slime papillae, and unfertilized archegonia, becomes meristematic and causes the axial tissue to grow up around the developing sporophyte. This axial growth is usually limited and may be responsible for forming only about the basal third of the shoot-calyptra as indicated by the position of the elevated accessory structures (Figs. 322, 324b, 326a, 329a, 335a-c, 341-343). The unfertilized archegonia and their associated stuctures which lie beyond the range of the meristmatic zone of receptacular tissue remain on the gynoecial disc posterior to the shoot-calyptra (Fig. 338b). Therefore, the position of the archegonium which is fertilized will also affect the ornamentation of the calyptra. For example, if the fertilized archegonium lies near the periphery of the terminal disc, the axial meristematic zone beneath it will likely include an inner reduced bract, thus elevating this structure onto the calyptra. In contrast, if the fertilized archegonium is located centrally on the disc, the inner reduced bracts would not be elevated unless the meristematic axial zone was large enough to extend to the periphery of the disc and underlie the inner bracts. Following cessation of axial growth, meristematic activity begins at the apex of the archegonial venter. The cellular collapse evident at the apex of the shoot-calyptra in Figure 329b, is a response of the small, delicate meristematic cells of this region to the SEM preparation technique. The mature shoot-calyptrae are cylindrical with 3-4-stratose walls, at least near the apex and in the basal third of the shoot-calyptra, due to the developmental involvement of the stem tissue (Figs. 320, 322,323, 341-343).

In H. intermedium and some individuals of H. andinum, the reduced inner bracts which have been elevated within the basal one-fifth or one-third of the shoot-calyptra undergo a period of secondary beneath it (Fig. 311b). A small ring of stem tissue which is generally located just below the fertilized archegonium and its closely juxtaposed bractlets, slime papillae, and unfertilized archegonia, becomes meristematic and causes the axial tissue to grow up around the developing sporophyte. This axial growth is usually limited and may be responsible for forming only about the basal third of the shoot-calyptra as indicated by the position of the elevated accessory structures (Figs. 322, 324b, 326a, 329a, 335a-c, 341-343). The unfertilized archegonia and their associated stuctures which lie beyond the range of the meristematic zone of receptacular tissue remain on the gynoecial disc posterior to the shoot-calyptra (Fig. 338b). Therefore, the position of the archegonium which is fertilized will also affect the ornamentation of the calyptra. For example, if the fertilized archegonium lies near the periphery of the terminal disc, the axial meristematic zone beneath it will likely include an inner reduced bract, thus elevating this structure onto the calyptra. In contrast, if the fertilized archegonium is located centrally on the disc, the inner reduced bracts would not be elevated unless the meristematic axial zone was large enough to extend to the periphery of the disc and underlie the inner bracts. Following cessation of axial growth, meristematic activity begins at the apex of the archegonial venter. The cellular collapse evident at the apex of the shoot-calyptra in Figure 329b, is a response of the small, delicate meristematic cells of this region to the SEM preparation technique. The mature shoot-calyptrae are cylindrical with 3-4-stratose walls, at least near the apex and in the basal third of the shoot-calyptra, due to the developmental involvement of the stem tissue (Figs. 320, 322, 323, 341-343).

In *H. intermedium* and some individuals of *H. andinum*, the reduced inner bracts which have been elevated within the basal one-fifth or one-third of the shoot-calyptra undergo a period of secondary growth and become very long and lingulate (Figs.

lingulate (Figs. 327a, 328). Thus, in this taxon, the extent of stem tissue involvement in calyptral development is variable. The axial zone is apparently small and includes only the bractlets and slime papillae immediately adjacent to the fertilized archegonium as evidenced by the numerous archegonia and accessory structures which remain on the gynoecial disc. If axial meristematic activity ensues for only a brief time, the bractlets are elevated only into the basal one-fifth or one-third of the shoot-calyptra and are relatively uneffected by the meristematic activity in the archegonium. However, if the zone is active for a longer period of time, the bractlets and slime papillae are raised to greater heights on the calyptra. In this case, some of the elevated structures may be positioned within the zone of archegonial meristematic activity and hence, can be carried to the apex of the shoot-calyptra by divisions in this second zone.

The elevated, secondarily elongated inner bracts associated with some shoot-calyptrae of H. andinum have never been described. In fact, some debate has even existed as to whether the calyptra itself is smooth with sterile archegonia restricted to the base (Spruce 1885; Schuster 1967), or whether unfertilized archegonia are elevated over the surface of the shoot-calyptra (Fulford 1962). Although H. andinum does indeed produce a shoot-calyptra as indicated by Fulford (1962), the sterile archegonia are scattered only over the basal third of the shoot-calyptra. However, in some cases, slime papillae and bractlets may be elevated toward the apex of the protective structure.

Like H. andinum, the shoot-calyptrae of H. blumii are also somewhat variable. The most frequently occurring morphology is one in which unfertilized archegonia, slime hairs, and bractlets are elevated only onto the basal one-fifth or one-third of the shoot-calyptra, while the midregion of the shoot-calyptra is smooth and the apex is covered by numerous foliar outgrowths, each terminated by a slime cell (Figs. 323, 331-333, 335a-c). However, old calyptrae whose ornamented mouths have been destroyed by their emergent sporophytes are indistinguishable from those of H. andinum (compare Figs. 330 and 336). At least some of the apical foliar outgrowths are apparently not elevated from the gynoecium, but are secondarily formed during the period of meristematic activity in the archegonial venter. This seems to be the most logical conclusion since it is known that bractlets are few within the gynoecium, yet the terminal foliar outgrowths on the shoot-calyptra are quite numerous. In addition, serial longitudinal sections made through the apex of the shoot-calyptra reveal that these foliar structures are inserted deep within the shoot-calyptra wall and not on the surface as would be expected if these structures had simply been carried aloft (Fig. 323). Not all H. blumii individuals bear such foliar outgrowths on the apices of their shoot-calyptrae, however. In some cases, archegonia and bractlets may be carried aloft to various heights depending on the temporal and positional coincidence of the axial and archegonial meristematic activity and the apex remains devoid of foliar outgrowths (Fig. 332). The secondary production of accessory structures on the usually smooth mid-region of the shoot-calyptra results in the structure being ornamented over its entire surface.

The literature reflects this variability in H. blumii shoot-calyptra morphologies. Schiffner (1900), Fulford et al. (1958), and Schuster (1967) described the protective structure as having sterile archegonia and slime papillae scattered over its surface, while Goebel (1891, 1930) and D. H. Campbell (1920) reported no such ornamentation. All agreed that the protective structure was large and massive with walls up to six cells thick, but Fulford et al. (1958) were the first to determine that the structure was actually a shoot-calyptra. However, Fulford et al. (1958) interpreted the terminal elaborations of the shoot-calyptra as being grossly malformed archegonia rather than foliar structures.

The shoot-calyptra of H. mnioides (Figs. 337-343) seems to be produced by a more extensive zone of stem tissue beneath the fertilized archegonium and adjacent unfertilized archegonia, bractlets, and slime papillae than the previous species. Here too, if the cells of the venter begin to divide before the stem activity has been completed so that the accessory structures are inserted below the venter zone, then the final form of the shoot-calyptra is one in which the apical one-half to two-thirds is smooth and relatively thin, while the basal one-half to one-third is thicker and covered with unfertilized archegonia, slime papillae, slime hairs and bractlets (Figs. 311c, 338a, 338b, 339a, 339b, 343). However, if the meristematic activity of the stem occurs long enough to elevate the accessory structures into the zone of archegonial meristematic activity, then ceases to function, the growth of the venter will continue to lift the structures to the apex of the calyptra, leaving the midregion smooth (Figs. 340a, 340b). Once elevated, no secondary growth is apparent in the bractlets, slime hairs, slime papillae, or archegonia. Due to the contributions of stem tissue and archegonial tissue, the walls of the mature shoot-calyptra may be as much as 5-6 cells thick and rarely up to 10 stratose locally (Fig. 343).

The shoot-calyptrae of H. mnioides have been briefly described by Horikawa (1929), Schuster (1967), and Inoue (1974). Horikawa (1929) reported the structure to be linear-oblong, 7-9 mm long X 1.4 mm wide, with 5-6-stratose walls. Inoue (1974) illustrated one immature shoot-calyptra with archegonia and slime papillae more than half the distance to the apex and one that is smooth distally and bears archegonia in the basal third. The ontogenetic patterns described in this study explain the apparent inconsistencies in Inoue's report. Although Schuster (1967) stated that the shoot calyptra is smooth and lacks leafy outgrowths, he did describe "terete projections" or "swollen gibbosities", and occasionally unfertilized archegonia over the surface. The absence of foliar outgrowths results from the lack of secondary elaboration of the raised bractlets.

Schistochila-type perigynia are produced by H. gibbsiae (Figs. 344; 345-348), H. chilensis (Figs. 349-351), H. indicum (Figs. 354, 355), H. grollei (Figs. 356a, 356b), and probably H. kashyapii through the activity of a meristematic zone of stem tissue which is located beneath the gynoecium (Figs. 311d, 311e). The extensiveness of this meristematic zone is highly variable in H. gibbsiae, and therefore, the types, number, and positioning of the accessory structures on the perigynia vary among individuals. In a few cases, the meristematic zone of axial tissue is small and is localized beneath the fertilized archegonium, a few of its adjacent unfertilized archegonia, slime papillae, and rarely a bractlet (Fig. 311d). Therefore, when meristematic activity commences, most of the unfertilized archegonia and intra-gynoecial structures are left on the gynoecial disc, while others are elevated only onto the basal fifth to third of the perigynium. Sometimes with more extensive meristematic activity a few archegonia and slime papillae are raised to the apex of the perigynium and occasionally, a bractlet is also elevated to about the midpoint of the perigynium (Figs. 346a, 346b, 347).

Although this form of perigynium, which is initiated via a restricted zone of meristematic stem activity, is of limited occurrence in H. gibbsiae populations, it is the predominant and perhaps the exclusive form found in H. chilensis (Figs. 349-351), H. indicum (Figs. 354, 355), and probably in H. grollei (Figs. 356a, 356b) and H. kashyapii. Only a single protective structure was seen in the available H. grollei specimens. Although the distal portion of the protective structure had been destroyed, raised archegonia and lingulate bractlets occur on the basal fifth of the structure (Figs. 356a, 356b). No mention of the sporophyte protective stucture of H. grollei exists in the protologue (Kumar and Udar 1977), but the overall appearance of the one present in the type material is reminiscent of the perigynia of *H. chilensis*, *H. indicum*, and the less elaborated form found in *H. gibbsiae*.

No sporophyte protective structures are present on the type specimens of *H. kashyapii* which were examined in this investigation. However, Udar and Kumar (1982) described this structure as a massive "shoot calyptra" with unfertilized archegonia and a few foliar outgrowths distributed over its entire surface.

The most frequently occurring form of perigynium in *H. gibbsiae* populations is one in which virtually the entire receptacular zone beneath the gynoecium becomes meristematic (Fig. 311e). Although some peripheral unfertilized archegonia, slime papillae, slime hairs, and bractlets remain on the disc, most are elevated with the growing stem tissue and are ultimately dispersed over the entire surface of the perigynium (Figs. 344, 348, 352, 353). The elevated structures are secondarily elaborated. The bractlets grow into extensive foliar structures via apical cell activity (Figs. 348, 353a, 353b). The slime papillae divide and become slime hairs, while the mucilage hairs elongate and branch (Figs. 352a, 352b). The unfertilized archegonia often become enlarged and grotesquely malformed (Fig. 352a). The mature perigynia are cylindrical and have walls which are ca. 6-9 (-12) cells thick (Fig. 344).

The published descriptions of the sporophyte protective structures of *H. gibbsiae* generalize it as one which is extensively covered with elaborated foliar outgrowths, slime hairs, and archegonia (Stephani 1924; E. O. Campbell 1959; Schuster 1967; Allison and Child 1975). However, the examination of numerous *H. gibbsiae* collections reveals the morphological variability inherent in these structures. Although perigynia which are extensively covered by elaborate foliar outgrowths, slime hairs, and archegonia are the predominant forms found in New Zealand populations, individuals occur in the same populations, albeit less frequently, which possess perigynia which are constructed from smaller axial meristematic zones, i.e., archegonia and slime papillae may be raised, but, with the exception of a rare bract, the perigynia are devoid of foliar outgrowths.

The descriptions of the sporophyte protective structures of _H. chilensis_ (Schuster 1971) and _H. indicum_ (Udar and S. Chandra 1961, 1965; Mehra and Kumar 1990) suggest perigynia such as those less frequently produced by _H. gibbsiae_. Schuster (1971) described _H. chilensis_ as forming a shoot-calyptra which is devoid of foliar outgrowths, but which usually has a few sterile archegonia carried aloft. _Haplomitrium indicum_ has been reported to form a "shoot calyptra" which is cylindrical, 5-7 mm long and 1.8-2.0 mm wide, with massive walls up to 10 cells thick locally, and which bears numerous archegonia at its base and often also towards the apex (Udar and S. Chandra 1961, 1965; Mehra and Kumar 1990).

Information on the sporophyte protective structure of _H. dentatum_ is very sparce. No fertilized female shoots were available for study and Kumar and Udar (1976) stated only that the "shoot calyptra" is 3.63-4.1 mm long and devoid of any outgrowths or unfertilized archegonia. This lax description suggests that these structures were not meticulously studied and creates uncertainty as to whether or not this structure is in fact a shoot-calyptra in the true developmental sense.

In addition to possessing shoot- or thallus-derived sporophyte protective structures, perianths or pseudoperianths commonly occur within the Jungermanniales and Metzgeriales, respectively. Within these lines, these "leaf"-derived structures are hypothesized to represent the most primitive modes of sporophyte protection. In fact, within these groups, the formation of massive shoot- or thallus-derived protective structures, namely, perigynia, is accompanied by the reduction of the perianth or pseudoperianth (Schuster 1966, 1984a).

So, in each of these groups, the primitive condition in the evolutionary progression of sporophyte protective structures is the "leaf"-derived structure (perianth or pseudoperianth). The addition of stem or thallus tissue along with the "gradual" suppression of perianths or pseudoperianths in forming a shoot-calyptra is more advanced. The complete suppression of the perianth or pseudoperianth and the creation of a massive perigynium represents the most derived condition.

Schuster (1984a) considered *Haplomitrium* to be "an exceedingly unspecialized genus" and maintained that the lack of a perianth in this taxon is primitive. In any case, the fact that *Haplomitrium* lacks a perianth and shows no evidence of ever having had such a structure, separates it from the Jungermanniales and Metzgeriales.

As first pointed out by Fulford et al. (1958), the presence of a derived form of protective structure such as the shoot-calyptra of *H. blumii* is somewhat inconsistent with the concept of *Haplomitrium* being ancestral. Indeed, both "primitive" (true calyptrae) and advanced forms of sporophyte protective structures (shoot-calyptrae and perigynia) are represented within the genus. As illustrated by Figures 311a-e, the starting point of the *Haplomitrium* line is the simple condition of forming a protective structure exclusively from archegonial tissue, i.e., a true calyptra (Fig. 311a). A slight elaboration of the process which provides additional protection for the young sporophyte adds a small amount of stem tissue to the calyptra to form a shoot-calyptra (Fig. 311b). Enlargement of the axial meristematic zone and an increase in the length of time the zone is active, plus divisions in the archegonial venter to "carry" the stem tissue higher make the protective structure more massive overall and increased its protective capabilities (Fig. 311c). Ultimately, the zone of axial meristematic activity became so extensive and was active for such a long period of time, that divisions in the apical region were completely eliminated (Figs. 311d, 311e).

Occasionally, more than one archegonium on a shoot is fertilized and multiple protective structures and sporophytes result. Usually no more than two such structures develop, but in one instance, Schuster (1967) observed three sporophytes developing on a single *H. gibbsiae* shoot. The potential for multiple sporophyte development indicates a breakdown of the system which suppresses multiple embryo development, and may also represent a regression to an ancestral expression (Schuster 1984b).

The Sporophyte

Gross Morphology

Sporophytes are known from *Steereomitrium*, and all species of *Haplomitrium* except *H. monoicum* (Engel 1981), *H. ovalifolium* (Schuster 1971), *H. grollei* (Kumar and Udar 1977), *H. kashyapii* (Udar and Kumar 1982) and *H. giganteum* (Stephani 1922; Grolle 1963; Schuster 1967). The mature sporophytes of *Haplomitrium* and *Steereomitrium* are identical to those found in the Jungermanniopsida and *Monoclea*, in being composed of a foot, an elongate seta, and a capsule which dehisces along specific lines of weakness. Very little morphological variation exists among the sporophytes of the different species of *Haplomitrium*, or between those of *Haplomitrium* and *Steereomitrium* (Table IX; Figs. 357-379). The foot is obconoidal and is embedded in the gametophyte to just above the first cycle of bracts (Fig. 359). When elongated, the seta is hyaline and may be up to 5 cm long and ca. 0.2-1.8 mm in diameter depending on the species (Fig. 358). The mature capsule is dark brown or reddish-brown and cylindrical in shape (Fig. 360). In some species such as *H. intermedium*, *H. gibbsiae*, *H. blumii*, and *H. andinum*, a short protrusion, or beak, tips the apex

(Fig. 361). The capsule dimensions range from 1.2-3.0 mm long X 0.4-0.8 mm in diameter in small plants such as H. intermedium and H. hookeri, to approximately 3.0-5.5 mm long X 0.8-1.5 mm in diameter in larger plants such as H. blumii and H. andinum (Table IX). Although these capsules are relatively large, they are not nearly as massive as those of Schistochila, which holds the distinction of producing the largest capsules (ca. 5-7 cm long) of any Jungermanniidae (Schuster 1984a).

According to Schuster (1984a), capsule shape is phylogenetically important, with spherical or nearly spherical capsules representing the most primitive condition and the ellipsoidal-cylindrical or cylindrical forms being derived. This latter morphology is found uniformly throughout Haplomitrium and Steereomitrium, and is also common in the Pallaviciniaceae (except Moerckia), Verdoorniaceae, Makinoaceae, and in the Jungermannialian representatives of Jackiellaceae, Radulineae, and Personiellineae. In addition, Monoclea also possesses ellipsoidal-cylindrical capsules, while most other Marchantioid taxa have obovoid to spherical capsules (Schuster 1984a).

Anatomy and Ontogeny

Although little is known about the sequence of divisions which produce the young sporophyte of Haplomitrium, it is hypothesized that the epibasal cell yields the sporophyte proper, i.e., the capsule, seta, and foot, while the hypobasal cell produces only the haustorium (Campbell 1959; Crandall-Stotler 1981; Schuster 1984a; Mehra and Kumar 1990). As the young Haplomitrium sporophyte develops, the foot, seta, and capsule become distinct and take on their ultimate forms. The foot is obconoidal and is composed centrally of quadrate cells which measure approximately 20-28 µm X 26-40 µm, and grades peripherally to much smaller cells which are about

TABLE IX. SPOROPHYTIC CHARACTERS OF HAPLOMITRIUM AND STEEREOMITRIUM[a]

	Seta (LXW; mm)	Capsule (LXW; mm)	Elaters No. Spirals	Elaters LXW (µm)	Spores Diam (µm)	Spores Sculpture	No. Dehis Lines
H. hookeri	3-4 enlarged central rows; 6.0-13.0 X 0.3- 0.4	1.2-3.0 X 0.6-0.8	1; or 2, 1 at ends	324-552 X 8-9; 174-180 X 11-16	28-36	vcl, limited fusion; large interstices	1, 2, or 4; c or inc
H. intermedium	gen; 2.5- 5.0 X 0.2- 0.3	2.0-2.5 X 0.4-0.6	2; rarely 3 in mid-region	420-492 X 6-9	21-28	vm densely ridged; small interstices	ND
H. gibbsiae	gen; 10.0-30.0 X 0.4- 0.5	3.0-4.0 X 0.5-0.8	1; or 2, 1 at ends	300-729 X 8-13	26-39	vc, smooth or rough; tetrads, diads or single	4, rarely 2, 3 or 5; inc

TABLE IX. CONTINUED

	Seta (LXW; mm)	Capsule (LXW; mm)	Elaters No. Spirals	Elaters LXW (µm)	Spores Diam (µm)	Spores Sculpture	No. Dehis Lines
H. chilensis[d]	ND	5.0-5.5 X 1.4-1.5	2 or 3; 1 at ends	ND	32-35	tbc; tetrads, diads or single	ND
H. indicum[e]	gen; 56.6 X 0.8- 1.8	5.0-5.9 X 1.0-1.8	2 or 3	400-800 X 16-18	21-22	papillose	1; inc
H. dentatum[f]	ND	ND	ND	ND	18-22	papillose; diads or single	ND
H. mnioides	hollow; 12.0-29.0 X 0.8- 1.3	3.0-4.0 X 1.0-1.5	2	342-415 X 7-10	25-35	tbc, small inter- stices	1

TABLE IX. CONTINUED

	Seta (LXW; mm)	Capsule (LXW; mm)	Elaters No. Spirals	Elaters LXW (µm)	Spores Diam (µm)	Spores Sculpture	No. Dehis Lines
H. blumii	gen; 18.0-30.0 X 0.8- 1.3	3.7-5.1 X 1.3-1.5	2	580-675 X 7-9	19-32	loose anastomosing ridges; large interstices	1[g]
H. andinum	gen; 11.2-32.5 X 0.9- 1.3	3.0-4.0 X 0.5-1.0	2	355-700 X 6-11	12-23	like H. blumii	1[h]
S. minutum	ND; 11.0-15.0 X 0.3- 0.4	1.1-1.7 X 0.7-0.8	1, 2 at ends	340-591 X 6-13	23-32	vcl, limited fusion; small interstices	4[i]; 2

[a]sporophytes unknown for H. ovalifolium, H. monoicum, H. indicum, H. grollei, H. kashyapii and H. giganteum; [b]Worley 1969 & Schuster 1967; [c]Campbell 1959; [d]Schuster 1971; [e]Mehra and Kumar 1990; [f]Kumar & Udar 1976; [g]Schuster 1967; [h]Spruce 1885; [i]Campbell 1987; c, complete; Dehis, Dehiscence; gen, generalized; inc, incomplete; LXW, length X medial width; ND, no data; tbc, tuberculate; vc, verrucate; vcl, verruculate; vm, vermiculose.

14-16 µm X 7-12 µm as seen in longitudinal view (Figs. 362, 363a). The curved outer walls of these peripheral cells give the foot a somewhat crenulate outline in sectioned preparations. The adjacent 3-4 rows of gametophyte cells are also reduced in size and appear compressed. The occurrence of a stained matrix between the two is suggestive of transfer tissue as described in other hepatics (Ligrone and Gambardella 1988). Unfortunately, ultrastructural details of this region were unavailable due to the paucity of live material.

Prior to elongation, the seta is nonarticulate and is composed of small, square, chlorophyllous cells with relatively thin, but elastic walls (Fig. 363b). Elongation of the seta is presumably accomplished through the absorption of water by the cells and the consequential rapid expansion of each seta cell wall (Thomas 1980). At this stage, the cells are hyaline and highly vacuolated with only a small amount of peripheral cytoplasm. In cross section, the typical seta of Haplomitrium is of Douin's (1908) generalized type, i.e., it is composed of numerous rows of polygonal cells surrounded by a single peripheral row of tiny (ca. 12.1-15.6 µm X 9.8-13.7 µm) quadrate epidermal cells (Fig. 364). The cells are arranged in a noncruciate pattern so that the original quadrant division sequence of the embryo is no longer visible. Embryonic quadrant origins are similarly masked in the setae of most Metzgeriales, i.e., the setae are generalized (Douin 1908; Schuster 1966, 1984a). In H. hookeri, the central portion of the seta is composed of approximately 3-4 rows of enlarged, polygonal cells surrounded by ca. 4-5 rows of smaller cells and an epidermal row of even smaller quadrate cells (Fig. 365). The central cells are almost twice as large as the peripheral rows of cells, i.e., the central cells are about 39.2-47.1 X 23.5-31.4 µm while the surrounding rows are about 23.5-27.5 X 12.5-19.5 µm.

The immature seta of H. mnioides also has a distinct central zone of 3-4 rows of hyaline cells surrounded by 4-5 rows of chlorophyllous cells and 1 row of epidermal cells (Fig. 366). However, with the elongation and maturation of the seta, the central cells are either unable to expand to the same extent as the surrounding cells, or they are programmed to degenerate. In any case, the elongated H. mnioides seta is hollow with 4-5 persistent rows of polygonal cells and a single row of quadrate epidermal cells (Figs. 367, 368). Frequently, Pellia Raddi and Trichocolea Dumort. bear hollow setae which are formed when the internal cells rupture due to their inabilities to expand to the same degree as the peripheral cells (Schuster 1966).

The central zones of hyaline cells in the setae of H. hookeri and H. mnioides are superficially suggestive of conducting tissue. No pitted end walls were found in these cells, but extensive study was prevented by the scarcity of sporophytic material. No hepatics are known to have conducting cells in their setae, although sporophytic hydroids are present in some mosses (Hébant 1977). More intensive studies are needed to determine the nature of these central cells, since the existence of conducting tissue or precursory conducting cells in the setae of Haplomitrium would be of great phylogenetic significance.

The capsule walls of Haplomitrium and Steereomitrium are unistratose throughout most their lengths, but may be 2-3 stratose at the bases and apices (Figs. 359, 369a, 369b, 379). In species with beaked capsules, the apical portion may, in fact, be up to 6-stratose (Fig. 370). Although E. O. Campbell (1987) reported the capsule wall of S. minutum as being 3 to occasionally 4 cells thick, scanning electron microscopy clearly shows that at least the median portion of the capsule wall is unistratose (Fig. 379). Apparently, the capsule sections illustrated by Campbell (1987) were made near the apex or base of the capsule.

Haplomitrium is not alone in bearing capsules that are unistratose. Members of the Marchantiidae also possess capsules with walls which are unistratose except at the apices where they are bistratose

(Crandall-Stotler 1981; Schuster 1984a). In contrast, the capsules of most representatives of the Jungermanniidae are 2-8-stratose (Schuster 1966, 1984a).

With the exception of the smaller, quadrate cells comprising the extreme apex and base of the Haplomitrium capsule (Figs. 369b, 370), the cells are long rectangular (Figs. 371, 372) and measure about 46.2-98.4 µm long X 19.8-35.2 µm wide in surface view. Nuclei, chloroplasts with starch grains, mitochondria, dictyosomes, ER, ribosomes, vacuoles, oil bodies, and lipid droplets are present in the young capsule wall cells, but as the capsule matures, their vacuolar volumes increase and the organelles begin to degenerate (Figs. 380, 381).

Each cell of the unistratose capsule valves is traversed by a single, longitudinal annular thickening band (Figs. 269a, 369b, 371, 378). This band extends down the middle of the outer and inner tangential walls and across the transverse radial walls to form a complete ring around the cell. The longitudinal radial walls remain unmodified. The band is formed toward the interior of the cell and therefore does not actually constitute a projection on the outer wall surface. The bands are clearly visible with the light microscope because of their pigmentation which ranges from yellow-gold to orange-red at the radial edges of the cell where the radial and longitudinal portions of the thickening meet (Fig. 372). Occasionally, a thickening may split and a portion of the cell may have two incomplete bands, but generally, they are uniformly single.

When the inner surface of the capsule wall is viewed from directly above with the scanning electron microscope, the annular thickenings appear as dark bands (Fig. 372). However, when the inner surface is viewed at an angle, each thickening band appears as a bulge running along the midline of each cell (Fig. 373). This is due to the fact that the thickened band remains firm, while the rest of the inner surface of the cell wall collapses as a result of the fixation procedure. In contrast, the thickening bands are not visible when the external walls of the capsule cells are viewed using SEM because the external cell walls possess additional layers of materials and thus, do not collapse during the SEM preparation procedure (Fig. 374). The thickened outer cell wall with its internal thickening band is apparent with the transmission electron microscope (Fig. 387).

The materials which form the band appear to be deposited on the inner surface of the respective cell walls by Golgi vesicles which pass through the plasma membrane (Fig. 382). Initially, a thin layer of material is deposited over the entire inner surface of the wall (Figs. 380, 382). The plasmodesmata that traverse the capsule cell walls appear to be plugged by the deposition of this additional material (Fig. 383). Compared to the fibrous middle lamella and the electron dense cell wall, the thin zone of thickening material appears granular and more electron opaque (Fig. 382). Following this initial general deposition, additional wall material is heavily deposited along the mid-line of the two opposite, transverse, radial walls and along the long axes of the external and internal tangential walls of the cell to form the single continuous annular band (Figs. 381, 384-386). As this band increases in thickness, it becomes densely osmiophilic and reaches a width of about 4.0-6.0 µm (Figs. 387-389).

As can be seen in both light and TEM preparations, the annular bands of the capsule wall cells which lie end to end are constructed alternately on conterminous cells. In adjacent cells the thickening of one cell forms slightly behind the other and the "younger" thickening is deposited slightly off center so that the radial thickenings of two cells are not "back-to-back" (Figs. 371, 384, 385, 388).

The paucity of living sporophytic material prohibited performing histochemical tests on the capsule wall thickening, so its chemical composition remains a mystery. However, Downey and Basile (1989) have extracted lignothioglycolic acid from mature sporophytes of *Pellia epiphylla* (L.) Corda. Perhaps a similar hydroxyphenyl lignin may also be a constituent of the capsule wall thickenings of *Haplomitrium*.

The capsule wall thickenings found among the Jungermanniidae may be semiannular, nodular, reticulate, or sheetlike, while in the Marchantiidae, thickening patterns range from the extensive bar-type thickenings on three sides of the capsule wall cells of *Monoclea*, to a complete lack of thickenings in many other marchantioid taxa (Schuster 1966, 1984a; Crandall-Stotler 1981). Among the hepatics, only *Haplomitrium* possesses single, longitudinal, annular thickenings. These are found in all species of *Haplomitrium* and are unique to it. In addition to the thickness of the capsule wall, Campbell (1987) used the capsule wall thickening pattern as a criterion for establishing the genus *Steereomitrium*. She stated that the outer capsule walls of *S. minutum* bear trigone thickenings at the angles and the inner cells have thickened brown radial walls. However, as shown in Figure 378, each cell of the unistratose capsule wall actually bears a single, longitudinal, annular band identical to that found in *Haplomitrium*. The multistratose capsule wall and distinct thickening patterns reported by Campbell (1987) are apparently the result of misinterpretations. The walls of the capsule cells and the longitudinal annular thickenings are similar in color and may have been interpreted as cell walls instead of thickenings. In this case, when one focuses through the capsule wall using the light microscope, the "bottom" band may be mistaken as the cell wall of a second row of capsule cells. The nodule-like thickenings described by Campbell (1987) likely stemmed from the nodular appearance created at the points where the longitudinal bands of adjacent cells meet. In any case, the capsule of *Steereomitrium* is in fact identical in form and anatomy to that of *Haplomitrium* (Figs. 375-379).

Elaters

The spores and elaters contained within the capsule arise from the archesporium (D. H. Campbell 1920; E. O. Campbell 1959). Each archesporial cell divides to give rise to an elaterocyte and a sporocyte. The sporocyte undergoes a series of mitotic divisions before meiosis occurs, while no further divisions occur in the elaterocyte (Fig. 390). The latter cell merely elongates, its cellular constituents degenerate, and spiral thickenings arise to produce the mature elater (Figs. 391-402a). Although all the typical vegetative organelles are initially present in the elater, they rapidly degenerate leaving only the large starch grains (0.7-1.0 µm long X 0.4-0.7 µm wide), huge lipid droplets (0.09-1.3 µm in diameter), and membrane fragments recognizable (Figs. 402a, 402b, 403). These too ultimately degenerate (Figs. 404, 405).

The pattern of deposition and ultrastructural appearance of the spiral thickenings are identical with those of the capsule wall thickenings. Initially, a thin zone of granular, more or less electron opaque material is deposited over the entire length of the inner surface of the elater cell wall (Fig. 402b), while continued deposition in selected areas creates the spiral thickening band (Figs. 402a, 402b, 403). It should be noted that in the elaters, as in the capsule wall cells, the thickenings are really internal deposits, not external protrusions of the wall as might be hypothesized from SEM micrographs (Figs. 391-401). As the elaters dry, in nature as well as in SEM preparations, the non-thickened wall areas collapse inwards so that the bands, in fact, appear to project outwards. At the light microscope level, the bands, although internal, are discernable due to their distinctive pigmentation.

As the thicknesses of the spiral bands increase, the granular consistency of the material is masked and the thickenings become densely osmiophilic and somewhat fibrous in appearance (Figs. 404, 405). In

H. mnioides, external globular structures were seen adhering to the outer walls of the elaters (Figs. 402a, 402b, 403). These structures, which lack bounding membranes, resemble lipid droplets. In their SEM study of spores and elaters, Taylor et al. (1974) noted "prominent flap-like" ornamentations on the elaters of H. gibbsiae. These are possibly homologous with the globules shown adhering to the H. mnioides elaters in Figures 402a, 402b, and 403.

The mature elaters of Haplomitrium are typically very long (to 800 µm), attenuate, and about 6-18 µm in diameter at the midregion (Table IX; Figs. 391-399). The elaters may be 1- or 2-spiralled throughout, or bispiralled with 1-spiralled tips, or rarely trispiral in the midregion. In some cases, a single thickening pattern may be species specific, but in other cases, elaters with various spiral patterns may be intermixed within the same capsule (Table IX). Haplomitrium intermedium, H. hookeri, and H. gibbsiae display the greatest variability in elater form, while the elater morphologies of H. mnioides, H. blumii and H. andinum are more uniform.

Haplomitrium hookeri (Figures 391-393). Both elaters that are bispiralled with 1-spiralled apices (Fig. 391), as well as those bearing only a single spiral (Fig. 393) may be found in a single capsule of H. hookeri. Occasionally, elaters with bluntly branched apices also occur (Fig. 392). In addition, Siessegger (1966), Schuster (1967), and Udar and Srivastava (1983) have observed a few H. hookeri elaters with 3-spiralled midregions, but these appear to be rare.

The unispiralled elaters are shorter and stouter than those with 2 spirals (Table IX) and were found adhering to the apices of the valves of open capsules where they are intermixed with bispiralled elaters with 1-spiralled tips. Similar monospiral apical elaters were also noted in H. hookeri by Gottsche (1843) and Pearson (1902). In contrast, Macvicar (1926), Frye and Clark (1943) and Schuster (1967) reported elaters bearing a single spiral attached to the base of the capsule, while Udar and Srivastava (1983) described elaters with 1 or 2 spirals being located at the apex of the capsule and short, stout elaters with annular or 1-2 spirals positioned at the base of the capsule. These contrasting reports imply that the location of the short, monospiral elaters varies among individuals. The descriptions of groups of elaters being attached to the apex or base of the capsule suggests the presence of an elaterophore. However, serial longitudinal sections through an unopened _H. hookeri_ capsule revealed only a few short elaters positioned at the base of the capsule, and no distinct elaterophore (Fig. 369b), Similarly, no apical elaterophore was found (Fig. 369a).

Haplomitrium intermedium (Figure 394). The elaters of this taxon are similar in size to those of _H. hookeri_, but are shorter and more slender than those of _H. gibbsiae_ (Table IX). Most _H. intermedium_ elaters bear two loose spirals which merge into a single spiral at the ends (Fig. 394). Rare elaters also exist in which a short portion of the bispiral band splits into three. This phenomenon was also reported by Berrie (1962), while Schuster (1967) described only bispiral elaters with single spirals at the tips.

Haplomitrium gibbsiae and Haplomitrium chilensis (Figure 395). Like _H. hookeri_, the elaters of _H. gibbsiae_ may be bispiralled with 1-spiralled apices, or may be unispiralled throughout. However, unlike those of _H. hookeri_, both types of _H. gibbsiae_ elaters tend to be very long and tightly spiralled (Table IX; Fig. 395).

The variability in elater thickening patterns exhibited by _H. gibbsiae_ is reflected in published descriptions. Schuster (1967) reported bispiralled elaters with 1-spiralled apices, as well as elaters which are 1-spiralled throughout, while Hässel de Menéndez (1975) described only bispiralled elaters with unispiralled tips. Stephani (1924), E. O. Campbell (1959) and Allison and Child (1975) observed only bispiral elaters and Taylor et al. (1974) discussed elaters with 1 tight spiral.

Like *H. gibbsiae*, the elaters of *H. chilensis* are reportedly bispiral medianly with a single spiral at the apices, or they may be trispiral medianly with a single spiral at the apices (Schuster 1971). No other information regarding the elaters of *H. chilensis* is provided by Schuster (1971).

Haplomitrium indicum. Although no *H. indicum* sporophytes were available for examination in this examination, the elaters of *H. indicum* have been described by Mehra and Kumar (1990). According to these workers, the elaters are 400-800 µm long, 16.5-18.0 µm broad at the widest portion, and bear 2-3 spiral thickening bands. Occasional branched elaters were also reported by Mehra and Kumar (1990).

Haplomitrium mnioides (Figure 396). The least variability in elater structure occurs in *H. mnioides*, *H. blumii*, and *H. andinum*. All of these species produce elaters which are bispiral throughout (Fig. 396). The elaters of *H. mnioides* are relatively short and bear two loose spirals (Table IX; Fig. 396). This same morphology was reported for *H. mnioides* by Horikawa (1929) and Schuster (1967), although Schuster (1967) also described rare elaters with 1-spiralled apices.

Haplomitrium blumii (Figure 397). The elaters of *H. blumii* are uniformly long and have two tight spirals (Table IX; Fig. 397). This description concurs with those of Nees (1830), Campbell (1920), and Schuster (1967).

Haplomitrium andinum (Figures 398-399). This species exhibits a wider range in elater lengths, but like *H. mnioides*, its elaters have two loose spirals (Table IX; Figs. 398-399). However, both Spruce

(1885) and Fulford (1962) reported elaters with one or two spirals. According to Spruce (1885), like *H. hookeri*, the elaters at the apex of the *H. andinum* capsule are short and bear single spirals. Although only bispiralled elaters were observed in this investigation, elaters from only two capsules were examined due to the scarcity of available sporophytes. So, it is not improbable that both 1- and 2-spiralled elaters may occur in some individuals of *H. andinum*, since this is not uncommon for species of *Haplomitrium*.

Steereomitrium minutum (Figures 400-401). Although the dimensions of the elaters of *Steereomitrium* are similar to those of *Haplomitrium*, their spiral thickening pattern is quite different (Table IX). The elaters of *S. minutum* have one spiral medianly which splits into two at the apices (Figs. 400-401). Like *H. hookeri*, the elaters tend to adhere to the apices of the valves of open capsules. The elater description provided by E. O. Campbell (1987) differs in some respects from the observations presented here. She found the elaters to be 300-400 µm long and only 5 µm broad at their maximum width with an irregular thickening band which is bispiral in some places and unispiral in others, or in the form of rings. These discrepancies are perplexing since the elaters observed in this study were taken from the isotype material. In any case, the thickening pattern which characterizes the elaters of *Steereomitrium* is unlike any known in *Haplomitrium* elaters.

Spores

Although the precise details of sporogenesis in *Haplomitrium* are unknown, spore wall development has been described for *H. hookeri* by Brown and Lemmon (1986, 1987) and Brown et al. (1986). This developmental process is most similar to that found in *Pallavicinia lyellii* (Brown and Lemmon 1986, 1987) in that both of these taxa have callosic exine precursors present in meiotic prophase and a mature wall composed of an outer sculptured layer of tripartite platelets and an inner multilaminar layer. According to Brown and Lemmon (1986, 1987) and Brown et al. (1986), the prophase sporocyte of H. hookeri has deeply lobed cytoplasm and a fibrillar sporocyte wall. Hence, the sporocytes are similar to those of the Jungermanniopsida as opposed to being spherical as in the Marchantiopsida (Schuster 1966; Neidhart 1979; Crandall-Stotler 1984). Localized callose containing extracellular material is deposited within the spore mother cell wall in a specific pattern which reflects the ornamentation pattern of the mature exine (Brown and Lemmon 1986, 1987). By the time tetrads are formed, following cytokinesis, these exine precursors are well developed around the spores, except at the proximal surface where the spores are contiguous. Next, tripartite lamellae appear in the exine precursors, the foveolate layer is initiated by deposition of an electron dense material external to the plasma membrane between the cytoplasmic projections, an inner multilaminar layer forms through the aggregation of tripartite lamellae, and finally, an intine is deposited external to the plasmalemma as the outer exine continues to thicken (Brown and Lemmon 1986).

Haplomitrium mnioides spores at this late tetrad stage of development were observed in this study (Figs. 406-408). The cytoplasm of each young spore of the tetrad is dense with a central nucleus, numerous chloroplasts with large starch grains which measure about 1.8-2.3 µm long X 0.7-1.1 µm wide and multiple large lipid droplets with diameters of ca. 1.6-2.1 µm (Fig. 406). Exine precursors were no longer visible at this late stage, but the sporocyte wall as well as the distinctly layered tetrad wall were present. A more or less fibrillar intine lies just external to the plasmalemma (Fig. 407). This wall is approximately 0.49-0.53 µm thick. A thin, multilaminar inner exine layer composed of alternating dark and light bands lies to the outside of the intine, while the sculptured layer of trilaminar lamellae (the future sexine) caps the lower lying lamellae of the foveolate layer (Fig. 407). The tripartite lamellae contain the sporopollenin deposits (Heckman 1970; Brown et al. 1986). The spores are still held in tetrads when the sculpture is copiously deposited over the entire distal surface of the young spores, but is extremely limited on the proximal surface where the spores of the tetrad meet (Figs. 406, 408).

Like that of H. hookeri, the sporocyte wall of H. mnioides is composed of three distinct layers (Fig. 407). At this late tetrad stage, the innermost layer is flocculent and encompasses the tripartite lamellae of the outer exine. The middle layer of the sporocyte wall is finely granular and slightly more electron dense, while the outer layer is fibrous and about 0.16-0.19 µm thick. Thus, it is thicker than the central layer which measures only about 0.08-0.12 µm.

Sections through the mature spores of H. mnioides and H. blumii are shown in Figures 409-412, and 413a and 413b, respectively. As the spores mature, the sexine develops an aereolate appearance (Figs. 412, 413b), which according to Brown and Lemmon (1986) is due to acetolysis of some of the matrix. What then remains in the mature ornamentation is presumably mostly sporopollenin. A large volume of the cytoplasm is still occupied by aggregations of lipid droplets (Figs. 409, 410). Plastids with distinct thylakoids and starch grains are also evident (Fig. 410). A large nucleus containing a central nucleolus and karyosome is centrally positioned in the cell (Fig. 410). The most prominent features of the spore walls are the electron dense sporopollenin sculpture of the outer exine (sexine) and the opaque intine (Fig. 409, 411-413). The foveolate layer of H. mnioides is rather low, rising only about 0.11-0.14 µm above the multilaminar inner exine layer, while the tuberculate sporopollenin sculptures are about 0.78-1.1 µm tall (Figs. 411, 412). Miyoshi (1967) found a similar tuberculate sculpture on the spores of H. mnioides. In longitudinal section, the sexine of H. blumii is often branched and rises about 1.0-1.4 µm above the nexine (Fig. 413a).

The mature spores of Haplomitrium range in size from (15-) 18-23 µm in H. andinum to 26-35 (-39) µm in H. gibbsiae (Table IX) and 32-35 µm in H. chilensis (Schuster 1971). The patterns of sporoderm sculpture are quite ornate and are species specific (Table IX; Figs. 414-424). Thus, the spores of Haplomitrium are most similar to those of some members of the Metzgeriales not only in the pattern of spore wall development, but also in the spore size and the extent of ornamentation (Heckman 1970; Inoue and Hibino 1984; Schuster 1984a; Steinkamp and Doyle 1984; Brown and Lemmon 1986, 1987).

The spores of H. hookeri, S. minutum, H. mnioides, H. gibbsiae, and H. chilensis are similar in size and in exine sculpture, with the ornamentation composed of wart-like or rod-like projections (Table IX; Figs. 414-420). In contrast, the sporoderm patterns of H. intermedium, H. blumii, and H. andinum are composed of ridges (Table IX; Figs. 421-424).

Haplomitrium hookeri and Steereomitrium minutum (Figures 414-416). The spores of H. hookeri and S. minutum are virtually indistinguishable. Both are between 23 and 32 µm in diameter and are verruculate with the verruculae widely spaced on a nexine which appears relatively low and smooth (Figs. 414-416). The verruculae are often broadened distally to form an irregularly-shaped cap. These wart-like projections may occur singly or sometimes fuse with one another to form abbreviated ridges (Figs. 414, 415b, 416b). The dimensions of the verruculae and interstices of these two species are almost identical. The interstices between verruculae in both H. hookeri and S. minutum are about 0.8-1.8 µm. The verruculae of H. hookeri have diameters of approximately 1.7-2.2 µm and are about 0.9-1.1 µm tall. In S. minutum these projections have diameters of about 1.6-2.2 µm and approximately 0.9-1.2 µm tall. The spores of H. hookeri do differ from those of S. minutum in one respect. Some of the expanded apices of the H. hookeri verruculae have minute, central apertures in the projections (Fig. 415b). These apertures were not seen in the verruculae of S. minutum (Fig. 416b). However, since the apertures in H. hookeri are probably formed through acetolysis, their presence or absence is not of great taxonomic consequence.

Although most descriptions of the spores of H. hookeri are based on light microscope observations, the spores were observed using electron microscopy by Udar and Srivastava (1983, 1984), Brown and Lemmon (1986, 1987), and Brown et al. (1986). The description presented here concurs with their general observations.

The spores of S. minutum were also examined using scanning electron microscopy by E. O. Campbell (1987). She described the spores as being ornamented by dome-shaped projections, but the micrograph in the protologue shows a spore which is identical with those described in this investigation.

Haplomitrium mnioides (Figures 417a, 417b). The spores of H. mnioides are only slightly larger than those of H. hookeri and S. minutum, but are densely tuberculate. The sporopollenin projections are in the form of small, rod-like structures, which, unlike H. hookeri and S. minutum have parallel walls and are not dilated distally (Fig. 417a, 417b). The tubercules are never fused and each is about 0.78-1.1 µm tall and approximately 0.61-0.80 µm in diameter. They are closely spaced with interstices of only about 0.4-0.7 µm.

Haplomitrium gibbsiae (Figures 418-420). The spores of H. gibbsiae tend to remain in tetrads or diads until very late in development (Fig. 418). Thus, in sporophytes in which the seta has elongated and the capsule has browned, the spores, which are fully ornamented, are still held in tetrads. The only other members of the genus with spores that tend to remain in tetrads or diads until this late stage in development are *H. chilensis* (Schuster 1971), *H. dentatum* (Kumar and Udar 1976), *H. indicum* (Mehra and Kumar 1990), and occasionally *H. hookeri* (Fig. 414) (Udar and Srivastava 1984).

The spores of *H. gibbsiae* are of average size for the genus (Table IX). The sporoderm is coarsely verrucate with large interstices and a coarse foveolate layer. The verrucae may be irregularly roughened (Figs. 418) or smooth (Fig. 419, 420) and are approximately 1.8-3.0 µm tall with diameters of about 2.0-3.0 µm. These projections are about 1.7-3.1 µm apart. Occasionally, small wart-like projections dot the interstices between the large verrucae (Fig. 418).

Hässel de Menéndez (1975) described the rod-like projections of the exine of *H. gibbsiae* as being 2.0-2.5 µm in height and 2.0-4.0 µm wide. Taylor et al. (1974) described rod-like projections with straight, smooth sides, while Schuster (1967) described either smooth or roughened, obovoid to subclavate, strongly elevated "papillae". Thus, it is highly probable that the dimensions of the verrucae as well as their roughened or smooth appearance is variable among individuals of this species.

Schuster (1971) relied heavily on spore characters to distinguish *H. gibbsiae* and *H. chilensis*. He described the spores of *H. chilensis* as being 32-35 µm in diameter with low, numerous, but distantly spaced, small tubercules. Schuster (1971) noted that the spores of *H. chilensis* are larger than those of *H. gibbsiae* and bear papillae which are under 1.5 µm tall, thus being lower and smaller than the typical projections of *H. gibbsiae*. Hässel de Menéndez (1975) challenged this by reporting *H. gibbsiae* spores with diameters of 29-36 µm. Her measurements are confirmed by the results of this study (Table IX), thus nullifying spore size as a useful character for distinguishing between *H. gibbsiae* and *H. chilensis*. The heights of the sporoderm projections do differ between the two species, but, considering the variability in the heights of the verrucae within _H. gibbsiae_, this appears to be an individual character as opposed to a specific trait.

Haplomitrium indicum. No spores of this species were seen in this investigation, however, Mehra and Kumar (1990) have provided descriptions of these cells. The spores of _H. indicum_ reportedly tend to remain in diads or tetrads until late in development, but may also be single. They are yellow-brown, finely papillose, and have diameters of about 21.6-22.43 µm (Mehra and Kumar 1990).

Haplomitrium dentatum. No spores of this species were available for examination, but Kumar and Udar (1976) have provided a brief description of these cells. As previously mentioned, the spores of _H. dentatum_ tend to remain in diads until late in development, but may also be single. They are relatively small with diameters of about 18.9-21.6 µm and are finely papillose (Kumar and Udar 1976). Udar and Srivastava (1983, Fig. 11, p. 41) published a scanning electron micrograph of one of these spores, but it is of poor quality. The spore is collapsed and the photograph is blurred, making the details of sporoderm sculpture impossible to determine.

Haplomitrium intermedium (Figures 421a, 421b). The spores of this taxon are relatively small and are covered by a densely ridged sporoderm (Table IX). Each ridge is approximately 0.35-0.70 µm broad and is vermiform (Figs. 421a, 421b). The vermiculae are quite extensive, being curved or bent in places to form irregular patterns. Occasionally, a few of the worm-like ridges fuse with one another forming loops (Fig. 421b). This fusion, along with the extensiveness of the ridges, and the very small interstices give the impression that, in some areas, the ridges are traversing over one another (Fig. 421a, 421b).

The only other descriptions of H. intermedium spores have been based on light microscope observations. Berrie (1962) described the spores as being 21-28 µm in diameter with a finely punctate sculpture. This description certainly does not agree with the pattern seen here. Schuster (1967), on the other hand, reported the spores as having diameters of 20-23 µm and being covered with small, close, irregular ridges which are often branching or anastomosing which is in accordance with these observations.

Haplomitrium blumii (Figures 422-423). The diameters of H. blumii spores are similar to those of H. intermedium and the sporoderm sculpture is also composed of relatively short ridges which anastomose. However, in H. blumii the ridges are widely spaced (Figs. 422-423). Each ridge is about 0.49-0.58 µm broad. Sometimes, the ridges are not completely anastomosing in which case one end of a ridge tends to bend back on itself to produce incomplete J-shaped ornamentations (Figs. 422a, 422b). The interstices between the ridges are very large and measure about 1.1-2.8 µm. The lamellae of the nexine are slightly raised and reticulate (Fig. 422a, 422b).

All other descriptions of H. blumii spores are based on light microscope observations. Campbell (1920) described the spores as bearing numerous, short, blunt projections. Schuster (1967) also interpreted the projections as being short, blunt papillae, but added that these papillae are partly fused into short ridges.

Haplomitrium andinum (Figures 424a, 424b). This taxon produces spores which tend to be slightly smaller than those of H. blumii, being only about (15-) 18-23 µm in diameter. As in H. blumii, however, the exine is sculptured into distantly spaced short ridges which may anastomose, positioned above a reticulate foveolate layer (Figs. 424a, 424b). The sculptured elements are widely spaced with the interstices being about 0.7-1.5 µm. Some of the ridges are short and bent in the middle to various degrees. Others are longer with one end curved to form a J-shape or rarely, a small loop (Fig. 424b).

Spruce (1885) and Fulford (1962) have provided the only other descriptions of H. andinum spores, and these were based on light microscope observations. Spruce (1885) noted that the spores are minute, barely exceeding the diameters of the elaters, and bear verruculose-striate ornamentations. Fulford (1962) reported the spores to be 18 µm in diameter with a striate-verruculose exine.

The exine sculpturing patterns of the spores provide evidence that the species of Haplomitrium are all very closely related, yet are different enough to demonstrate species distinctions. In addition, the similarities in the spore morphologies of H. hookeri and S. minutum suggest that these taxa are very closely related. Spore morphology is considered to be a good taxonomic character both at the species level and at higher levels of the hierarchy depending on the group involved because spores are believed to be less variable than many other parts of the plant (Clarke 1979).

The sculpture patterns within the genus appear to be based on simple patterns involving wart- or rod-like ornamentation with increased complexity attained through varying degrees of fusion to yield ridged patterns. A similar condition was found among the spores of the Schistochilaceae of South America (Schuster and Engel 1977). In this family, Schuster and Engel (1977) proposed an evolutionary progression from the primitively simple papillose to tuberculate patterning, to the imperfect aggregation of papillae into short vermiculate lines, and ultimately, into anastomosing networks. Although a distinct morphological progression based on the fusion of exine elements is clear in Haplomitrium, the evolutionary progression of species based on sporoderm features is not necessarily linear. It seems more likely that, within Haplomitrium, at least two lines of spore morphologies have diverged from a sporoderm type like that found in H. hookeri, i.e., one which shows both single wart- or rod-like projections as well as limited fusion into abbreviated ridges. From this pattern, spores which lack the fusion of exine elements, such as those of H. mnioides and H. gibbsiae could emerge. In a different line, fusion may have been intensified to yield patterns like those of H. andinum, H. blumii, and H. intermedium spores. Spore size also fits this scenerio in that H. hookeri spores are more or less intermediate in size for the genus, i.e., 28-33 µm in diameter, while those of H. mnioides, H. gibbsiae, and H. chilensis fall within the upper limits of this range (26-35 µm and 28-35 µm, respectively) and those of H. andinum, H. blumii, and H. intermedium are at the lower limits (15-23 µm, 22-30 µm, and 23-38 µm, respectively).

While the dimensions of the elaters of Haplomitrium most resemble those of the Marchantiales (Schuster 1984a), the pattern of spore wall deposition as well as the range of spore sizes and the patterns of sporoderm sculpture appear to be most similar to those of the Metzgeriales, particularly the Pallaviciniaceae. The micrographs of Aneura pinguis (L.) Dum., Riccardia kodamae Mizut. and Hatt., and Makinoa crispata (Steph.) Miyake spores presented by Inoue and Hibino (1984) show ornamentation patterns similar to those of Haplomitrium. Aneura pinguis and R. kodamae spores are similar to those of H. hookeri, but show slightly more fusion of the verruculae into J-shaped ridges (Inoue and Hibino 1984, p. 188, fig. 6, p. 189, fig. 2). The spores of M. crispata have irregularly anastomosing vermiculose ridges (Inoue and Hibino 1984, p. 187, fig. 1) which are similar to those found on H. intermedium spores. Therefore, in spore sculpturing, the spores of M. crispata fall somewhere between those of H. intermedium and H. blumii.

Capsule Dehiscence

The capsules of Haplomitrium open along predetermined lines. Dehiscence may be along 1, 2, 4, or rarely 3 or 5 lateral sutures which may open completely or incompletely depending on the species or individual involved (Figs. 425-427). The pattern of capsule dehiscence is directly related to the pattern of thickenings which characterize the capsule wall cells. While the typical thickening pattern of the capsule wall cells is that of a longitudinal annular band, the special cells comprising the dehiscence lines possess transverse thickening rods (Figs. 425, 426b, 426c, 428). So, as the capsule dries, the longitudinal, radial walls collapse inwards, while the special cells bearing transverse rods do not. The pull generated on both sides of the dehiscence cell row by the decreasing widths of the surrounding cells ultimately causes the capsule to tear along one side of the dehiscence line. The special row of dehiscence cells remain attached to one of the valve margins (Fig. 428). The variable number of final lines of dehiscence depends on both the number of special cell rows differentiated in the capsule, as well as the number that actually are physically torn during the process (Fig. 426a, 427). The lack of apical tearing which occurs in some species is probably at least partially correlated with the presence of polystratose capsule apices (Fig. 426a, 426b).

The patterns of dehiscence are summarized for known species in Table IX. Dehiscence patterns are most variable in H. hookeri and H. gibbsiae, and are most uniform in H. mnioides, H. blumii, and H. andinum. The patterns are unknown in H. intermedium (Berrie 1962; Schuster 1967) and H. dentatum (Kumar and Udar 1976).

Most herbarium specimens of H. hookeri in which dehiscent capsules were present revealed that they had opened completely along two opposite, lateral lines of weakness to create two blunt valves which are completely separated at the apex (Fig. 425). In fact, this mode of dehiscence is demonstrated by the neotype of *H. hookeri*. Other capsules were seen which had opened along four lines, either completely to form four individual valves, or incompletely so the valves remain attached at the apex. In addition, at least one collection made by Hooker in 1812 (*Hooker s.n.* [BM]) contained a capsule which had burst along 5 lines, 3 of which are united at the apices.

Apparently, although four lines of weakness occur in the capsules of *H. hookeri*, in some instances, some of the sutures, or portions of them, may be nonfunctional (Worley 1969). Schuster (1984a) cited examples of Jungermannialian taxa with imperfectly functional lines of dehiscence as well as some species of hepatics in which accessory suture lines form in the capsules. This latter phenomenon could explain the presence of five lines in *H. hookeri*.

Dehiscence in *H. hookeri* has been attributed to 2 or 4 lines of weakness by Carrington (1874) and Frye and Clark (1943), while Gottsche (1843), Evans (1917), and Worley (1969) reported that the capsules open completely or incompletely along four lines of weakness. In addition, Pearson (1902) and Frye and Clark (1943) noted that dehiscence may be restricted to just one line of weakness.

According to Udar and Srivastava (1983), the capsules of Indian *H. hookeri* plants usually open along a single longitudinal slit which originates near the middle of the capsule and progresses toward the apex and base. The capsule walls then roll back to release the contents of the capsule. They also noted that the Indian capsules occasionally open incompletely along two opposite lateral lines. These workers concluded that a regional distinction exists between the mode of dehiscence in capsules of European and American *H. hookeri* plants and Indian populations in that the former dehisce into four valves which remain united at the apices and bases of the capsules. However, this conclusion is not valid since European

H. hookeri capsules may also dehisce along 1 or 2 lines of weakness, as well as 4 complete or incomplete sutures.

Similar dehiscence patterns occur in *H. gibbsiae*. According to Campbell (1959), the capsule usually opens into 4 valves, but occasionally into 2, 3, or 5 valves which remain united at the apex and the base. As is the case in *H. hookeri*, multiple, sometimes incomplete dehiscence lines as seen in Figures 426a-c, are responsible for these phenomena.

Dehiscence in *H. mnioides*, *H. blumii*, and *H. andinum* is generally along a single lateral suture which allows the capsule walls to fold back on themselves and expose the spores and elaters. Therefore, dehiscence in these taxa is like the "spoon-type" of dehiscence described for *Monoclea* (Campbell 1954). This pattern has been observed in *H. mnioides* in this investigation and has also been described by Schuster (1967). A single dehiscence line has also been reported in *H. blumii* by Nees (1830), Andreas (1899), D. H. Campbell (1920), and Schuster (1967), and for *H. andinum* by Spruce (1885).

Campbell (1987) described the capsule of *Steereomitrium minutum* as opening from the apex toward the base along four lines to create four valves. However, one capsule included in the isotype collection has opened completely along two opposite lateral lines of weakness to form two blunt valves just as in *H. hookeri* (Fig. 377).

While some species of *Haplomitrium* dehisce along 2, 3, or 4 lines of weakness in a manner similar to many members of the Jungermanniidae, other *Haplomitrium* species dehisce along a single suture such as occurs in the Marchantiidae. *Haplomitrium* also shares other systematically significant traits with these two hepatic groups. For instance, it is similar to members of the Jungermanniidae, particularly the Metzgeriales, in seta anatomy, capsule shape, spore size, spore morphology, and sporoderm development and sculpture. In capsule wall thickness and elater length, *Haplomitrium* is most like

Monoclea which also possesses an elongating seta, and a cylindrical-shaped, unistratose capsule which dehisces along a single suture. However, the mechanism of dehiscence in *Monoclea* probably differs from that of *Haplomitrium* due to its different capsule wall thickening pattern. In any case, the single, longitudinal annular thickenings which traverse the cells of the unistratose capsule wall of *Haplomitrium* are unique and distinguish *Haplomitrium* from all other bryophyte taxa.

Sporeling Development

Ontogenetic Pattern

Historically, attempts to determine the pattern of sporeling development of *Haplomitrium* have been mostly unsuccessful, in that the spores rarely germinated, or if they did, the protonemata usually died early in development. Gottsche (1843) tried in vain to germinate *H. hookeri* spores on moistened sand. Goebel's (1891) germination experiments also failed, and according to him, no one had observed spore germination in *Haplomitrium* as of 1891. This led to the belief that some special factor was necessary for germination to occur. Then, in 1959, E. O. Campbell successfully germinated *H. gibbsiae* spores on andesitic clay. The resulting protonemata failed to develop beyond a 7-celled stage, however. A few years after Campbell's attempt, Nehira (1961, 1966) germinated *H. mnioides* spores on half strength Knop's solution. But, as was the case with *H. gibbsiae*, protonemal growth ceased at the cell mass stage. Only Yang (1967) and Yang et al. (1968) were able to grow the sporelings of *H. mnioides* and *H. blumii*, respectively, to maturity by transferring the sporelings, which had become cell masses, from a liquid 50% Hoagland solution to a semisolid 50% Hoagland + 1% agar medium. However, their descriptions of the ontogenetic patterns were without specific developmental details of the four stages reported, which were based on the general shape of the sporelings.

In this study, sporeling ontogeny was investigated in both H. *mnioides* (Figs. 429-439; 448-471) and *H. blumii* (Figs. 440-447). Mature spores of both taxa were sown on five types of agar media (Beneke [1903], Hurey-Py [1948], Bold [1949], Hatcher [1965], and Basile's F_1 [1975]) and placed in an environmentally controlled chamber. Germination was observed on all media within 32 days for *H. mnioides* and 70 days for *H. blumii*. The sporelings were placed in a depression slide and manipulated so each protonema could be observed from multiple angles (Bartholomew 1985). The observed early stages of ontogeny were identical for both taxa (compare Figs. 429-439 and 440-447). The spores, which are originally 25.0-35.0 µm in diameter in *H. mnioides* (Fig. 417a) and 19.0-32.0 µm in diameter in *H. blumii* (Figs. 422a, 423), imbibe water, become greener, and swell to diameters of 39.6-44.0 µm and 35.4-42.2 µm, respectively. As each spore expands, the physical stresses placed on the spore wall cause it to stretch and finally rupture at the trilete mark, the weakest point, uncovering a portion of the spore (Fig. 429). Since all divisions occur outside the spore wall, the sporelings are truely exosporic.

The exosporic nature of *H. mnioides* was first recognized by Nehira in 1961. Later, he illustrated the sporoderm covering the first dividing wall of the spore and added that the exospore remained on part of the multicelled protonema (Nehira 1966). Yang (1966, 1967) and Yang et al. (1968) were unable to determine when the spore wall was torn in either *H. mnioides* or *H. blumii*. Likewise, Campbell (1959) seemed uncertain of the pattern in *H. gibbsiae*, stating that sometimes the spore wall breaks open at the 2- or 3-celled stage, while other times a hyaline mucilage hair emerges first.

The confusion regarding the spore wall dehiscence pattern may be due to difficulties in determining the extent of the spore wall. The thin nature of the stretched sporoderm sometimes makes its limits hard to define. The amount of the sporeling covered by sporoderm also varies slightly according to the angle at which the protonema is viewed.

The emergent spore is first partitioned transversely to form a basal cell and slightly smaller terminal cell (Figs. 430, 440). The basal cell, which is dense with plastids and is covered by sporoderm remnants, remains undivided while four apical quadrants are formed in the terminal cell as follows. The cell is divided into two cells by a curved vertical wall (Figs. 431a, 431b, 441). Each daughter cell is then divided by a second vertical wall, which in each case extends perpendicular to the first division wall (Figs. 432a, 432b) to form four quadrants in apical view (Figs. 433a-c, 442). Shortly after quadrant formation, transverse divisions cut the four cells sequentially to produce two tiers, each composed of four cells as can be seen in lateral view (Figs. 434a-c, 435, 443). Each cell of the two tiers is then traversed one at a time by a vertical wall to form a protonema consisting of two tiers of terminal cells, each composed of eight cells, subtended by an undivided basal cell, still partially covered by the sporoderm (Figs. 436, 437a, 437b, 444a, 444b).

Although quadrant and tier formation has not been described by previous workers, some H. mnioides protonemata illustrated by Nehira (1961, Figs. 5-7, p. 187) and Yang (1966, Plate II, Figs. 4-6 and 11, p. 16) suggest the existence of this sequence. For example, the apical quadrant shown in Figures 433c and 434c are identical to Yang's (1966) Figure 4. Similarly depicted are lateral views representing the formation of double tiers (compare Figs. 436-439 with Yang 1966, Figs. 5-6, and 11). In the figure legend Yang (1966) classifies Figures 5-6 and 11 as stages in cell mass formation, overlooking the details of the pattern. The intricacies of the pattern became evident in this investigation only when the protonemata were manipulated using the technique outlined by Bartholomew (1985) which allowed the cells to be viewed from numerous angles.

After formation of the two tiers, generalized divisions occur in all cells including the basal cell, so no recognizable pattern of development is discernable. These generalized divisions are responsible for the formation of globose cell masses (Figs. 438a, 438b, 445a, 445b, 446a, 446b) For at least the next 60 days, divisions continued and the size of the protonemata increased, but no distinctive developmental patterns were observed (Figs. 439a, 439b, 447). At the time of this writing, the H. blumii sporelings had not developed beyond this stage of development. A similar massive stage has been reported for all three species of Haplomitrium in which sporeling ontogeny has been studied (Campbell 1959; Nehira 1961, 1966; Yang 1966, 1967; Yang et al. 1968).

The next stage of development is characterized by the production of numerous, leafless, cylindrical axes arising from all surfaces of the massive protonema (Figs. 448, 449). In some cases, protonemata in the early period of shoot development appear triangular in outline, especially if one shoot dominates and grows to a greater length than the others (Figs. 448, 449). This stage likely corresponds with the "triangular-thalloid stage" cited by Yang (1967), while his "rhizomatous stage" probably corresponds to the later period when the shoots have reached much greater lengths and are more conspicuous.

The cylindrical shoots are built through repeated tier formation via quadrants which form in cylindrical shoot initials. No apical cell is involved. Four quadrants are formed in one of the protonemal cells by three divisions as previously described (Fig. 450). One of the four quadrant cells is then partitioned by a curved wall which connects its two radial walls. In apical view, this yields a single central cell, surrounded by four peripheral cells (Figs. 451-453). In surface view, the central cell often appears somewhat triangular in outline (Fig. 452a). This arrangement also appears in the development of _H. hookeri_ gemmae (Furuki 1986). The fate of the central cell was not determined in gemmae ontogeny, but in _Haplomitrium_ sporelings, it swells (Fig. 452b) and functions as a cylindrical shoot initial. Following an anticlinal division in one of the peripheral cells to yield six peripheral cells (Fig. 453), four new quadrants are formed in the swollen central initial by three walls (Fig. 454). In this way, the length of the shoot is increased by one tier. A new cylindrical shoot initial is then delimited in one of the quadrants by a fourth wall and the pattern is repeated. The elongation process continues in this manner as multiple tiers are formed through repeated cylindrical shoot initial and quadrant formation.

Concomitant with the increase in shoot length is an increase in girth from the apex to the base of the cylindrical shoots. The diameter of each tier is determined by a sequence of vertical and longitudinal divisions within the peripheral row of cells, i.e., the epidermal cells. Immediately after a tier is formed, cells are added to the core of the cylinder by periclinal divisions in the peripheral row of cells followed by subsequent divisions in the inner cells themselves. Concomitant anticlinal divisions in the peripheral cells increase the shoot circumference to keep pace with the growing diameter of the tier brought about by the dividing inner cells (Figs. 455a-d). The divisions occur in a fixed pattern and produce distinct cell arrangements within the tiers. In a progressive sequence from the apical to the basal tiers of a shoot, the fundamental cell arrangements are as follows: 4 epidermal cells around 1 central cell (Figs. 451, 452), 6 epidermal cells around 1 inner cell (Figs. 453, 455a), 8 epidermal cells around 4 inner cells (Fig. 455b), 10 epidermal cells around 8 inner cells (Fig. 455c), 12 epidermal cells around 8 inner cells (Fig. 455d), 14 epidermal cells around 8 inner cells, and 16 epidermal cells around 8 inner cells.

Slime cells are generated from epidermal cells and occur at numerous locations along the cylindrical axes (Figs. 455c, 456). An epidermal cell of a tier swells and is cut into two unequal daughter cells. The inner, larger daughter cell remains an epidermal cell, while the smaller cell elongates to form a unicellular slime papilla. In addition to those of the cylindrical shoot, certain epidermal cells of the massive protonema may enlarge and become active unicellular slime cells (Fig. 457). So much mucilage is exuded by these slime cells that the protonemata stick together in clusters. This clustering was cited as a distinctive feature of the sporeling ontogeny of H. blumii by Yang et al. (1968). However, this phenomenon is merely a consequence of copious slime production and, in light of the slimy nature of Haplomitrium, is probably characteristic of the sporelings of all species of the genus.

Oil bodies initially become recognizable in the cells of the shoots in the form of a single, glistening spherical body located in the center of each cell. As development proceeds, the number of spherical oil bodies per cell increases (Fig. 458). Eventually, different shapes arise, so by the juvenile stage, the cells contain numerous, adult-like oil bodies (Fig. 459).

Ultimately, a tetrahedral apical cell is created at the apex of each cylindrical shoot when a cylindrical shoot initial begins to segment as an adult apical cell (Figs. 460, 461a, 461b). This generative cell is slightly sunken in the shoot tip and in surface view, appears as an equilateral triangle with each side measuring approximately 22 µm in length. As derivatives are cut parallel to each of these three faces in a spiral sequence (Figs. 461a, 461b, 462), the dorsal cutting face is slightly reduced in length to approximately 17.5 µm (Fig. 461a). Unicellular slime papillae overarch and protect the apical cell as it segments (Figs. 462a, 462b, 464). The merophytes develop as in the adult, i.e., the central cell of the 3-celled merophyte yields a leaf while the other two cells give rise to the stem (Figs. 462, 464). Three rows of leaves are thus produced which coincide with the three segmenting faces of the apical cell. The leaves of the dorsal row are transversely inserted and slightly smaller in size than the obliquely inserted lateral rows (Fig. 463).

Leaf development is as described in the adult plant by Bartholomew (1987). The initial growth is through the activity of a wedge-shaped initial (Fig. 462) which segments to produce a triangular-shaped leaf (Fig. 464), followed by basal divisions in the lamina which extend the leaf to its final form (Fig. 465). Also, scales, which are formed from divisions in the cells at the base of the slime papillae before the onset of leaf production, can be found posterior to the leaves just as in the adult plant (Fig. 463).

Concomitant with this apical cell activity is the differentiation of the central conducting strand. A short distance below the apical cell, the elongate, thin-walled cells of the hydroids begin to form (Figs. 466, 467).

Interestingly, erect leafy axes and not the rhizome, are the first shoots produced. Subsequent exogenous branching from the leafy shoots gives rise to more leafy shoots and the rhizomatous system (Figs. 468-470). Branching may occur at numerous points along the axis, so a single leafy shoot may produce multiple branches (Figs. 463, 469, 470). Leafless stolons are similarly generated through branching from the rhizome. As the juveniles continue to age, the cells of the massive protonema gradually degenerate. Thus, a miniature *H. mnioides* plant is formed, complete with a rhizomatous system, erect leafy shoots, and stolons (Fig. 471).

Pattern Comparisons and Phylogenetic Implications

The limited descriptions of sporeling ontogeny in *H. mnioides* presented by Nehira (1961, 1966) and the general descriptions outlined by Yang (1966, 1967) agree well with the pattern described in this study despite differing environmental conditions. In addition, the early stages of H. gibbsiae (Campbell 1959), and the pattern outlined for H. blumii (Yang et al. 1968) appear to be like that of H. mnioides. Presumably, then, the fundamental pattern of sporeling development as described for H. mnioides is constant for the species and probably for the genus.

The sporeling ontogeny of Haplomitrium was initially classified by Nehira (1961), as belonging to Fulford's (1956) Nardia type which is characterized by the exosporic development of a globose, ovoid, or cylindrical protonema. Nehira (1961) suggested that this type of pattern could indicate a phylogenetic link with the Jungermanniales, since the Nardia type is often observed in most genera of leafy hepatics. In 1966, Nehira removed Haplomitrium from the Nardia type and placed it, along with Megaceros Campb., into the newly created Calobryum type. Yang (1967) also noticed the resemblance of the young Haplomitrium protonemata with those of the Anthocerotales, but added that the leafy gametophyte and "triangular thalloid stage" were reminiscent of Sphagnum protonemata.

I feel the morphological resemblance between the protonemata of Haplomitrium and the anthocerotes is purely superficial and a close relationship between the two is not supported by developmental evidence. The globose protonemata of Megaceros form exclusively through generalized divisions (Nehira 1963; Renzaglia 1978). There is no quadrant or tier formation leading to the massive stage in the hornworts as there is in Haplomitrium. Likewise, a comparison of the protonemal development of Haplomitrium with that of Sphagnum as described by Nishida (1978) reveals no similarities.

Quadrant development and the formation of two tiers, each of four cells, is a well documented stage in the sporeling ontogeny of the Marchantiales (Inoue 1960; Nehira 1983). However, this stage is preceded by germ tube and filament production and followed by the formation of a cell plate, events which are not reminiscent of the pattern found in Haplomitrium.

The quadrant and tier formation that characterize the sporeling pattern of Haplomitrium are further reflected in the early stages of other thalloid hepatics, such as Fossombronia and Monoclea. The early stages of sporeling ontogeny in Monoclea forsteri Hook. were described by Proskauer (1951) and later in more detail by E. O. Campbell (1954). According to Campbell (1954), the spore wall stretches but remains unbroken as the spore swells, but she does not say when the wall is ruptured and this cannot be determined from her illustrations. In contrast, Proskauer (1951) stated that the sporoderm stretches and is finally ruptured with the swelling of the spore, making the germination pattern exosporic, just as in Haplomitrium. Campbell (1954) goes on to describe the production of four apical quadrants by the same divisional sequence as occurs in Haplomitrium. At this point, the patterns apparently deviate, however. According to Campbell (1954), an apical cell with two cutting faces is delimited in Monoclea by an "inclined wall in one of the cells at the quadrant or three-celled stage". The other cells of the quadrant divide a few times, but further development is through the activity of the apical cell. Also, unlike Haplomitrium, rhizoids are formed in Monoclea protonemata at the 1- or 2-celled stage (Proskauer 1951; Campbell 1954) and the sporeling apical cell is converted from one with two cutting faces to a wedge-shaped cell (Campbell 1954).

Early stages of sporeling ontogeny in the Metzgerialian taxon Fossombronia are even more similar to those of Haplomitrium. According to Renzaglia and Bartholomew (1985), Fossombronia is exosporic and, after the first transverse division of the spore, immediately forms four apical quadrants. However, as transverse walls delimit the two terminal tiers in Fossombronia, a tetrahedral apical cell is created by one additional division in the apical quadrant.

Although a tetrahedral apical cell is also formed from four quadrants in Haplomitrium, this does not occur until much later in ontogeny. Segmentation of the tetrahedral apical cell, and merophyte development to form an axis bearing three rows of leaves in which one leaf is reduced in size is also similar in the two patterns. However, this sequence of leaf generation does not continue in Fossombronia, for the tetrahedral apical cell is transformed into a lenticular initial as the plant adopts a prostrate growth form (Renzaglia and Bartholomew 1985).

Takakia has long been considered closely related to Haplomitrium and although no Takakia sporelings are known, its regenerants superficially resemble the sporelings of Haplomitrium (Crandall-Stotler, unpublished data). The regenerants are massive with four quadrants of cells present on various surfaces of the cell mass. However, before any phylogenetic inferences are made based on the sporeling/regenerant ontogenies of these taxa, the detailed ontogeny of the Takakia regenerants must be ascertained.

Recently, Murray (1988) suggested the possibility of a phylogenetic relationship among the Andreaeopsida, Takakia, and Haplomitrium. Indeed, the protonemal stage of Andreaeobryum Steere & Murray, as documented by Murray (1988), is similar to that of Haplomitrium in that numerous terete filaments form from a massive protonema before apical cell formation occurs. The single cell which terminates each Andreaeobryum cylinder differs from the cylinder initial of Haplomitrium in shape and size and it does not appear to be cut into a quadrant of four cells as occurs in the central initial of Haplomitrium. Each Andreaeobryum cylinder is constructed of tiers, however, and as illustrated by Murray (1988, pls. 1; 5 and 2; 9), transverse sections through these axes show a single peripheral row of approximately 9-12 cells surrounding a central zone of 7-8 cells. This internal pattern is reminiscent of that found in transections of Haplomitrium cylinders. Also, as in Haplomitrium, an apical cell forms in each cylindrical shoot of Andreaeobryum resulting in the production of numerous leafy shoots from a single protonema. Decisive differences in the sporeling pattern of Andreaeobryum when compared with that of Haplomitrium include endosporic and precocious germination, the production of germ tubes, caulonemata, chloronemata, rhizoids, and beaked mucilage cells in the former.

The multiseriate filamentous, or cylindrical form is a frequently observed building block in the sporeling and regenerant ontogeny of numerous bryophyte taxa whether multiple axes are produced from a single protonemata as in Haplomitrium and Andreaeobryum sporelings (Murray 1988) or Takakia regenerants (Crandall-Stotler, personal communication), or a single multiseriate filament is incorporated in the developmental scheme. The latter condition occurs in Marchantialean protonemata (Inoue 1960; Nehira 1983), numerous Metzgerialean protonemata such as Fossombronia (Bartholomew 1985; Renzaglia and Bartholomew 1985), Pellia (Bartholomew 1985), Makinoa Miyake (Nehira 1966, 1983; Bartholomew 1985), sporelings of the Pallaviciniaceae (Nehira 1983; Bartholomew 1985, 1986), and the anthocerotes Dendroceros japonica Steph. (Nehira 1963, 1983) and Anthoceros L. (Nehira 1983). The incorporation of a cylindrical stage in the developmental pattern is not necessarily indicative of an intimate phylogenetic relationship among these taxa. As evidenced by their frequent occurrence and structural commonality in the young stages of numerous cryptogams, the construction of cylinders is probably the most efficient ontogenetic design available to the developing organisms.

Although some stages of Haplomitrium's sporeling ontogeny are similar to other patterns, the complete sequence is unparalleled in any other known bryophyte sporeling pattern. It is unique in its repeated quadrant and tier formation without apical cell involvement, in the production of multiple shoots from a single massive protonema, and its lack of rhizoids. Based on known sporeling ontogenies, the similarities that exist between Haplomitrium and Fossombronia suggest that Haplomitrium's closest affinities lie with the Metzgeriales.

Experimental and Culture Studies

Spore Culture Experiments

The developmental rates for H. mnioides sporelings cited in the literature vary widely because different environmental conditions were utilized in each investigation (Nehira 1961, 1966; Yang 1966, 1967). Variations in temperature, pH, and light regimes are known to effect the growth rates of bryophytes (Bopp 1983). Therefore, the developmental rates found in this study cannot be compared with those reported in the literature for H. mnioides sporelings. In this investigation, however, light quality, light intensity, photoperiod, and temperature conditions were identical for each replicate. The only variable was the type of medium utilized, making a comparison of media effects possible. Beneke (1903), Hurey-Py (1948), Bold (1949), Hatcher (1965), and Basile's F^1 (1975) media comprised the five macronutrient recipes tested. As illustrated in Table X, these media differ in mineral concentrations, pH, and the type of nitrogen source.

Comparisons were made only for H. mnioides sporelings since the H. blumii protonemata had not yet reached the juvenile stage of development at the time of this writing. The rate and pattern of spore germination for H. mnioides were identical on all media tested, i.e., germination was exosporic and commenced within approximately 32 days after the spores were sown. However, the germinated spores on the Bold medium showed no further development and soon died.

The fundamental pattern of sporeling development was identical on the remaining media, although only

TABLE X. COMPOSITION OF MACRONUTRIENT MEDIA

Medium	Salts (g/l)										
	NH_4NO_3	$(NH_4)_2SO_4$	$Ca(NO_3)_2$	$CaCl_2$	$MgSO_4$	KH_2PO_4	K_2HPO_4	NaCl	NaH_2SO_4	$FeCl_3$	pH
Beneke	0.2	-	-	0.1	0.1	0.1	-	-	-	0.004	5.03
Bold	-	-	-	0.25	0.075	0.175	0.075	0.025	-	-	4.80
F[1]	-	0.5	-	0.5	0.25	-	-	-	-	-	6.34
Hatcher	-	-	0.26	-	0.735	-	-	0.165	0.20	-	4.80
Hurey-Py	-	-	0.5	-	0.125	0.125	-	-	-	0.01	4.93

the protonemata on the Hatcher and Beneke media produced leafy gametophores. Those on the Hurey-py medium progressed only to the cell mass stage and death soon followed. Those on F^1 progressed to the cylindrical shoot stage but then began to deteriorate. These juvenile gametophytes were a pale, yellow-green color as opposed to the healthy, deep green sporelings on the Beneke and Hatcher media. The cells of the F^1 juveniles were highly vacuolated with relatively large, but few chloroplasts. The oil bodies appeared to be in the process of breaking down. In fact, some groups of cells beneath the meristematic tips and leaves of the shoots were empty and already dead. At this time, the juveniles were transferred to Hatcher medium where they began to grow and develop normally (Fig. 472).

The rates of ontogeny differed slightly, with the protonemata on the Hurey-Py medium developing most slowly, reaching the cell mass stage in 60 days, while those on the F^1, Beneke, and Hatcher media required only 50 days to reach this same stage. Protonemata with shoots were observed on Beneke medium within 108 days of cell mass formation, while none were seen until 329 days after cell mass formation on the F^1 and Hatcher media. The sporelings on the Beneke medium continued to mature more rapidly, with juvenile plants observed 370 days after the spores were sown, while those on F^1 and Hatcher media required 422 days to reach an equivalent state of maturity.

Yang (1967) tested the growth of H. mnioides sporelings on 50% Hoagland, Knop, Beneke, and 50% Hoagland + gibberillic acid (GA) 10^{-5}, 10^{-6}, and 10^{-7} M media. He found the 50% Hoagland medium to be the most satisfactory, stating that Beneke was good at the start but declined later. However, the Beneke medium was apparently utilized without trace elements, since, in 1968, Yang reported satisfactory growth on Beneke medium supplemented with Hoagland trace elements. Likewise, he reported normal growth on 50% Hoagland + 10^{-6} M indole acetic acid (IAA) (Yang 1968). Apparently, H. mnioides sporelings can be grown to maturity on numerous types of media and under various light and temperature regimes.

Of the H. mnioides cultures which reached maturity, those grown on Hatcher medium display the most normal morphologies, i.e., they are most similar to field grown H. mnioides plants. Like all other H. mnioides plants maintained in culture, the juveniles on the Hatcher medium possess erect leafy shoots as opposed to the prostrate aerial shoots of field grown plants (Fig. 471). However, the leafy shoots are similar to those grown in nature in that the leaves are entire, subrotund, moderately spaced, and are produced in three ranks with one row slightly reduced in size (Table XI; Figs. 471, 473). In addition, after 26 months in culture and numerous transfers onto fresh Hatcher medium, some of these shoots became fertile, producing numerous antheridia which are organized in a terminal rosette (Fig. 474).

Twenty-six month old sporelings grown on Beneke medium appeared to grow well and were a healthy-looking grass green color, but were very abnormal. Their aerial shoots were long, generally unbranched, and bore three rows of extremely reduced leaves (Figs. 475a, 475b, 476). Some of these leaves consist of no more than 3, 4, or 5 cells, are triangular in outline, and usually are terminated by a single slime cell (Fig. 475b). Others are somewhat subrotund, but still much smaller (ca. 208-780 µm long X 258-780 µm wide) than the subrotund leaves of adult field grown plants (Figs. 475b, 476). The "scales" positioned at the base of the shoot are reduced to 2-celled slime papillae (Fig. 477). These morphologies suggest that the ontogenetic sequences which yield "normal" leaves and scales under natural conditions are suppressed at a very early stage in the Beneke medium. The irregular divisions which normally would cut the basal cells of the slime papillae to produce scales do not occur. Thus, the slime papillae remain unmodified. Similarly, in many cases, the divisions in the basal cells of the laminae of the young leaves which would normally extend the leaves to

TABLE XI. EFFECTS OF L-PROLINE AND DIFFERENT LIGHT CONDITIONS ON *HAPLOMITRIUM MNIOIDES* JUVENILES

Treatment[a]	Shoot Length (mm)	Shoot Width (μm)	Internode Length (mm)	Lateral Leaf (LXW;μm)	Dorsal Leaf (LXW;μm)
Control[bc]	2.6-3.1	232-421	0.3-0.6	342-990 X 144-414	180-324 X 99-171
L-proline[bc] (10^{-4} M)	1.9-3.3	168-342	0.3-0.8	115-129 X 98-100	94-108 X 87- 92
Low Light[bd]	5.4-8.5	183-260	1.0-1.5	230-254 X 113-1217	100-135 X 72-102
High Light[be]	3.4-8.3	224-350	0.5-0.6	148-241 X 115-128	71- 84 X 61- 70
Clear Filter[bf]	3.2-6.1	180-210	0.3-0.4	171-267 X 90-105	117-160 X 62- 98
Blue Filter[g]	2.0-2.6	143-207	NA	NA	NA
Red Filter[h]	1.8-2.1	140-200	NA	NA	NA
Incandescent Light[i]	4.5-5.0	150-212	NA	NA	NA

[a]all grown on Hatcher medium + Nystatin; [b]blue spectrum = 1.2 $\mu W/cm^2/nm$, red spectrum = 0.8 $\mu W/cm^2/nm$; [c]1929.9 lux; [d]861.1 lux; [e]3002.2 lux; [f]3216.0 lux; [g]740.2 lux, blue spectrum = 0.56 $\mu W/cm^2/nm$, red spectrum = 0; [h]534.6 lux, red spectrum = 0.69 $\mu W/cm^2/nm$, blue spectrum = 0; [i]3910.4 lux, far-red spectrum = 11.0 $\mu W/cm^2/nm$, blue spectrum = 2.1 $\mu W/cm^2/nm$, red spectrum = 8.5 $\mu W/cm^2/nm$; LXW, length X width; NA, not applicable.

their ultimate forms are also suppressed. Therefore, the leaves remain at the triangular stage of development. In other cases, the divisions of the basal leaf lamina are merely limited as opposed to being completely suppressed, and some cells do elongate and divide to build a subrotund leaf. However, the divisions are not extensive enough to extend the leaf to the full dimensions found in field grown individuals.

Branching is similarly repressed in the Beneke juveniles. The branch initials appear to segment for only a short time before being suppressed leaving only a minute raised area on the shoot surface (Fig. 476). No elongate branches were observed in the distal portions of the leafy shoots.

The plants involved in this study were repeatedly transferred to fresh medium during their 26 month growing period. Therefore, the phenovariations exhibited must be due to some factor in the medium and not to a depletion of specific macronutrients or a change in pH. Perhaps, these morphological variations may be associated with nitrogen availability. Basile (1975) noted that while some bryophytes show similar developmental patterns on both nitrate-media and nitrate and ammonium media, others may develop differently on the two types. Although the fundamental pattern of sporeling development in *Haplomitrium* was unaltered by the different media, slight morphological variations did exist. The sporelings did not grow well and were not healthy on the F^1 medium in which nitrogen is supplied only as ammonium. In contrast, nitrogen is supplied only as nitrate in Hatcher medium, which supported the best and most normal sporeling growth. Both nitrate and ammonium contribute nitrogen in Beneke medium, on which leaf ontogeny and terminal branching was suppressed. As in Hatcher medium, nitrogen is supplied only as nitrate in both Hurey-Py and Bold media. However, these media generally have lower concentrations of the other essential elements than does Hatcher medium and therefore, may be incapable of supplying the quantities of nutrients necessary to maintain the growth of the Haplomitrium sporelings. In addition, the Bold medium, which was unable to support growth beyond germination, is the only one of the five media tested which contains HPO_4, and has a pH above 6.0 (Table X).

Yang (1968) maintained cultured H. mnioides gametophytes which had been grown from spores on 50% Hoagland's medium for three years. The aberrant gametophytes were reportedly slow growing and produced no gametangia. Yang (1968) found that cultured plants were generally taller, more slender, with abnormally long internodes, and more frequent terminal branching as compared with field grown adults.

The juvenile H. mnioides plants cultured from spores are not unique in producing erect leafy shoots. Adult H. mnioides plants, collected in nature and maintained in a growth chamber on their natural substrate over moistened vermiculite for at least three months, also become erect and isophyllous. Originally, hydroxyproline studies were planned in order to determine whether an erect, isophyllous growth form (the theoretically primitive habit [Fulford 1965; Schuster 1966; Basile 1974]) could be induced in prostrate, anisophyllous H. mnioides plants by the exogenous addition of hydroxy-l-proline. Hydroxyproline is exclusively located in the structural proteins of the primary cell wall and is important for cell extension (Lamport 1965). Walls with little hydroxyproline have lower tensile strengths than do walls which are rich in hydroxyprolines (Lamport 1965). Hydroxy-l-proline is an antagonist to proline-hydroxyproline metabolism (Steward et al. 1958; Basile 1979). Therefore, Basile (1979) hypothesized that the exogenous addition of hydroxy-l-proline blocks this pathway and prohibits the production of hydroxyproline proteins. Therefore, wall elasticity increases. In contrast, the exogenous addition of proline not only allows the pathway to proceed, but increases the amount of hydroxyproline protein incorporated into the wall. Thus, wall elasticity decreases. Basile (1967, 1968, 1974, 1979) reported that the exogenous addition of hydroxy-l-proline to various anisophyllous members of the Jungermanniales allowed the ventral face of the apical cell to lengthen. Thus, the plants become isophyllous. The exogenous addition of L-proline has the opposite effect in that it prohibits the cutting faces of the apical cell from elongating and, thereby, anisophylly is maintained (Basile 1968, 1974). However, the H. mnioides plants adopted an erect, isophyllous growth form without the addition of any chemicals, when the plants, on their natural substrate, were placed on moistened vermiculite in an environmental chamber. Since the hydroxyproline phenovariation is already expressed by all the plants in culture, it is impossible to test the effects of hydroxy-l-proline. However, in an attempt to induce a reversion to the natural growth form, and thus, assess the involvement of naturally produced hydroxyproline, juvenile H. mnioides plants were treated with the hydroxproline antagonist, L-proline.

Often, the key factor in inducing morphological changes in phenovariants is light, since temperature usually effects growth rate, not ontogeny (Bopp 1983; Bartholomew 1985). Therefore, different light intensities and wavelengths were also tested in an attempt to ascertain the environmental factor responsible for the aberrant isophyllous growth form. The plants were maintained under these experimental conditions for three months, then observed using light and scanning electron microscopy. The results of these observations are summarized in Table XI. Unfortunately, attempts to study the fine structure of the juvenile plants failed due to fixation difficulties. Despite repeated attempts, the cell structure was never adequately preserved. Perhaps the copious amounts of mucilage which encompass the young plants acts as a barrier to fixative penetration. In any case, no information was obtained regarding the cytological differences among the treated plants.

None of the treatments utilized induced a complete return to the prostrate habit typical of vegetative _H. mnioides_ plants. The leafy shoots were all erect or at best, obliquely ascending (Figs. 478-485). The juveniles treated with L-proline are similar to the control plants (those on Hatcher medium) in shoot length, width, and internodal lengths. However, all three rows of leaves produced by the L-proline plants are transversely inserted and are more similar to one another in size, i.e., the shoots are more isophyllous (Fig. 478; Table XI).

The juveniles which received a low light intensity (861.1 lux) as well as those which received a high light intensity (3002.2 lux) all produced longer shoots than did the control plants (1929.6 lux) (Table XI; Figs. 479, 480a, 480b). The low light plants also produced the longest internodes of any of the treated plants (Table XI). Although differences in light intensities did not alter leaf shape, it did affect leaf size and insertion. The leaves of both low light and high light plants are more reduced in size than are those of the control plants (Table XI; Figs. 479, 480a, 480b). The plants which received low light produced two rows of succubously inserted leaves, and a third row of transversely inserted leaves (Fig. 479). In contrast, the plants exposed to the high light intensity bear three rows of transversely inserted leaves, but, like the control and low-light plants, one row of leaves remains reduced in size (Figs. 480a, 480b).

The results of the light quality experiments are inconclusive since the colored filters used, not only altered the wavelengths, but also the intensities of the light. The plants which received full-spectrum light (clear filter) produced relatively long leafy shoots with internodes of average lengths (Table XI; Fig. 481). The leaves of these plants are typically subrotund and anisophyllous, but all three rows of leaves are transversely inserted (Fig. 481; Table XI). As shown in Table XI, the shoot lengths, leaf sizes, and leaf insertions of the full-spectrum plants are similar to those of the high light intensity juveniles. This is to be expected since these plants also received full light intensity of over 3000 lux.

The plants which received wavelengths high in the red spectrum as well as plants which received mostly blue spectrum wavelengths are similar to those grown under low light intensity (both the red and blue filters decreased the intensities to 740.2 and 524.6 lux, respectively) (Figs. 482-484; Table XI). The shoots of the plants which were grown under either red or blue wavelengths are stunted and leaves are absent. Only slime papillae cover the cylindrical axes (Fig. 484). Since the production of short axes is evident in the plants which were exposed to 861.1 lux, it is likely that this trait is correlated with light intensity. However, it is impossible to ascertain whether the lack of leaves is induced by light quality, intensity, or a combination of both factors.

The juveniles which were exposed to the incandescent light high in far-red wavelengths came the closest of all the treated plants to producing prostrate shoots. The majority of these shoots are obliquely ascending, although some are erect (Fig. 485a). Most of the shoots lack leaves and are covered by slime papillae or slime hairs, but one axis was found which had a single cycle of leaves (Figs. 485a, 485b). It was impossible to determine whether these leaves were produced while the plants were exposed to the far-red wavelengths, or whether they were already present when the juveniles were transferred to the experimental conditions and further growth did not occur. However, since the other shoots were leafless, the latter explanation is most likely.

The results of these experiments suggest that the most normal growth of the *H. mnioides* juveniles occurs on Hatcher medium, under relatively low light intensities, i.e., between 860 and 2000 lux, and with full-spectrum light which contains at least some far-red wavelengths. Light quality appears to play a greater role in determining the erect stature of the leafy shoots than do hydroxyprolines. Although it was postulated that L-proline would exaggerate the anisophyllous expression obtained on Hatcher media, it actually induced greater isophylly.

Culture-Induced Adult Phenovariations

Adult Haplomitrium plants obtained from the field and maintained in culture on a macronutrient agar medium or on their original substrate over moistened vermiculite often display various phenovariations (Figs. 486-507). In general, these plants become taller, more slender, have longer internodes, and often exhibit more frequent terminal branching than do field grown plants. These general modifications were also noted in cultured H. hookeri plants by Fulford and Diller (1956) and Sharma et al. (1960), and in H. mnioides by Yang (1968). Various culture-induced aberrations occur depending on the species. For example, after about eight months in culture on vermiculite, the leafy, vegetative shoots of H. gibbsiae had become tall and slender with distantly spaced, bifid or crescent-shaped leaves, some of which were slightly reduced in size (Figs. 486-488). Slime hairs also became apparent along the stem (Fig. 489). After about 17 months in culture some leafy shoots were distictly anisophyllous with one reduced row of transversely inserted leaves and two rows of succubously inserted leaves which are elliptical to subrotund, and entire with finely dentate margins (Fig. 490). In addition to the modifications of the leafy shoots, the rhizomatous systems are very abbreviated in cultured plants. Only a single, long, slender, stolon descends from each leafy shoot (Fig. 486).

As previously mentioned, H. mnioides plants become erect and nearly isophyllous after being exposed to culture conditions for about three months. In addition, after approximately six months in culture, the leaves become bifid or highly dissected with long lateral projections forming along the margins both on vegetative and fertile plants (Figs. 491-493). These dissected leaves are similar to those observed in the eight month old H. gibbsiae plants. Similar culture-induced bifid and incised leaves have been reported in H. mnioides plants grown on MSK-3 agar by Kitagawa (1984) and by Fulford and Diller (1956) on H. hookeri plants maintained on Beneke medium.

In addition to the production of dissected leaves, numerous other types of aberrant growths occur along the stems, in leaf axils, and on the ventral surfaces of the basal laminae of the leaves (Figs. 494-499). Slime hairs (Fig. 494, 495), irregularly shaped scales often tipped by slime cells (Fig. 496), as well as numerous cylindrical, multicellular outgrowths, some bearing slime papillae and some without (Figs. 497-499) are produced not only in H. mnioides, but also in H. intermedium, H. gibbsiae, and H. hookeri. Fulford and Diller (1956) and Sharma et al. (1960) described similar outgrowths produced by H. hookeri in their culture studies which involved glucose and amino acid treatments. Some of the meristematic aberrations observed on H. hookeri by these workers included "ridges or wings" on the ventral side of leaves, small "piles" of cells on leaves and stems, and "pendant pouches" on the leaves (Fulford and Diller 1956; Sharma et al. 1960).

Terminal branching also occurs more frequently in cultured plants than in those grown in their native environments. Haplomitrium gibbsiae and H. hookeri both commonly produce an elongate terminal branch at a 45° angle from the main shoot, which lends a dichotomously forked appearance to the plant (Figs. 500, 501). Branching was most abundant in H. hookeri plants grown on Hatcher medium supplemented with kinetin (Figs. 502a, 502b). Here, a leafless branch emerges from the axil of at least one leaf in virtually every series of merophytes (Fig. 502a). The kinetin, which induces cellular divisions, apparently suppresses the dormancy periods of the branch apical cells and induces segmentation and derivative development in each branch apical cell. Although leafless, the branches are covered by slime papillae, mucilage hairs, and minute scales with terminal slime cells (Fig. 502b). These branches appear to develop like the cylindrical sporeling axes rather than like branches of adult plants, i.e., one of the epidermal cells of the stem functions as a quadrant forming initial. The branch is then formed by repetitive quadrant formation from this initial cell until a shoot initial is ultimately delimited.

At the time of this writing, the H. blumii plants which have been maintained on vermiculite for eight months show no leafy shoot modifications. However, the stolon apices which are exposed to light have become green, branched, and leafy (Fig. 95). Longitudinal sections through the apex of one of these leafy stolons reveals a tetrahedral apical cell which measures 19.8 µm wide X 48.4 µm deep. The apical derivatives develop in the manner typical for leafy Haplomitrium shoots, i.e., one leaf is produced per merophyte. The fact that stolons have the potential to become secondarily leafy when exposed to light shows that this axis is essentially like a leafy shoot, which, in the dark suppresses leaf development.

Gametangia were produced in culture by H. hookeri, H. mnioides, and H. gibbsiae plants. The H. hookeri plants grown on Hatcher medium produced archegonia in the same manner as described by Fulford and Diller (1956) for H. hookeri plants maintained on Beneke medium. The archegonia are produced at the tip of the stem and become intercalary between leaves through continued apical segmentation. Fulford and Diller (1956) reported that an irregular lamella secondarily formed on each row of archegonial neck cells on plants maintained on both Beneke agar and liquid Beneke + tryptone media. No such aberrant outgrowths existed on the archegonia produced by the H. hookeri plants maintained on the Hatcher medium in this study.

The H. mnioides plants maintained over moistened vermiculite produced both antheridia and archegonia.

The gametangia were formed in apical rosettes just as in field grown plants (Fig. 215). Even plants grown from spores to maturity on Hatcher medium produced antheridia in this manner (Fig. 474).

The *H. gibbsiae* plants on vermiculite formed antheridia, but, these did not form in the manner characteristic for the species. Instead of being produced in a terminal rosette, the antheridia are distributed along the shoot, usually in pairs (Fig. 507). This antheridial arrangement is reminiscent of that found in *Steereomitrium minutum* and some individuals of *H. hookeri*. The absence of the apical rosette in cultured *H. gibbsiae* males appears to be due to the rapid elongation of the internodes between the apical spirals which produce the antheridia.

Although regenerants were reported as early as 1894 from *Haplomitrium* plants maintained under artificial conditions (Schostakowitsch 1894), asexual reproduction by specialized cells was thought to be absent in natural populations until the discovery of gemmae on *H. hookeri* plants collected in the Yatsugatake Mountains of Japan (Furuki 1986). Furuki (1986) described the gemmae of *H. hookeri* as developing from the stem epidermal cells near the apices of young plants and then being distributed over the entire stem surface. Gemma formation begins through periclinal divisions in the epidermal cells with the outer cell giving rise to a 2-4-celled filament. A globose gemma is then produced through repeated divisions from the terminal cell of the filament (termed the apical cell by Furuki [1986]). Furuki's (1986) illustrations show that four quadrants are created in the terminal cell of the filament and later a single central cell is delimited, thus suggesting a pattern similar to that which occurs during the sporeling ontogeny of *Haplomitrium*. The mature gemma is stalked, multicellular, globose, and bears a few 1-2-celled mucilage hairs (Furuki 1986). Furuki (1986) noted that this type of gemma is rare among hepatics.

Before Furuki's (1986) discovery of gemmae, in nature, *Haplomitrium* was thought to multiply vegetatively solely through the production of new branches from the rhizomatous system as old portions of the plant died off (Gottsche 1843; Lilienfeld 1911; Fulford and Diller 1956; Schuster 1967; Udar and Srivastava 1983). Udar and Srivastava (1983) suggested that the spores of *Haplomitrium* are often an ineffective means of reproduction and thus, the rapid branching from the rhizome appears to be the only means of multiplication for the plants.

Regenerants were observed on the leaves of *H. intermedium* and *H. mnioides* plants from vermiculite cultures and *H. hookeri* plants from Hatcher medium (Figs. 503-506). These propagules form from cells of both the leaf margin and lamina. The young *H. intermedium* regenerant begins as a sessile, multicellular globose mass of cells (Fig. 503), which then grows into an elongate, slender, cylindrical filament composed of multiple tiers of four rows of cells (Fig. 504a). The growth of this cylinder is by an initial which appears somewhat triangular in surface view (Fig. 504b). The presence of four cells peripheral to the initial suggests that the filament is built in a manner similar to that responsible for cylindrical shoot construction in *Haplomitrium* sporelings, i.e., the initial swells and is cut by three divisions to form four quadrants of cells, a new initial is delimited in one of the quadrants, and the process is repeated to build the leafless filament.

The young regenerants of *H. hookeri* and *H. mnioides* differ from those of *H. intermedium* in that they are stalked, more massive, and delimit a tetrahedral apical cell at a much younger stage of development (Figs. 505, 506). One and two-celled slime papillae are produced from the apical derivatives which overarch the segmenting apical cell and are displaced along the axis as the regenerant grows (Fig. 506). Only these young, leafless stages of development were observed. Schostakowitsch (1894) and Inoue (1964) observed regenerants in *H. hookeri* and *H. mnioides*, respectively, with early developmental stages similar to those reported here. According to Schostokowitsch (1894) and Inoue (1964), regeneration usually involves 4-8 adjacent leaf cells, but occasionally only a single leaf cell may be involved. These cells divide to form a short stalk which is 2-4 cells high and supports a globose, multicellular regenerant. One cell of the regenerant becomes an apical cell with three cutting faces which appears equilateral in surface view (Schostokowitsch 1894). Segmentation from this apical cell continues for a long time with the segments producing only slime papillae, but ultimately a leafy shoot is produced (Schostokowitsch 1894; Inoue 1964).

Therefore, under artificial conditions, multiplication is achieved not only by means of the rhizomatous system acting as a perennating structure, but also through regeneration from leaf cells. In addition, the phenotypic plasticity inherent in the genus is emphasized by artificial culture conditions. This is most obvious in plant size, leafy shoot stature and symmetry, increased branching from leafy shoots, leaf size, shape, and even insertion (but not basal field thickness), and internodal lengths. In some species, such as *H. gibbsiae*, gametangial positioning is also affected by artificial conditions suggesting that this trait is not as genetically fixed as one might hypothesize from a study of herbarium specimens. Many of the culture-induced characters described here are also evident in some field grown individuals, but these traits are the rarity, not the norm, and usually are not expressed to the extreme that they are under artificial conditions. Apparently, the interactions of nutrient media and artificial light and temperature regimes cause these infrequently expressed phenotypes to be manifested more intensely and uniformly in cultured individuals.

TAXONOMIC TREATMENT

In an attempt to reflect evolution, the classification of *Haplomitrium* among the higher taxonomic categories has varied according to the changing phylogenetic concepts regarding the origin of bryophyte groups and the amount of divergence which has occurred between the groups. For instance, those who considered the bryophytes as a monophyletic group, namely, a single division, with the single class Hepaticopsida (= Hepaticae) to encompass all hepatics (with the anthocerotes excluded), placed *Haplomitrium* in the subclass Jungermanniidae (Schuster 1953, 1966, 1967, 1972, 1984b; Schljakov 1972), which subsequently contained the Haplomitriales, Jungermanniales, and Metzgeriales. These three basic units have also been recognized by Evans (1939) and Schljakov (1972), but as suborders and superorders, respectively. In contrast, the bryophytes have also been perceived as a polyphyletic group represented by three divisions (Hepatophyta, Bryophyta, Anthocerotophyta) (Stotler and Crandall-Stotler 1977; Crandall-Stotler 1981, 1984). This necessitated that the basic hepatic groupings be elevated to the class level as opposed to the ordinal level. Thus, in order to complement the Jungermanniopsida and Marchantiopsida, Stotler and Crandall-Stotler (1977) established the class Haplomitriopsida which included the sole order Haplomitriales and family Haplomitriaceae. This classification maintained the consistency of the phylogenetic system and simultaneously implied the uniqueness of *Haplomitrium* with regard to other hepatics.

Regardless of the classification system employed, the uniqueness of *Haplomitrium* has been widely accepted and reflected by its placement in the order Haplomitriales and family Haplomitriaceae (Goebel 1891; D. H. Campbell 1920; Buch 1936; Evans 1939; Schuster 1966, 1967, 1972, 1984b; Grolle 1964; Schljakov 1972; Hässel de Menéndez 1975; Stotler and

Crandall-Stotler 1977). However, its ties to the Mezgeriales were not always recognized. The suggestion of such an affinity by D. H. Campbell (1920), Schuster (1966, 1967), and Crandall-Stotler (1981) placed even greater emphasis on the re-evaluation of the systematic status of Haplomitrium. Indeed, the results of this investigation show that Haplomitrium is very closely related to the Metzgeriales, while still differing in some aspects. A summary of the traits which are shared either entirely or in part by Haplomitrium and the Metzgeriales follow: 1) apical organization and the pattern of apical derivative development; 2) leaf ontogeny; 3) spore wall development; 4) the presence of a central conducting strand of pitted hydroids; 5) exogenous branching in which the branch primordium remains dormant for a period of time before the branch apical cell is delimited; 6) sporophyte protective structures derived from thallus and/or archegonial tissue; 7) massive, generalized sporophyte setae; 8) relative spore size and extent of sporoderm ornamentation; 9) long, slender, attenuate elaters. The gap between Haplomitrium and the Jungermanniales is emphasized by their divergent patterns of apical derivative development, leaf ontogenies, plant symmetry in anisophyllous taxa, branching modalities, and the lack of a perianth in Haplomitrium. In addition, Haplomitrium is distinguished from all other bryophyte taxa by its unistratose capsule wall with unique annular, longitudinal thickening bands, the identical patterns of early antheridial and archegonial ontogenies, the 4-seriate construction of the archegonial necks, its unique Haplomitrium-type spermatozoid ultrastructure, the unique pattern of sporeling development, a leafless, rhizomatous system with descending, hyaline stolons and the complete absence of rhizoids.

As pointed out by Schuster (1984b), a classification system in which Haplomitrium is placed into its own class while the Jungermanniales and Metzgeriales are grouped together in the

Jungermanniopsida implies that the Metzgeriales and Jungermanniales are more closely related to one another than are the Metzgeriales and Haplomitriales. Thus, Schuster (1984b) rejected the organization presented by Stotler and Crandall-Stotler (1977) in favor of his previously designed system in which the hepatic orders Haplomitriales (= Calobryales), Metzgeriales, and Jungermanniales were all grouped within the subclass Jungermanniidae, class Hepaticopsida (= Hepaticae). However, his scheme treats the Haplomitriales, Metzgeriales, and Jungermanniales as equal entities of the Jungermanniidae, and therefore, provides no means of determining whether Haplomitrium is more closely related to the Metzgeriales or to the Jungermanniales. In addition, this system is inadequate in that it does not recognize the apparent polyphyletic origin of the bryophytes. Therefore, a scheme is needed which reflects the polyphyletic nature of the bryophytes and which better reflects the relationships of Haplomitrium among the hepatics. This is best accomplished by discarding the class Haplomitriopsida and transferring Haplomitrium to the Jungermanniopsida with the Haplomitriales and Metzgeriales grouped together into a newly created subclass, the Metzgeriidae, with only the Jungermanniales retained in the Jungermanniidae. The placement of the Haplomitriales and Metzgeriales in the same subclass emphasizes the close relationship between these taxa, while simultaneously divorcing Haplomitrium from the Jungermanniales. Therefore, Haplomitrium is classified as follows: division Hepatophyta, class Jungermanniopsida, subclass Metzgeriidae, order Haplomitriales, family Haplomitriaceae.

METZGERIIDAE BARTHOLOMEW-BEGAN
Phytologia, 69: 464-466 (1990).

Diagnosis - Plants thalloid or leafy, leaf or thallus wing development from a single wedge-shaped cell of a

3-celled apical derivative, trigones absent, conducting strand of pitted hydroids sometimes present; branches furcate or monopodial, usually of exogenous origin with a dormant, multicellular primordial stage preceding elongation; gametangia variable in position, rarely acrogynous, sporophyte never enclosed by a leaf-derived perianth, spores (15-) 19-40 µm, with the spore wall development often involving preprophase, callosic, exine precursors; elaters long, slender, attenuate.

HAPLOMITRIALES BUCH EX SCHLJAKOV
Bot. Zhurn., 57: 499 (1972).

Calobryales D. H. Campb. Ann. Bot., 34: 12 (1920) (nom. nud.). Haplomitriales Buch Suomen Maksasammalet, 116 (1936) (nom. nud.). Calobryales D. H. Campb. emend. Schust. Nova Hedw., 13: 5 (1967) (nom. nud.). Calobryales D. H. Campb. emend. Schust. ex Hamlin Rec. Dom. Mus., 7: 315 (1972).

Diagnosis - Plants with a prostrate, leafless rhizome with descending, hyaline, leafless stolons; rhizoids absent. Leafy shoots erect, 3-ranked, isophyllous or anisophyllous with the reduced row of leaves, dorsal. Antheridia and archegonia with a common pattern of development at least in early ontogenetic stages; antheridial stalk 4-seriate; archegonial neck 4-seriate. Sporophyte protective structure a shoot-calyptra, a true calyptra, or a perigynium; perianth absent. Sporophyte capsule wall unistratose with each cell traversed by a single, annular, longitudinal thickening band. Spores (15-) 19-33 (-36) µm in diameter.

HAPLOMITRIACEAE DĚDEČEK

Archiv. Přírod. Proskoum Czech., 5: 68 (1883).

Calobryaceen Goebel Ann. Bot. Gard. Buitenzorg, 9: 21 (1891). Haplomitrioideae Engler Syll. Pflanzenf., 2: 50 (1898). Calobryaceae Cavers New Phytologist, 9: 222 (1910). Scalieae C. Jens. Denmarks Mosser, 1: 25; 60 (1915). Haplomitriaceae Verdoorn Manual Bryol., 422 (1932).

Since the Haplomitriales is a monotypic order, its diagnosis will also circumscribe the Haplomitriaceae.

Variation and Differentiation - As discussed by Schuster (1967), *Haplomitrium* shares superficial characters with *Takakia*, such as the development of a highly branched rhizomatous system, the lack of rhizoids, a central conducting strand of hydroids, large, isolated gametangia in scattered arrangements, and copious secretions of mucilage. In fact, *Haplomitrium* and *Takakia* have long been considered to be closely related and have even been classified in separate suborders, the Calobryiinae and Takakiinae Schust., respectively of the Haplomitriales (Schuster 1967). However, as suggested by Crandall-Stotler (1986) and Crandall-Stotler and Bozzola (1988, 1989) and confirmed in this investigation, *Takakia* gametophytes differ from *Haplomitrium* in many ultrastructural, ontogenetic, and morphological characteristics including the apical organization and the pattern of derivative development, leaf development, hydroid ontogeny, antheridial ontogeny (McFarland et al. 1989), slime cell morphology, plastid ultrastructure and number, plasmodesmatal ultrastructure, presence versus absence of oil bodies, and chromosome number. The recent discovery of moss type sporophytes in *Takakia*, in fact indicate a relatively large phylogenetic gap between the taxa (Smith 1990).

Murray (1988) discussed a possible shared ancestry between *Haplomitrium* (which she classified in the class Haplomitriopsida) and the Andreaeopsida through *Takakia* (which she considered as comprising the Takakiopsida). However, except for a very slight, probably superficial, resemblance in the production of multiple, cylindrical axes from a massive protonema during sporeling ontogeny, *Haplomitrium* and *Andreaeobryum* share no characters.

Calobryum Nees, *Steereomitrium* E. O. Campb., and the fossil genus *Gessella* Pouls. have also been classified as members of the Haplomitriaceae (Goebel 1891; D. H. Campbell 1920; Poulsen 1974; E. O. Campbell 1987). However, *Steereomitrium* is herein synonymized with *Haplomitrium*, and *Calobryum* and *Haplomitrium* were united in 1963 (Schuster 1963). *Calobryum* was originally thought to be acrogynous (Goebel 1891; D. H. Campbell 1920) in contrast to the acrogynous *Haplomitrium* (Gottsche 1843; Leitgeb 1875; Lilienfeld 1911). However, E. O. Campbell (1959) showed that at least one species of *Calobryum* (*C. gibbsiae* Steph.) is anacrogynous rather than acrogynous as previously reported. Based on this finding and his own observations, Schuster (1963) proposed that the supposed anacrogynous/acrogynous boundary was invalid and united the taxa into the single genus *Haplomitrium*.

In 1974, Poulsen discovered *Gessella* at Slagelse no. 1 in western Sealand, Denmark. The fossils date to the Permian period of the Paleozoic. According to Poulsen (1974), this taxon is like *Haplomitrium* in being rhizomatous and bearing radial sterile shoots possessing three rows of isophyllous, transversely inserted, unistratose leaves. Also, the capsules are cylindrical with a unistratose wall, and the spores are coarsely verrucose. It differs from *Haplomitrium* in bearing very long, ligulate to lanceolate or sublanceolate leaves, with a small median denticle on a distal extremity. With the inclusion of this genus, then, the family Haplomitriaceae contains two genera -- *Haplomitrium* and the fossil genus *Gessella*.

HAPLOMITRIUM NEES

Naturg. Eur. Leberm., 1: 109 (1833) (nom. cons.).

Scalius S. F. Gray Nat. Arr. Brit. Pl., 1: 704 (1821) (nom. rej.), non Scalia Sims Bot. Mag., 956 (1806) (orth. var.: Scalia S. F. Gray mut. Lindb. Hep. Hib. lect. 532 [1875]); Grolle (1983). Scaliusa O. Kuntze Rev. Gen. Pl., 839 (1891). Mniopsis Dum. Comm. Bot., 114 (1822), non Mniopsis Martius Nov. Gen. et Sp., 1: 3 (1822) (orth. var.: Mniopis Dum. An. Fam., [1829] [nom. inval.]); Grolle (1983).

Type: Jungermannia hookeri Smith [= Haplomitrium hookeri (Smith) Nees] [BM].

Calobryum Nees in G., L. & N., Syn. Hep., 507 (1846). Cladobryum Nees ex Endlicher Gen. Pl., 1337 (1840) (nom. inval.); Grolle (1983). Rhopalanthus Lindb. Manip. Musc. II., 390 (1874) (orth. var.: Rhossalanthus Lindb. mut. Aust. Bull. Torrey Bot. Club., 6: 21 [1875]); Grolle (1983). Steereomitrium E.O. Campb. Mem. New York Bot. Gard., 45: 569 (1987) (syn. nov.).

Description - Plants rhizomatous with erect, prostrate, or obliquely ascending, radial leafy shoots and subterranean stolon systems, with rhizoids absent on all axes. Gametophyte axes of three types, subterranean rhizomes, stolons, or aerial leafy shoots; growth via tetrahedral apical cells; central strands of thin-walled, elongate hydroids present. Rhizomes branched, chlorophyllous, bearing 1- to 2-celled slime papillae, leaves lacking; apex of the rhizomatous axis turning upward and growing into an aerial leafy shoot following a period of prostrate growth. Stolons unbranched (branched), hyaline, positively geotropic, possessing 1- to 2-celled slime cells, leaves lacking.

Leafy shoots erect, obliquely ascending, or prostrate, unbranched or rarely with exogenous, lateral basal or distal terminal branching, chlorophyllous. Leaves three-ranked, usually decurrent, ranging from subrotund to ovate or rhomboidal to lingulate-triangular, entire to shallowly or deeply lobed, bearing 1- to 2-celled slime papillae at least on the margins. Symmetry isophyllous or anisophyllous with the smaller leaves comprising the dorsal row in anisophyllous representatives. Dorsal leaves always inserted transversely, lateral leaves inserted transversely, subtransversely, or succubously depending on symmetry. Basal leaves smaller, different in shape, and more distantly spaced than those of the main body.

Dioicous or monoicous. Antheridial and archegonial ontogeny identical in early developmental stages. Mature gametangia massive, axillary or supra-axillary, scattered along the stem in undifferentiated or slightly enlarged bracts, aggregated into loose terminal clusters, positioned at alternating swollen nodes, aggregated into well-defined terminal rosettes on swollen discs, or free-standing on the stem. Slime papillae, and/or slime hairs and minute bractlets often intermixed among the gametangia. Plants acrogynous or anacrogynous, occasionally intermixed within single populations of a species. Spermatozoid blepharoplast of the Haplomitrium-type, characterized by an extremely broad spline (more than 57 microtubules) widening distally in two stages, an open spline aperture (like the basal bodies) located left-of-center, and an elongated curved, drop-shaped lamellar strip (Carothers and Rushing 1988).

Sporophyte protective structure consisting of a shoot-calyptra, true calyptra, or a perigynium; perianth absent. Sporophyte consisting of a foot, elongating seta and capsule. Foot obconical. Seta massive. Capsule cylindrical; capsule wall unistratose becoming multistratose at the extreme apex and base, all cells possessing a single longitudinal, annular thickening band; dehiscence variable, occurring along 4 incomplete or complete lines, 2 complete lines, or by a single suture. Spores (15-)

19-33 (-36) µm in diameter. Elaters very long and attenuate, 350-700 µm long X 7.0-10.0 µm in diameter at the broadest portion. Spore:elater ratio 3:1.

Sporeling development of the *Calobryum* type with a massive protonema producing multiple leafless shoots via repeated quadrant and tier formation without apical cell involvement, with leafy shoot production involving a tetrahedral apical cell, with rhizoids lacking.

Chromosome number n = 9 [*H. blumii* (Lorbeer 1934), *H. hookeri* (Lorbeer 1934; Berrie 1959), *H. mnioides* (Yang 1966; Tatuno 1941), *H. dentatum*, and *H. indicum* (Mehra and Kumar 1980)].

Historical Account - The genus *Haplomitrium* was originally described by Nees von Esenbeck in 1833. This name has since been conserved over the earlier, illegitimate generic homonyms *Scalius* S. Gray and *Mniopsis* Dum., along with their orthographic variants *Scalia* S. Gray mut. Lindb. and *Mniopis* Dum. (Grolle 1983). Nees (1833) distinguished the genus from all others based on its overall stature, its possession of a sac-shaped calyptra, and its lack of a perianth. The name *Haplomitrium* is derived from the Greek *aploos*, simple + *mitrion*, little cap (Frye and Clark 1943), or from *haplos*, one + *metra*, headband or girdle (Bold et al. 1980) and apparently refors to the shoot calyptra serving as the sole protective structure for the sporophyte (Frye and Clark 1943).

Engel (1981) recognized 13 species of *Haplomitrium*. Since then, one new species has been described (Udar and Kumar 1982) and one has been synonymized (Hässel de Menéndez 1975). Of the 13 described species, 7 are here retained, while 6 were found to be taxonomically equivalent with pre-existing elements of the genus. In 1985, Gao and Li described a new variety of *H. mnioides*, but this taxon is not retained here. However, a new variety of *H. hookeri* is described based on *Steereomitrium minutum*. Thus, *Haplomitrium* presently consists of seven species and one infraspecific taxon.

Variation and Differentiation - Although E. O. Campbell (1987) allied Steereomitrium with Haplomitrium based on general habit, the absence of rhizoids, and a marked tendency toward isophylly, she maintained that Steereomitrium was distinct from Haplomitrium due to its possession of a true calyptra subtended by a sheath of crowded, immature leaves, a 3-4-stratose capsule wall in which the external cells have trigonous thickenings and the inner cells bear uniformly thickened brown radial walls, and because the plants possess the potential for asexual reproduction through gemmae. However, these characters are not valid. True calyptrae subtended by clusters of bracts are also known from H. hookeri and H. monoicum, and gemmae have been reported from natural populations of H. hookeri (Furuki 1986). As discussed previously, the reported multistratose capsule wall and distinct thickening patterns are apparently the result of misinterpretations, for the capsule wall is unistratose with each cell bearing a longitudinal, annular thickening just as in Haplomitrium. Thus, Steereomitrium displays no characteristics which are distinct from those of Haplomitrium and is, therefore, here synonymized.

Distribution and Ecology - Haplomitrium displays a disjunct distribution incorporating both trans-oceanic and trans-continental ranges (Fig. 508) (Schuster 1966, 1967; Engel 1981; Chandra et al. 1987). Although the widest ranging species (H. hookeri) extends throughout the northern hemisphere, the center of greatest species diversity is the southern hemisphere, specifically Australasia (Fig. 508). In fact, three species (H. ovalifolium, H. monoicum, and H. intermedium) are endemic to Australasia.

The genus exhibits a broad altitudinal range which extends from sea level to approximately 3550 m.s.m. Within the range of a particular species, the taxon is very local and usually occurs as individual plants or in small tufts (Schuster 1967). Soil, wood, and/or rock may serve as the substrate for *Haplomitrium*, but soil habitats are the most common with all except two species restricted to this substrate -- *H. blumii* which also occurs on wood and rock and *H. mnioides* which occurs on both soil and rock.

Structural Features Useful for Subgeneric and Interspecific Differentiation

The results of this morphogenetic study suggest that many characters which traditionally have been used as distinguishing traits for *Haplomitrium* taxa are, in fact, highly plastic and are not adequate systematic markers. These characters include plant size, leaf distribution, the number of leaf cycles produced on a shoot, leaf size, leaf shape, and in some cases, even shoot symmetry and leaf insertion. Obviously, then, very few vegetative characters are stable enough to be taxonomically dependable. One such useful vegetative trait appears to be leaf thickness, since basal polystraty is fundamentally unmodified by artificial culture conditions (at least in *H. hookeri*, *H. gibbsiae*, *H. intermedium*, and *H. mnioides*). However, although the basal leaf thicknesses are relatively constant, the extent and position of the basal zones are variable among individuals of a population of a species, so leaf thickness must be considered cautiously. Similarly, oil body characters may provide taxonomic inference if utilized with the realization that many characteristics of these organelles vary with the age and health of the cells which contain them.

Ephemeral structures provide the most reliable taxonomic information since they are exposed to environmental pressures for only a short period of time and therefore, are least vulnerable to environmental modification. Such structures include sporophytes, sporophyte protective structures, and gametangia. Sporophyte characters are likely the most stable since the sporophyte is protected from environmental pressures by gametophytic tissue for most of its existence. However, many sporophytic characters, such as capsule morphology, capsule wall thickening patterns, dehiscence, and foot morphology, vary little throughout the genus and thus, are useless for differentiating among infrageneric taxa. In contrast, spore sculpturing patterns, spore sizes, and in some cases, elater thickening patterns and seta anatomy have proved to be important differential characters, particularly at the species level. In fact, the species for which spore architecture is known can be distinguished purely by their exospore sculpturing patterns. Sporophyte protective structure ontogeny is also a crucial systematic character. Like many of the sporophytic traits, gametangial morphology is also relatively uniform throughout the genus, except for minor modifications in archegonial neck and antheridial stalk lengths, and in the color of the antheridia. However, this latter character varies with antheridial age and can only be useful when dealing with living material. In contrast, gametangial positioning is shown to be an important taxonomic character, even though its expression is no doubt influenced more by environmental conditions than are sporophytic traits or the type of sporophyte protective structures. Whether or not a species possesses the potential for apical proliferation from fertile shoots is also important, but this potential is often difficult to determine with certainty. In light of the evidence observed, all the systematic and phylogenetic decisions made in this study are based upon the characters discussed above, with sporophytic, sporophyte protective structure, and gametangial characters weighted most heavily.

Subdivisions of the Genus and Relationships Among Taxa

Haplomitrium has been interpreted by Schuster (1967) as being composed of three subgenera, namely, Haplomitrium, Calobryum (Nees) Schust. and Archibryum Schust. The subgenus Haplomitrium, typified by H. hookeri, is defined by Schuster (1967) as being composed of plants which are leafy throughout, with leaves which are unistratose throughout and subequal in size from the base to the apex of the stem. The oil bodies are numerous, minute, and homogeneous. The gametangia are normally scattered along the stem and the antheridia are yellow to pale orange (Schuster 1967). The species comprising the subgenus Archibryum (represented only by H. intermedium according to Schuster [1967]) are similar to those of the subgenus Haplomitrium, but are distinguished by their frequent terminal branching, the axillary origin of the stolons, and the possession of oil bodies which are large, granular and few per cell (Schuster 1967). In contrast, Schuster (1967) described the subgenus Calobryum as being characterized by plants in which the leaves are distally clustered and enlarged. The oil bodies are numerous, minute, and homogenous, and the gametangia are normally aggregated onto more or less definite, swollen receptacles as opposed to being scattered. The antheridia of this group may be ivory-white or orange. According to Schuster (1967), this subgenus, typified by H. blumii, also includes H. gibbsiae, H. indicum, H. mnioides, H. andinum, and H. giganteum.

Engel (1981) recognized only the subgenera Haplomitrium and Calobryum. He felt that the discovery of H. monoicum and H. ovalifolium reduced the degree of isolation attributed to the subgenus Archibryum by Schuster (1967), and instead, regarded this group as a section of the subgenus Haplomitrium. In addition, he created the section Protohaplomitrium Engel of the subgenus Haplomitrium for H. monoicum which he characterized by the presence of entire leaves each bearing a border of thick-walled cells, the existence of a striolate-papillose cuticle, the absence of slime papillae on all plant parts, the monoicous habit of the plants, and the formation of ivory-white antheridia with extremely reduced stalks comprised of only 1-2 tiers of cells. However, in light of the results of this investigation, a distinct section for *H. monoicum* does not seem warranted. Entire leaves characterize numerous species and leaves with borders of thick-walled cells also occur in *H. intermedium*. Slime papillae are present (at least occasionally) on the leaf margins of *H. monoicum*, and the occurrence of a striolate-papillose cuticle was not verified in this study. The monoicous habit and reduced, white antheridia serve as good species characters, but are not sufficient to define a separate section.

Therefore, following this investigation, I recognize only the subgenera *Haplomitrium* and *Calobryum*, with the former subgenus divided into the sections *Haplomitrium* and *Archibryum*. In addition, the characters which define each group are revised so that they represent the most stable and workable characters possible. Except for *H. monoicum* and *H. gibbsiae*, the species included in each subgenus remain as discussed by Schuster (1967) and Engel (1981). For the reasons discussed above, the section *Protohaplomitrium* is eliminated and *H. monoicum*, being most similar to *H. hookeri*, is placed in the subgenus *Haplomitrium* section *Haplomitrium*. The plasticity in symmetry, insertion, leaf characters, and particularly the newly discovered variability in gametangial positioning in *H. gibbsiae* suggest that this species, like *H. intermedium*, is intermediate between the subgenus *Calobryum* and the subgenus *Haplomitrium*. Based on this premise, the characters of the section *Archibryum* are altered and its limits are expanded to allow inclusion of *H. gibbsiae*.

KEYS TO THE TAXA OF *HAPLOMITRIUM*

KEY TO THE SUBGENERA AND SECTIONS OF *HAPLOMITRIUM*

1. Leafy shoots isophyllous or slightly anisophyllous with all leaves transversely inserted or with some subtransverse or succubous; gametangia scattered along the stem or in loose terminal clusters with undifferentiated bracts 2

1. Leafy shoots anisophyllous with the leaves of one row inserted transversely and two rows inserted succubously; gametangia produced on a swollen receptacle to form a terminal rosette with enlarged, differentiated bracts . . . . subg. *Calobryum*

    2. Sporophyte protective structure derived exclusively from archegonial tissue (= true calyptra) . . . . subg. *Haplomitrium* sect. *Haplomitrium*

    2. Sporophyte protective structure derived from a combination of archegonial and stem tissue (= shoot-calyptra) or exclusively from stem tissue (= perigynium) . . . . subg. *Haplomitrium* sect. *Archibryum*

KEY TO THE SPECIES OF THE SUBGENUS *HAPLOMITRIUM* SECTION *HAPLOMITRIUM*

1. Plants dioicous; median leaf cells 35-50 µm X 25-35 µm; antheridia gold or orange, with long stalks of 3-7 tiers of cells 2

1. Plants monoicous; median leaf cells 50-80 μm X 50-70 μm; antheridia white, with short stalks of 1-2 tiers of cells . . . . . . 2. *H. monoicum*

2. Male shoots with distal leaves or bracts in a perfect triquetrous arrangement 3

2. Male shoots with an imperfect spiral of distal leaves or bracts . . . 3. *H. ovalifolium*

3. Spores covered by verruculae which often possess a central hole; elaters 2-spiralled becoming 1-spiralled at the tips, or 1-spiralled throughout . . . . 1a. *H. hookeri* var. *hookeri*

3. Spores covered by verruculae lacking central holes; elaters 1-spiralled becoming 2-spiralled at the tips . . . . 1b. *H. hookeri* var. *minutum*

KEY TO THE SPECIES OF THE SUBGENUS *HAPLOMITRIUM* SECTION *ARCHIBRYUM*

1. Plants with a shoot-calyptra; gametangia in ill-defined, terminal rosettes, at alternating swollen nodes, scattered distally in axils of leaves or bracts, or free-standing on stems; spores single, covered with dense vermiculose ridges . . . . . . 4. *H. intermedium*

1. Plants with a perigynium; gametangia on swollen discs in well-defined, terminal rosettes, rarely scattered along the stem; spores in tetrads, diads, or single with distant, roughened or smooth verrucae . . . . 5. *H. gibbsiae*

KEY TO THE SPECIES OF THE SUBGENUS CALOBRYUM

1. Vegetative shoots prostrate, fertile shoots erect; spores tuberculate . . . 6. *H. mnioides*

1. Vegetative and fertile shoots erect; spores ridged . . . . 7. *H. blumii*

VEGETATIVE KEY TO THE SPECIES OF HAPLOMITRIUM

1. Terminal branching common; stolons originating from leaf axils of the aerial shoots or from rhizomes, stolons branched; oil bodies 6-24 µm, few per cell (<15) . . . . 4. *H. intermedium*

1. Terminal branching rare; stolons originating exclusively from rhizomes, stolons simple; oil bodies 2-14 µm; numerous per cell (>20) 2

   2. Leaves with a distinct 1-2-celled border of enlarged cells . . . . . 3. *H. monoicum*

   2. Leaves without a distinct border of enlarged cells 3

      3. Leaves varying from entire to shallowly or deeply lobed on the same shoot . . 4

      3. Leaves always entire on the same shoot 6

         4. Leaves with an extensive polystratose basal field . . . . 5. *H. gibbsiae*

         4. Leaves with a small bistratose zone restricted to the extreme base . . 5

5. Leaves (at least some) divided almost completely to the base . . . . 2. *H. ovalifolium*

5. Leaves never divided completely to the base . . . . 1. *H. hookeri*

6. Leafy shoots prostrate . . . . 6. *H. mnioides*

6. Leafy shoots erect . . . . 7. *H. blumii*

SUBGENUS *HAPLOMITRIUM*

Diagnosis - Leafy shoots erect, leafy throughout or with the leaves clustered terminally; isophyllous with all leaves transversely inserted or slightly anisophyllous with the lateral leaves transverse, subtransverse, or succubous. Leaves entire or shallowly to deeply 2- or 3-lobed. Dioicous or monoicous. Gametangia scattered along the shoot, arranged in loose terminal clusters, or intercalated at nodes. Apical proliferation common. Sporophyte protective structure a true calyptra, shoot-calyptra, or perigynium. Capsule dehiscence variable, along 1-(2-3) -4 (-5) complete or incomplete dehiscence lines. Elaters short and stout or long attenuate, 2 (-3)-spiralled with 1-spiralled tips or 1-spiralled throughout within a single capsule.

SUBGENUS *HAPLOMITRIUM* SECTION *HAPLOMITRIUM*

subg. *Haplomitrium* sect. *Protohaplomitrium* Engel *Ann. Missouri Bot. Gard.*, *68*: 674 (1981) (syn. nov.).

Diagnosis - Leaves unistratose, becoming bistratose at the base. Dioicous or monoicous. Gametangia scattered along the shoot or in terminal clusters, gametangial production never coupled with stem dilation. Sporophyte protective structure a true calyptra.

1. HAPLOMITRIUM HOOKERI (SMITH) NEES
Naturg. Eur. Leberm., 1: 111 (1833).

Jungermannia hookeri Smith in Sowerby's Engl. Bot., 36, pl. 2555 (1813). Scalius hookeri (Smith) S. F. Gray Nat. Arr. Brit. Pl., 1: 705 (1821). Mniopsis hookeri (Smith) Dumort. Comm. Bot., 114 (1822). Lejeunea hookeri (Smith) Spreng. Syst. Veg., 4: 234 (1827). Gymnomitrium hookeri (Smith) Corda in Opiz Beitr., 651 (1829). Scalia hookeri (Smith) Lindb. Musci Scand., 10 (1879). Scaliusa hookeri (Smith) O. Kuntze Rev. Gen. Pl., 839 (1891). Haplomitrium hookeri subsp. montanum Schiffn. Krit. Bemerk. über die eur. Leberm., 27: 22 (1941). Mniopsis acutifolia Dumort. Syll. Jung. Eur., 75 (1831). Haplomitrium cordae Nees Naturg. Eur. Leberm., 1: 112 (1833).

Type: England, New Forest, "Two capsules: one burst at the sides only a peculiar character of this J. forming an ellipse crowned by a crown of filaments.", Hooker s. n. (BM, neo [Grolle 1976: 87]).

Historical Account - Haplomitrium hookeri was first discovered by Charles Lyell on December 21, 1812, growing in a ditch near the road from Cadnum to Poultons in New Forest, Hampshire, England (Smith 1813). Specimens were given to Smith, who published the original description in English Botany. He named the taxon Jungermannia hookeri in honor of Sir Wm. Jackson Hooker. Although Hooker apparently preferred the name J. proteifolia (Gottsche 1843; Lindberg 1885), he agreed to Lyell's suggestion that Smith use *hookeri* as the specific epithet. Gray (1821) transferred the species to *Scalia*, and Nees von Esenbeck (1833) provided the new combination of *H. hookeri*.

1a. *HAPLOMITRIUM HOOKERI* VAR. *HOOKERI*

Diagnosis - Stolons originating from the rhizome, unbranched. Leafy shoots erect, isophyllous or slightly anisophyllous; terminal branching rare. Leaves ranging from ovate to orbicular or rhomboidal, entire or shallowly to deeply lobed, bistratose only at the extreme base; median leaf cells 38-48 µm long X 28-32 µm wide. Oil bodies minute 2.0-4.0 µm, homogenous to finely granular, 12-26 per cell. Dioicous. Gametangia axillary or supra-axillary in loose terminal clusters, in scattered undifferentiated or slightly modified bracts or free-standing along the axis. Apical proliferation present. Antheridia yellow-orange, with the stalks 3-4 tiers high, 69-100 µm long. Sporophyte protective stucture a true calyptra. Sporophyte seta with a central zone of enlarged cells. Capsule dehiscence along (1-) 2-4 (-5) complete or incomplete sutures. Spores 28-33 (-36) µm in diameter, occasionally retained in tetrads until late in development; the exine verruculate with limited fusion of the verruculae; each verrucula occasionally with a central aperture. Elaters 2-spiralled becoming 1-spiralled at the tips or 1-spiralled throughout; some remaining attached to the valve apices of opened capsules.
(Figures 1-16, 69, 70, 78, 79, 81-88, 103-105, 107-109, 135, 138, 141, 145, 156, 160, 165-167, 170, 172, 174-180, 262, 277, 291, 295-298, 306, 312-315, 319, 357, 360, 365, 369, 372-373, 391-393, 414, 415, 425, 428, 501, 502, 505, 506.)

1b. HAPLOMITRIUM HOOKERI (SMITH) NEES VAR. MINUTUM (E. O. CAMPB.) BARTHOLOMEW-BEGAN, COMB. et. STAT. NOV.

Steereomitrium minutum E. O. Campb. Mem. N.Y. Bot. Gard., 45: 569 (1987) (syn. nov.).

Type: New Zealand: North Island, Performance trial plots. Turf Research Institute, nr. Palmerston North, "amongst Cotula maniototo Petri", Campbell 16015 (MPN, holo), (Campbell 1987); Campbell s. n. (NY, iso).

Diagnosis - Stolons originating from the rhizome, unbranched. Leafy shoots erect, isophyllous or slightly anisophyllous; terminal branching rare. Leaves ranging from ovate to orbicular or rhomboidal, entire or shallowly to deeply lobed, bistratose to rarely tristratose at the extreme base; median leaf cells 29-54 µm long X 26-36 µm wide. Oil bodies 2.0-4.0 µm, homogeneous to finely granular, 30-40 per cell. Dioicous. Gametangia axillary or supra-axillary in loose terminal clusters, in scattered undifferentiated or slightly enlarged bracts, or free-standing along the axis. Apical proliferation present. Antheridia gold, with the stalks 6-7 tiers high, 117-201 µm long. Sporophyte protective structure a true calyptra. Capsule dehiscence usually along 2 or 4 lines of weakness. Spores (23-) 26-32 µm in diameter, the exine verruculate with limited fusion of the verruculae; verruculae lacking central apertures. Elaters 1-spiralled becoming 2-spiralled at the tips; some remaining attached to the valve apices of opened capsules.
(Figures 146, 232-234, 370, 316, 376-379, 400, 401, 416.)

Variation and Differentiation - Haplomitrium hookeri is distinguished from all other Haplomitrium taxa by its unique spore morphology. It is the only Haplomitrium species with verruculate spore exines with limited fusion of the verruculae (Figs. 414-417).

Haplomitrium hookeri var. *hookeri* and *H. hookeri* var. *minutum* are virtually indistinguishable except for three minor definable characteristics: the spiral thickening patterns of the elaters, the structure of the verruculae which cover the spores and the length of the antheridial stalks (3-4 tiers in *H. hookeri* var. *hookeri* versus 6-7 tiers in *H. hookeri* var. *minutum*). The elaters are similar in size and shape, but those of *H. hookeri* var. *minutum* are unispiralled medianly and bispiralled at the apices. Most elaters of *H. hookeri* var. *hookeri* display the opposite pattern, i.e., they are bispiralled medianly and unispiralled at the tips. *Haplomitrium hookeri* var. *hookeri* also possesses an additional form of elater which is short, stout, and unispiralled throughout. This latter elater type was not seen in the limited material of *H. hookeri* var. *minutum* available. Like the elaters, the spores of each taxon are similar in shape, size, and sculpturing, but, the verruculae of *H. hookeri* var. *minutum* do not possess central apertures like those occasionally present in the verruculae of *H. hookeri* var. *hookeri*.

Distribution and Ecology - Of all *Haplomitrium* species, *H. hookeri* exhibits the widest distribution (Fig. 508). It ranges throughout the temperate zone of the northern hemisphere, and with the synonymizing of *Steereomitrium minutum*, is also known from the southern hemisphere. Chandra et al. (1987) theorized that *H. hookeri* migrated southward during the Pleistocene glaciation from a nearctic center of origin somewhere in Scandinavia to establish its current range in North America, Europe, India, and Japan, but this is unsubstantiated.

According to Worley (1969), *H. hookeri* is relatively common on the west coast of North America in the subalpine and alpine zones of the coastal mountains from Glacier Bay, Alaska, south to northern Washington. In the central portion of the United States it has been found by N. G. Miller in the Rocky Mountains near Denver, Colorado (Gradstein 1974). On the east coast it occurs on Mt. Washington, New Hampshire (Evans 1917), and at the base of Mt. Katahdin in Maine (Schuster 1966, 1967). In Greenland, it is known from only one site on the west coast at Ulkebrugt near Holsteinsborg (Böcher 1963). It's European range was summarized by Schuster (1967) as extending from the Carpathian Mountains, northward throughout Germany, Belgium, The Netherlands, Denmark, Great Britain and Ireland, to Norway, Sweden, Finland, and Spitzbergen. In Asia, *H. hookeri* occurs in Ghunsa, Nepal (*Long 16759*) and has been reported in India from Deoban in the Western Himalayas (Udar and Kumar 1977), and from Darjeeling in the Eastern Himalayas (Udar and V. Chandra 1965). In addition, Inoue (1971) collected *H. hookeri* in middle Japan on Mt. Tateyama.

The ecology of western North American representatives reportedly differs from that of the eastern North American, European and Asian plants (Worley 1969). The habitat of the latter group is restricted mainly to hygric areas such as sandy depressions, moist heath, and margins of ditches (Schuster 1967; Udar and V. Chandra 1965; Inoue 1971; Udar and Kumar 1977). Schuster (1967) points out that *H. hookeri* is not tolerant of intermittent dry periods. However, at least in its western North American range, *H. hookeri* has a relatively broad ecological tolerance and is not a strict hygrophyte (Worley 1969). The species occurs at the base of boulders in open alpine meadows in the southern part of its range, while in the northern part, it was found among heathers at cliff bases or on rocky slopes in mesic to somewhat dry, exposed sites on the crests of barren, alpine ridges (Worley 1969).

Numerous bryophytes are found in association with *H. hookeri* throughout the northern hemisphere. Its most common associates, at least in Eurasia and eastern North America, are *Riccardia* S. Gray, *Fossombronia*, *Pellia*, and *Aneura* Dum. (Evans 1917; Schuster 1967; Inoue 1971; Udar and Kumar 1977). In western North America its associates include

Isopterigium elegans (Brid.) Lindb., *Andreaea nivalis* Hook., *Gymnomitrion* Corda, *Calypogeia* Raddi, *Nardia* S. Gray, and occasionally *Moerkia blyttii* (Moerck.) Brockm. and *Takakia lepidozioides* (Worley 1969).

In the southern hemisphere, *H. hookeri* var. *minutum* was found growing as a weed in the *Cotula maniototo* Petri turfs of the bowling green at the Cashmere Bowling Club, Christchurch, New Zealand. The turfs were grown for performance trials in both open plots and in a greenhouse at the Turf Research Institute, Palmerston North, New Zealand (Campbell 1987). The manner in which the turfs were treated was not discussed by Campbell (1987), but the *C. maniototo* plants were thought to have originated from Dunback, Otago. There are no reported findings of this genus from that site, however. Therefore, the only *H. hookeri* var. *minutum* plants that have been studied are those grown under artificial conditions.

Specimens Examined - UNITED STATES: Alaska, Baranof Is., *Worley & Hamilton 9933* (H, G); Glacier Bay Natl. Monument, *Worley & Streuler 13027* (G); Mt. Olds, *Worley & Rawson 12542* (H). CANADA: British Columbia, Queen Charlotte Islands, *Schofield & Spence 83691* (F). SCOTLAND: Argyll, *Long 9956* (BM); Easterness, *Long 9737* (BM); Forres, *Dalby 1867* (BM); N. Harris, *Long 9407* (G); S. Harris, *Long 9509* (G); Peeblesshire Co., *Crundwell s. n.* (F); Mid Perth, *Long 9930* (BM, G), *Long 6932* (BM); W. Ross, *Long 12224* (BM). ENGLAND: New Forest, *Hooker s. n.* (neotype), *39*, *Lyell s. n.* (BM); s.l., *Hooker s. n.* (BM); Kinnordy, *Hooker s. n.* (BM), *Lyell s. n.* (MANCH); Cornwall, *Ralfs s. n.* (BM); Penzance, *Ralfs s. n.* (BM), *Curnow s. n.* (BM, MANCH); Yorkshire, *Ware s. n.* (MANCH). FINLAND: Helsinki, *Lindberg s. n.* (BM, G. MANCH). DENMARK: Jutland, *Jensen 1347* (AD, G), *Jensen 324* (BM, G, MANCH). EAST GERMANY: Brandenburg region, *Osterwald 1346* (AD). WEST GERMANY: Berlin, *Loeske s. n.* (BM, H); Hamburg, *Gourlie s. n.*, *Kohlmeyer s.n.* (BM). AUSTRIA: Vorlarlberg State, *Loitlesberger 1348*, *s. n.* (AD, G, H). CZECHOSLOVAKIA: Prague, *Lorbeer H56* (culture collection communicated from M. L. Sargent). POLAND: Carpathian Mts., Lilienfeld 19 (F). NEPAL: Ghunsa Khola, Long 16759 (E). JAPAN: Tattori, Faurie 392 (G); Honshu, Mt. Tateyama, Yamada & Yoshimura 969 (F, G, H). NEW ZEALAND: North Island, Palmerston North, Turf Research Institute, Campbell s. n. (isotype of Steereomitrium minutum, NY), Campbell s. n. (ABSH).

2. HAPLOMITRIUM OVALIFOLIUM SCHUST. Bryologist, 74: 131-143 (1971).

Type: New Zealand, Campbell Island, "soil over small ledges, vertical sea cliffs of Mt. Yves Villarceau, ca. 200-275 ft., above rookeries of the Rockhopper Penguin, at Penguin Bay", Schuster 70-3248, (Schuster 1971) (hb Schuster, holo).

Diagnosis - Stolons originating from the rhizome, unbranched. Leafy shoots erect, isophyllous or slightly anisophyllous; terminal branching rare. Leaves ranging from ovate to orbicular or rhomboidal, entire or shallowly to deeply lobed, bistratose only at the extreme base; median leaf cells 25-38 (-60) µm long X 23-36 µm wide. Oil bodies 2.5-5.0 µm, homogeneous to finely granular, 25-35 per cell. Dioicous. Gametangia axillary or supra-axillary in loose terminal clusters, in scattered undifferentiated or slightly modified bracts or free-standing along axis. Male shoots with a distal imperfect spiral of leaves or bracts. Apical proliferation present. Antheridia yellow, stalk lengths unknown. Sporophyte protective structure a true calyptra. Sporophytes unknown.

Historical Account - Haplomitrium ovalifolium is known only from the type collection which was unavailable for examination. Therefore, the above diagnosis is based entirely on the original description of the species by Schuster (1971). Although the type specimen is cited in his protologue as being housed at the New York Botanical Garden (NY), it is not among their holdings, nor is it among the Schuster collections received to date at the Field Museum of Natural History (F). According to Schuster (personal communication) the type specimen was not deposited at NY and is still among his personal collections.

Variation and Differentiation - *Haplomitrium ovalifolium* appears to be extremely similar, if not identical to *H. hookeri*. According to the published information for *H. ovalifolium* (Schuster 1971), the two taxa can be separated only by the production of an imperfect spiral of male bracts by *H. ovalifolium*. This character was not seen in *H. hookeri*, but its occurrence is not improbable considering the great plasticity inherent in the species, especially when influenced by different environmental conditions. Without examining *H. ovalifolium*, it is impossible to ascertain its taxonomic status with certainty. However, it seems highly probable that this species will prove to be synonymous with *H. hookeri*, especially in light of the fact that *H. hookeri* is now known from New Zealand.

Distribution and Ecology - *Haplomitrium ovalifolium* has been collected by Schuster (1971) on South Island and Campbell Island, New Zealand. It occurs on soil over small ledges on vertical cliffs (Schuster 1971). At its lower altitudes it is often associated with *Aneura pinguis* (L.) Dum., *Fossombronia*, and *Riccardia*, while in its subalpine range it occurs with *Phyllothallia nivicola* Hodgs., *Allisonia cockaynii* (Steph.) Schust., *Solenostoma* cf. *rufiflora* Colenso, *Balantiopsis* Mitt., and *Riccardia* (Schuster 1971).

3. HAPLOMITRIUM MONOICUM ENGEL
Ann. Missouri Bot. Gard., 68: 668-669 (1981).

Type: New Caledonia, Montagne des Sources reserve, "ca. 20 air-km NE of Noumea. On shaded bank. Alt. 750 m.", Crosby 14165 (F, holo).

Diagnosis - Stolons originating from the rhizome, unbranched. Leafy shoots erect, isophyllous or slightly anisophyllous; terminal branching absent. Leaves ranging from ovate to orbicular to ellipsoidal, entire, bordered by one or sometimes two rows of enlarged cells, with the basal field small, median, and bistratose; median leaf cells 50-74 µm long X 52-71 µm wide. Oil bodies unknown. Monoicous. Gametangia axillary or supra-axillary, in loose terminal clusters or in scattered undifferentiated or slightly enlarged bracts in the distal quarter of the shoot. Apical proliferation present. Antheridia white, with the stalks 2-3 tiers high, 19-24 µm long. Sporophyte protective structure a true calyptra. Sporophyte unknown.
(Figures 17, 18, 21-25, 181-184, 247, 260, 261, 301, 317, 318.)

Historical Account - In 1981, Marshall Crosby discovered in New Caledonia the only known monoicous species of *Haplomitrium*. The material was given to Engel who named the plant *H. monoicum*.

Variation and Differentiation - *Haplomitrium monoicum* appears most similar to *H. hookeri* and *H. ovalifolium* with which it shares similar habits, symmetry, gametangial loci, and calyptra ontogeny. It differs from these taxa in bearing entire leaves with borders composed of 1 or 2 rows of enlarged cells, in having small, median bistratose basal fields as opposed to being bistratose only in a small zone near the point of leaf insertion, and in possessing larger median leaf cells. *Haplomitrium intermedium* is the only other known *Haplomitrium* species which possesses similar leaf borders. In addition, the median leaf cell sizes in these two taxa are quite similar.

Haplomitrium monoicum differs from all other Haplomitrium taxa in being monoicous. Engel (1981) also listed as distinguishing species characters the absence of slime papillae and the production of white antheridia with extremely reduced stalks, composed of only 2-3 tiers of cells. However, with the aid of SEM, slime cells were found to be present. The reduced height of the antheridial stalks does appear to be unique to H. monoicum (see Table VI). Numerous other taxa (H. hookeri, H. intermedium, and H. mnioides) commonly possess antheridial stalks which are only 3-4 tiers tall, the shortest of these (seen in H. intermedium) being 42-72 µm high, whereas the stalks of H. monoicum are only 19-24 µm tall.

Distribution and Ecology - Haplomitrium monoicum is known only from its type locality in Montagne des Sources reserve, New Caledonia (Engel 1981) (Fig. 508). Apparently, the plant is very rare and extremely difficult to locate due to its tiny size and because it grows scattered rather than in mats (Kitagawa, personal communication). As documented by Engel (1981), the type specimens were found on a shaded, vertical clay bank of an intermittent stream in an Araucaria forest.

Specimens Examined - NEW CALEDONIA: Montagne des Sources reserve, Crosby 14165 (holotype, F), Kitagawa 21694 (hb. Kitagawa).

SUBGENUS HAPLOMITRIUM SECTION ARCHIBRYUM (SCHUST.) ENGEL (Ann. Missouri Bot. Gard., 68: 675 [1981]) EMEND. BARTHOLOMEW-BEGAN

subg. Archibryum Schust. Nova Hedw., 13: 28 (1967).

Diagnosis - Leaves unistratose, becoming 2-4 stratose at the base. Dioicous. Gametangia scattered along the shoot or localized on dilated discs in terminal rosettes or at alternating swollen nodes. Sporophyte protective structure a shoot-calyptra or perigynium.

4. HAPLOMITRIUM INTERMEDIUM Berrie Proc. Linn. Soc. N. S. W., 87: 191-195 (1962).

Type: Australia, New South Wales, Bilpin, Mt. Irvine Road, "wet cliff face, permanent dripping water, shade from herbs, etc.", Berrie 2/62/1 (SYD, holo).

Diagnosis - Stolons originating from the leaf axils of aerial shoots or from the rhizome, commonly branched. Leafy shoots erect, isophyllous or slightly anisophyllous; terminal branching common. Leaves ranging from ovate to orbicular to elliptical, usually entire, bordered by a row of enlarged cells, bistratose at the extreme base; median leaf cells 55-80 µm long X 40-50 µm wide. Oil bodies 6.0-24.0 µm, granular, 3-12 per cell. Dioicous. Gametangia axillary or supra-axillary, in scattered undifferentiated or enlarged bracts, in ill-defined terminal rosettes, at alternating swollen nodes or free-standing along shoot. Apical proliferation present or absent. Antheridia orange, with the stalks 3-4 tiers high, 46-72 µm long. Sporophyte protective structure a shoot-calyptra with elevated archegonia, bractlets, and slime papillae in the basal third; bractlets secondarily elongated. Sporophyte seta generalized. Capsule dehiscence unknown. Spores (21-) 23-28 µm in diameter, the exine of densely arranged vermiculose ridges. Elaters 2-spiralled

throughout, rarely 3-spiralled medianly.
(Figures 19, 20, 26-34, 75, 96, 97, 124, 143, 144, 155, 164, 185-191, 263, 307, 324, 325, 394, 421, 503, 504.)

Historical Account - Berrie discovered and named *H. intermedium* in 1962. The specific epithet *intermedium* was chosen to suggest that the species is intermediate between *Haplomitrium* and *Calobryum* in gametangial position, i.e., gametangia may be positioned apically, at nodes along the stem, or scattered without definate loci (Berrie 1962).

Variation and Differentiation - *Haplomitrium intermedium* is similar to *H. monoicum* in habit, symmetry, and the fact that gametangial positioning is variable. However, it differs in possessing the potential for stem dilation at the sites of gametangial production and in the possession of a shoot-calyptra as opposed to a true calyptra. *Haplomitrium intermedium* is unique among *Haplomitrium* species in its frequent terminal branching, unique oil bodies, axillary stolons with exogenous branching, and spores which are densely covered with vermiculose ridges.

Distribution and Ecology - *Haplomitrium intermedium* is known only from New South Wales, Australia (Fig. 508). According to Berrie (1962), it occurs on soil which has accumulated in irregularities on lightly shaded, wet cliff faces. It is often associated with *Symphyogyna* Nees & Mont., *Pallavicinia* S. Gray, and small mosses (Berrie 1962).

Specimens Examined - AUSTRALIA: New South Wales, Blue Mts., *Thiers 2979* (AD), Bilpin, *Berrie 2/62/1* (holotype), *3/62* (SYD), Adelina Falls, *Berrie 8/61* (SYD); Kuring-gai Chase, *Beard 1/59* (SYD); Bowen's Creek, *McKay 3951* (MUCV), *Selkirk s. n.* (ABSH).

5. HAPLOMITRIUM GIBBSIAE (STEPH.) SCHUST. J. Hattori Bot. Lab., 2: 25 (1963).

Calobryum gibbsiae Steph. Spec. Hep., 6: 76 (1917).

Type: New Zealand, N. Island, Te Aroha, Wairongami Mine, "Hab. on clay banks, Alt. 2000", L. S. Gibbs 1064 (G, lectotypus nov.).

Calobryum indicum Udar & Chandra Rev. Bryol. et Lichen., 33: 555 (1965) (syn. nov.). Haplomitrium indicum (Udar & Chandra) Schust. Hep. & Anth. N. A., 1: 638 (1966) (syn. nov.). Haplomitrium chilensis Schust. Bryologist, 74: 133 (1971); Hässel de Menéndez (1975). Calobryum dentatum Kumar & Udar J. Indian Bot. Soc., 55: 23 (1976) (syn. nov.). Haplomitrium dentatum (Kumar & Udar) Engel Ann. Missouri Bot. Gard., 68: 668 (1981) (syn. nov.). Haplomitrium grollei Kumar & Udar Geophytology, 7: 260 (1977) (syn. nov.). Haplomitrium kashyapii Udar & Kumar J. Indian Bot. Soc., 61: 73 (1982) (syn. nov.).

Diagnosis - Stolons originating from the rhizome, unbranched. Leafy shoots erect, isophyllous, rarely anisophyllous; terminal branching rare. Leaves rhomboidal or ranging from orbicular to ovate; basal field 2-4-stratose; median leaf cells 65-85 µm long X 45-65 µm wide. Oil bodies 2.0-9.0 µm, finely granular, 16-38 per cell. Dioicous. Gametangia in well-defined, terminal rosettes, rarely scattered along the stem. Apical proliferation present or absent. Antheridia white to yellow-orange, with the stalks 4-10 tiers high, 81-233 µm long. Sporophyte protective structure a perigynium bearing archegonia, bractlets, and slime papillae over entire surface, with bractlets absent or with one or two bractlets raised only to basal one-fifth to one-third; secondary elaboration of raised structures common. Sporophyte seta generalized. Capsule dehiscence along (2-3-) 4 (-5) incomplete lines. Spores 20-39 µm in diameter, usually remaining in tetrads or diads until late in development, exine coarsely verrucate, with the verrucae roughened or smooth, 1.7-3.2 µm high. Elaters 2-spiralled (3-spiralled), becoming 1-spiralled at the tips, or 1-spiralled throughout. (Figures 35-44, 71, 74, 80, 99, 100, 111, 112, 137, 139, 140, 147-149, 152, 153, 158, 169, 192-212, 236-239, 245, 252, 253, 255, 257, 259, 265, 266, 271, 273, 276, 278, 283, 286, 290, 292-294, 300, 302-304, 308, 309, 344-356, 370, 374, 375, 395, 418-420, 426, 486-490, 500, 507.)

Historical Account - Gibbs first collected the type specimen of *H. gibbsiae* in November, 1907, in a forest at Wairongamī Mine, Te Aroha, New Zealand (Gibbs 1911). She sent the specimens to Stephani (1924), who described the species and named it *Calobryum gibbsiae* in honor of its collector. The species was ultimately transferred to *Haplomitrium* by Schuster (1963).

Variation and Differentiation - *Haplomitrium gibbsiae* is similar to *H. intermedium* in both symmetry and habit, but differs in possessing leaves having extensive multistratose basal fields, in displaying less varibility in gametangial positioning, and in the production of a perigynium. The gametangia of this taxon are almost always produced on a swollen terminal disc surrounded by enlarged bracts. However, under certain environmental conditions, the gametangia may be dispersed along the stem. The perigynium has traditionally been described as being covered by foliar outgrowths as well as elaborated archegonia and mucilage hairs (Stephani 1917; E. O. Campbell 1959; Schuster 1967). However, examination of numerous collections from New Zealand shows that the ornamentation of this structure can be quite variable. Occasional individuals occur which possess perigynia which are virtually smooth and lack foliar outgrowths at least in the distal two-thirds, or which may rarely have an occasional bractlet carried midway to the apex, or may be covered exclusively by archegonia and/or slime papillae or mucilage hairs. In addition to this set of characters, H. gibbsiae possesses spores which tend to remain in tetrads or diads until late in ontogeny and are covered with large, roughened or smooth verrucae which are 1.7-3.2 µm high.

When Schuster (1971) named the South American species, H. chilensis, he recognized that it was closely allied to the New Zealand species H. gibbsiae stating, "... if only the gametophyte were under consideration it might easily be possible to consider this species identical with H. gibbsiae." According to Schuster (1971), the plants are distinguished by spore size, height of the verrucae of the spores, the lack of foliar outgrowths on the perigynium, the absence of apical proliferation, oil body size and number, and the presence of enlarged "stoloniform" outgrowths from the shoot bases. Based on her examinations of the type specimens of both H. chilensis and H. gibbsiae, along with observations of H. gibbsiae plants from Argentina (particularly material from Lago Escondido in Tierra del Fuego), Hässel de Menéndez (1975) disputed these differences and synonymized H. chilensis with H. gibbsiae. Similarly, the results of this investigation indicate that these two species are synonymous. The range of spore sizes and of exine verrucae heights observed for spores taken from New Zealand H. gibbsiae plants encompass the measurements cited by Schuster (1971) as being specific for H. chilensis. The same was found to be true concerning the range of oil body characters. As noted above, perigynia without foliar outgrowths such as those described by Schuster (1971) for H. chilensis also occur intermixed in New Zealand populations of H. gibbsiae plants which possess the more ornamented form of perigynia. Apical proliferation was not seen in fertile H. chilensis plants, but the potential for proliferation is present as transections through mature gametangial discs reveal tetrahedral apical cells. Both H. gibbsiae and H. chilensis bear short, leafless branches at the bases of leafy shoots. The extent to which these structures elongate is likely influenced by environmental conditions.

Like *H. chilensis*, the Indian taxa *H. indicum*, *H. dentatum*, *H. grollei*, and *H. kashyapii* are here considered synonymous with *H. gibbsiae*. Most of the features used to delimit these taxa are based on plant sizes, symmetries, leaf shape, or whether or not shoots are leafy throughout or have the leaves clustered terminally (Udar and S. Chandra 1961, 1965; Kumar and Udar 1976, 1977; Udar and Kumar 1982). However, as previously discussed, these characters are highly malleable in *Haplomitrium* and therefore, are not dependable species characters. Furthermore, the complete range of variability observed in the Indian endemics also occurs in *H. gibbsiae* populations from New Zealand.

Although gametangial positioning was also used to define the Indian species, the authors apparently failed to recognize the similarities with apically proliferating specimens of *H. gibbsiae*. Like the New Zealand populations of *H. gibbsiae*, the reproductive shoots of *H. indicum*, *H. dentatum*, and *H grollei* exclusively produce gametangia which are arranged in well-defined terminal discs and which commonly proliferate apically resulting in numerous "crops" or alternating nodes of gametangia. Male *H. kashyapii* plants and some female shoots also produce gametangia in similar terminal rosettes, while other female shoots produce archegonia which are scattered along the stem. In addition, Kumar (1987) has described individuals of a presumed undescribed species which produce antheridia which are scattered along the shoots, while female shoots possess archegonia which are aggregated into well-defined terminal rosettes. From his description, these plants also appear to be *H. gibbsiae* individuals. This displacement or scattering of the gametangia is identical to that which occurred in male *H. gibbsiae* plants maintained in culture over moistened vermiculite, demonstrating that the potential for the production of scattered gametangia exists in *H. gibbsiae* and that this condition can be induced by different environmental conditions.

In addition to gametangial positioning, Udar and Kumar (1982) also emphasized the anisophylly and succubous insertion of the lateral leaves as distinguishing characters of *H. kashyapii*. However, similar individuals also occur in New Zealand populations of *H. gibbsiae*, although less frequently than isophyllous or nearly isophyllous individuals with transverse or subtransverse leaf insertions (compare Figures 37 and 40).

The known sporophytic and calyptral characters of the Indian taxa also support their synonymy with *H. gibbsiae*. The perigynia, like those of *H. chilensis* are identical with the less elaborated forms found in *H. gibbsiae*. Of the above taxa described from the Indian subcontinent, spores have been reported only for *H. dentatum* and *H. indicum*, and, although slightly smaller than typical *H. gibbsiae* spores, they tend to remain in diads or tetrads until late in development (Kumar and Udar 1976; Mehra and Kumar 1990). *Haplomitrium gibbsiae* and occasionally *H. hookeri* are the only species known to possess this character.

Final analysis of all of these characters indicates that *H. chilensis*, *H. indicum*, *H. dentatum*, *H. grollei*, and *H. kashyapii* are either ecophenes or fixed genetic variants of *H. gibbsiae*. Each represents a small, local population of *H. gibbsiae* in which a particular phenotype, which is typically expressed less frequently in the widespread New Zealand *H. gibbsiae* populations, becomes the most frequently expressed condition due to some different environmental factor or factors or due to genetic drift. Thus, these taxa should have no taxonomic status and should be considered synonyms of *H. gibbsiae*. However, it must be remembered that only a few populations of *H. chilensis*, *H. indicum*, *H. dentatum*, *H. grollei*, and *H. kashyapii* currently are known, thus limiting our knowledge concerning the ranges of variability inherent in these populations. Perhaps more extensive collecting in southern South

America and the Himalayan regions of India will reveal individuals expressing the more typical H. gibbsiae phenotypes.

Distribution and Ecology - Haplomitrium gibbsiae is widely distributed throughout New Zealand, and with the synonymizing of H. chilensis, and H. indicum, H. dentatum, H. grollei, and H. kashyapii, also occurs in southern South America and India, respectively (Fig. 508). The species exhibits a broad ecological tolerance which is particularly evident in New Zealand where it is a pioneer on moist, disturbed sites, such as clay banks (Gibbs 1911; E. O. Campbell 1959), but also occurs in swampy, Sphagnum areas, and under alpine conditions on bare wet soil among small rocks (Schuster 1967). Although Schuster (1967) cited H. gibbsiae as occurring from sea level to at least 4200 ft in New Zealand, Allison and Child (1975) described it as a lowland species, ranging from sea level to 2000 ft.

In New Zealand, on North Island, Campbell (1959) found H. gibbsiae intermixed in mats dominated by Schistochila splachnophylla (Hook. & Tayl.) Steph. In Dunedin, on South Island, Allison and Child (1975) found it associated with Schistochila colensoana Steph. and Treubia lacunosa (Col.) Prosk. Numerous associations correlated with habitat were noted by Schuster (1967) including an association with Sphagnum and Lepidozia (Dum.) Dum. in swampy areas, with Trichotemnoma Schust., Kurzia v. Martins, Jamesoniella (Spruce) Carring., and Radula Dum. on wet cliff faces, and with Allisonia Herzog. and Marsupella Dum. in the snow tussock zone.

In South America, H. gibbsiae is known from Chile and Argentina (Schuster 1971; Hässel de Menéndez 1970, 1975). The plants occur at sea level and are often found in turfs on soil in shady, damp woods, often associated with Telaranea plumulosa (Lehm. & Lind.) Fulf. (Hässel de Menéndez 1975).

In India H. gibbsiae is known only from the Darjeeling region of the Eastern Himalayas. All of the Indian collections are from the Teesta Valley near Ghoom and share identical altitudinal ranges (1969-2121 m.s.m.) and ecologies. The plants reportedly grow on shaded, sandy soil over rocks in association with Jungermannia L. emend. Dum., Riccardia multifida (L.) S. Gray, Pogonatum P. -Beauv., and Aplozia (Dum.) Dum. (Udar and S. Chandra 1965; Kumar and Udar 1976, 1977; Udar and Kumar 1982).

Specimens Examined - CHILE: Prov. Valvivia, Rio Futa, Engel 10847 (F); Prov. Osorno, Refugio Antillanca, Engel 11513B, 11537 (F); Prov. Chiloe, Loncomilla, Engel 12133 (F); Prov. Magallanes, Cerro Pta. Arenas, Tsudgii & Yoshida 710 (H); Prov. Aisen, Ventisquero Solor, Seki 1245 (H). INDIA: Darjeeling, Ghoom, Udar & Kumar C-69/4 (holotype of H. dentatum), Udar & Kumar C72/1 (holotype of H. grollei), Udar & Kumar C70/5, Udar & Kumar C73/1 (holotype of H. kashyapii) (LWU). TASMANIA: Mt. Darwin, Engel 16590 (F); Mt. Wellington, Ratkowsky H996 (HO). NEW ZEALAND: Hari Hari westland, Allison H4729 (CHR); North Island - Te Aroha, Gibbs 1064 (lectotype, G); Wellington Dist., Akatarawa Mts., Brownsey H8727, H8740 (WELT), Fife 5737 (CHR), Sneddon s. n., Fife 8453 (ABSH); Rangitaiki River, Allison H568a, H568b, H2263 (CHR); Coromandel Peninsula, Allison H1443 (CHR), Moore 36875 (CHR); Southern Kaingaroa Plains, Allison H2264 (CHR); Mt. Ruapehu, Allison H2266 (CHR); South Island - Fox Glacier, Schuster 48359 (G); Mt. Arthur, Fife 8453 (CHR); Okira Gorge, Schuster s. n. (MUCV); Maungatua Range, Allison H845 (CHR); Bullet River, Allison H6050 (CHR); Akalone Bush, Allison H6684 (CHR); Howden-Mackenzie, Taylor 135749 (CHR); Fiordland, Burrell, Scott & Taylor 136036, Morice, Wylie & Taylor 135342 (CHR); Leith Valley, Allison, Scott & Taylor 135503 (MUCV), Scott 94, 766 (hb. Scott); Dunedin, Schuster & Taylor 136035 (CHR); Doubtful Sound, Taylor 135736 (CHR); Stewart Island - Halfmoon Bay, Adams H3444 (WELT), Thule Bay, Schuster, Scott & Taylor 135631 (CHR).

SUBGENUS CALOBRYUM (NEES) SCHUST.
Nova Hedw., 13: 40 (1967).

Diagnosis - Leafy shoots erect or prostrate, usually with leaves clustered terminally; anisophyllous, with a reduced dorsal row of leaves transversely inserted, with the lateral leaves succubous. Leaves entire, rarely shallowly lobed or dentate. Dioicous. Gametangia localized on a dilated terminal receptacle subtended by enlarged bracts forming a well-defined terminal rosette. Apical proliferation present or absent. Sporophyte protective structure a shoot-calyptra. Capsule dehiscence along one suture. Elaters long attenuate, uniformly bispiral throughout.

6. HAPLOMITRIUM MNIOIDES (LINDB.) SCHUST.
J. Hattori Bot. Lab., 26: 225 (1963).

Rhopalanthus mnioides Lindb. Manip. Musc., 2: 390 (1874). Calobryum mnioides (Lindb.) Steph. Spec. Hep., 1: 399 (1900). Haplomitrium mnioides var. delicatum Gao & Li Wuyi Sci. J., 5: 234 (1985) (syn. nov.). Scalia rotundifolia Mitt. Trans. Linn. Soc. Lond., Ser. 2, 3: 204 (1891). Haplomitrium rotundifolium (Mitt.) Steph. Bull. Herb. Boiss., 5: 79 (1897). Calobryum rotundifolium (Mitt.) Schiffn. Oesterr. Bot. Zeitschr., 49: 389 (1899).

Type: Japan, Exped. of U. S., (H-SOL, lectotypus nov.).

Diagnosis - Stolons originating from rhizome, unbranched. Leafy shoots prostrate to obliquely ascending when sterile, erect when fertile, anisophyllous, terminal branching rare. Leaves subrotund, entire; basal field 2-3-stratose; median leaf cells 55-95 µm long X 30-55 µm wide. Oil bodies 3.0-9.0 µm, finely granular, 28-50 per cell. Dioicous. Gametangia produced in well-defined terminal rosettes. Apical proliferation present.

Antheridia orange, with the stalks 3-4 tiers long, 197-248 µm long. Sporophyte protective structure a shoot-calyptra with elevated intra-gynoecial structures within the basal one-third or restricted to the basal one-third and the apex. Sporophyte seta hollow at maturity. Capsule dehiscence along 1 line. Spores (25-) 28-35 µm in diameter, the exine densely tuberculate. Elaters 2-spiralled throughout. (Figures 45-55, 73, 91-94, 101, 102, 106, 110, 113-123, 127-134, 136, 142, 150, 161, 162, 168, 171, 173, 213-220, 135, 240-244, 246, 148-251, 254, 256, 258, 264, 274, 275, 280, 282, 285, 287-289, 310, 337-343, 358, 366-368, 380-386, 388, 389, 396, 402, 403, 406-412, 417, 427, 429-439, 448-485, 491-499.)

Historical Account - _Haplomitrium mnioides_ was originally described from male specimens and named _Rhopalanthus mnioides_ by Lindberg (1874). In 1891, Mitten named _Scalia rotundifolia_ based on female specimens. Schiffner (1899) determined that these genera were taxonomically equivalent and proposed the new combination, _Calobryum rotundifolium_. Stephani (1900) retained the earlier, legitimate epithet and called the plant _C. mnioides_. The species was ultimately transferred to _Haplomitrium_ by Schuster (1963).

Variation and Differentiation - _Haplomitrium mnioides_ differs from all other species of _Haplomitrium_ in that its vegetative shoots are prostrate or obliquely ascending, while only the fertile shoots are erect. In addition, it is the only species of _Haplomitrium_ which possesses an elongate sporophyte seta which is hollow, and spores which are densely covered with tuberculae. Unlike the previously diagnosed species, but like _H. blumii_, _H. mnioides_ is strongly anisophyllous, its gametangia are exclusively produced in well-defined terminal rosettes, capsule dehiscence is typically along a single line of dehiscence, and only bispiralled elaters are known.

Recently, Gao and Li (1985) named the new variety H. mnioides var. delicatum from China. They apparently based their decision on the delicate texture of the plants and the fact that only 2-4 cycles of distal leaves are produced. As previously shown, however, leaf distribution and plant texture are easily altered by different environmental conditions. Therefore, this new variety is likely a phenovariant of H. mnioides and is not here given taxonomic status.

Distribution and Ecology - Haplomitrium mnioides is widespread throughout central Japan and extends southward through Taiwan (Fig. 508). The species exhibits an altitudinal range of 20-800 m.s.m. and grows on shaded soil or wet granitic rocks (Horikawa 1929; Schuster 1967; Inoue 1974). It has also been collected from Chongan Xian, and the Fujian, Guangdong, Guangxi, and Sichuan provinces of southern China (Gao and Li 1985) (Fig. 508). Here H. mnioides was found growing on wet rocks often mixed with Dumortiera hirsuta (Swartz) Reinw., Blume et Nees, Heteroscyphus Schiffn., and Philonotis Brid.

Specimens Examined - TAIWAN: Hwalien Co., Lai 2041 (F); Chiayi Co., Lai 10080 (F, H, G): Chiayi Hsien, Lin 19 (H); Taipei, Stephani 186 (G). JAPAN: s.l., Lindberg s. n. (lectotype, H); Kumemura, Kudaira s. n. (G); s.l., Miya s. n. (G); Honshu Island - Aki-cho, Hiroshima, Ando s. n. (H); Aki, Horikawa 62 (F, H); Mie Prefecture, Owase City, Inoue 9346 (TNS); s.l., Inoue 24763 (TNS); Kyushu Island - Kagoshima Pref., Suzukawa, Mizutani 10360 (F); s.l., Iwatsuki s. n. (AD); s.l., Saito s. n. (TNS); Shiratani, Mizutani 10782 (F); s.l., Noguchi & Imae 511 (G); s.l., Tagawa & Kitagawa 369 (hb. Kitagawa); s.l., Takaki & Mizutani 773 (F, H, G); Mt. Mochomu, Inoue & Gradstein 506 (F, H, G); Shiratani-Unsui-Kyo Valley, Iwatsuki 142 (H); Miyazaki Pref., Obi, Hattori 101 (F, H), Hattori 2547, Mizutani s. n. (TNS), Koponen 11214 (H), Mizutani s. n. (ABSH); Inotani Exp. Forest, Schofield, Iwatsuki &

Chuang 36776 (F); Mt. Osuzu, Inoue 23771 (TNS); Kaeda Valley, Inoue 24474 (TNS); Kumamoto Pref., Uyemura, Mayebara 457 (H, G); Shikoku Island - Kochi Pref., Kitagawa 16746 (hb. Kitagawa); Ryukyu Islands, Inoue 20923 (TNS); Hiroshima Pref., Isl. Miyazima Doei s. n., Furuki 7408 (ABSH).

7. HAPLOMITRIUM BLUMII (NEES) SCHUST. J. Hattori Bot. Lab., 26: 225 (1963).

Monoclea blumii Nees Hep. Javan., 2 (1830).
Calobryum blumii (Nees) Nees in G., L. & N., Syn. Hep., 507 (1846).

Type: [Java, Blume s. n.] (STR, lectotypus nov.).

Scalia andina Spruce Trans. proc. Bot. Soc. Edinburgh, 15: 532 (1885) Lectotypus (nov.): Spruce s. n. (syn. nov.).
(note:) Spruce s. n., Andes Quitenses, alt. 1300 m (BM) is selected from among two syntypes and here designated as the lectotype.
Scalia carnosula Mitt. Trans. Linn. Soc. Lond., Ser. II, 3: 204 (1891) (syn. nov.). Calobryum andinum Steph. Spec. Hep., 1: 400 (1900) (syn. nov.). Haplomitrium andinum (Steph.) Schust. J. Hattori Bot. Lab., 26: 225 (1963) (syn. nov.). Tylimanthus giganteus Steph. Spec. Hep., 6: 247 (1922) (syn. nov.). Calobryum giganteum (Steph.) Grolle J. Hattori Bot. Lab., 26: 5 (1963) (syn. nov.). Haplomitrium giganteum (Steph.) Grolle Oesterr. Bot. Zeitschr., 111: 191-192 (1964) (syn. nov.). Haplomitrium mnioides (Gottsche ms.) Bescherelle, J. Bot. (Morot), 7: 191 (1893) (syn. nov.).

Diagnosis - Stolons originating from the rhizome, unbranched. Leafy shoots erect, anisophyllous, terminal branching rare. Leaves subrotund to obovate, entire; basal field 2-4-stratose, extensive; median leaf cells 50-125 µm long X 40-80 µm wide. Oil bodies

3.0-15.0 µm, granular, 60-75 per cell. Dioicous. Gametangia on swollen discs of well-defined terminal rosettes. Apical proliferation absent. Antheridia orange, with the stalks 3-5 (-10) tiers high, 92-252 µm long. Sporophyte protective structure a shoot-calyptra with elevated intra-gynoecial structures within the basal one-third, restricted to the basal one-third and the apex, or over its entire length, secondary outgrowths and elaboration of raised structures common. Sporophyte seta generalized. Capsule dehiscence along one line. Spores (15-) 19-30 µm diameter, the exine loosely ridged with the ridges often anastomosing, with the interstices large. Elaters 2-spiralled throughout.
(Figures 56-67, 72, 89, 90, 95, 125, 126, 151, 154, 159, 163, 221-231, 267-269, 272, 279, 284, 299, 305, 320-323, 326-336, 359, 361-364, 371, 387, 390, 397-399, 404, 405, 413, 422-424, 440-447.)

Historical Account - Nees von Esenbeck (1830) first named this plant *Monoclea blumii* in honor of Blume who originally discovered the plant in Java. Nees (1830) categorized the species as a member of *Monoclea* because the capsule dehisces along a single suture. Later, when he created the new genus *Calobryum*, Nees (Gottsche et al. 1844-1847) transferred the species from *Monoclea* to *Calobryum*. In 1963, Schuster transferred the taxon to *Haplomitrium*.

Variation and Differentiation - Although similar to *H. mnioides* in shoot symmetry, leaf insertion, leaf shape, and leaf thickness, *H. blumii* differs from *H. mnioides* in that both its fertile and vegetative shoots are erect and apical proliferation is absent on fertile shoots. In addition, the spores of *H. blumii* are smaller than those of *H. mnioides* and are loosely ridged as opposed to being densely tuberculate.

A comparison of the species variability exhibited by *H. blumii* and that displayed by plants traditionally identified as *H. andinum* reveals that these plants are indistinguishable. The characters traditionally used to distinguish H. andinum, i.e., unistratose leaves, a smooth calyptra with raised structures restricted to the basal portion, and minute spore size (Spruce 1885; Fulford 1962; Schuster 1967), are not valid. Each lateral leaf and bract bears a bistratose zone of various size at its extreme base. In some cases, this zone may encompass only 2-3 cell rows, while in other cases, it may cover as many as the basal 5-7 rows of cells. So, like H. blumii, the leaves are not unistratose throughout. In addition, both taxa produce shoot-calyptrae which are similar to those of H. mnioides in that intra-gynoecial structures may be restricted to the basal one-fifth or one-third of the structure or may be carried to the apex depending on the size of the axial meristematic zone, its length of activity and the consequential positional and temporal coincidence of the raised structures with the zone of archegonial meristematic activity. Secondary elaboration of raised structures, particularly bractlets, occurs in both Latin American (H. andinum) and Asiatic populations (H. blumii). The only observable difference between these structures is that archegonia are not elevated beyond the basal one-third of the shoot-calyptra of H. andinum, while they may be carried higher on H. blumii shoot-calyptrae. As is the case with the perigynia of H. gibbsiae, the variability in the ornamentation of the shoot-calyptrae is due to the size of the apical meristematic zone responsible for the elevation of the intra-gynoecial structures. In other words, the only type of intra-gynoecial structures which will be elevated onto the shoot-calyptra are those which are located directly above the axial meristematic zone.

The spore characters of H. andinum and H. blumii also support their synonymy. The size ranges overlap (H. andinum spores = 15-23 µm, H. blumii spores = 19-30 µm) and the pattern of exine sculpturing of both is composed of widely spaced, often anastomosing ridges.

The unique growth regime which is responsible for the reduced form of the two most apical dorsal leaves found in most vegetative shoots of H. andinum, remains the only character which separates H. andinum and H. blumii. However, although the apical dorsal leaves on H. blumii vegetative shoots are not significantly smaller, they are more lingulate-triangular in outline than are the other dorsal leaves of the shoot. More importantly, some H. blumii plants from Luzon, Philippines possess individual shoots which display great variability in dorsal leaf size. While most of the dorsal leaves on a single leafy shoot are broader than the stem, a few reduced leaves which are clearly narrower than the stem may be interspersed at various heights along the dorsal row. These plants clearly link H. andinum and H. blumii.

The demonstrated variability in dorsal leaf morphologies among H. blumii (including H. andinum) shoots further serves to merge H. giganteum with H. blumii. The delayed maturation of the upper series of dorsal leaves which occurs in the vegetative shoots and many fertile shoots of H. andinum as well as the variously reduced dorsal leaves of the Luzon H. blumii plants demonstrate that the mechanisms responsible for the delay in dorsal leaf development as well as its resumption are controlled by some environmental factor or factors. It is likely that in the three existing shoots of H. giganteum, dorsal leaf development was delayed at or shortly after the triangular stage of leaf development (as is usually the case with the two most apical dorsal leaves of H. andinum shoots), and the mechanism which induces the resumption of dorsal leaf development was similarly inhibited, so that the leaves remained highly reduced. The H. blumii shoots from Luzon demonstrate that such inhibiting factors are expressed in the Philippines.

The only character besides the extremely reduced dorsal leaves which distinguishes H. giganteum from H. blumii is the apparent unistratose composition of the leaves. However, these leaves have not been sectioned and it is quite likely that, as is the case with all other species of the genus, the leaves possess at least a small bistratose zone near the point of insertion. Even so, this lone character would not be sufficient to support specific status. Therefore, *H. giganteum* is here synonymized with *H. blumii*.

Distribution and Ecology - *Haplomitrium blumii* occurs in Java, Sumatra, New Guinea (Schuster 1967; Piippo 1984), the Genting Highlands of Malaysia (*Furuki 384* [ABSH]), and Luzon and Mindanao of the Philippines (Fig. 508). The species is also known from Jowai, India (Udar et al. 1968) and the Guangdong province in southern China (Gao and Li 1985) (Fig. 508). Yang (1966) has reported *H. blumii* as also occurring in Ali-shan and Taiping Shan, Taiwan, but according to Lai (personal communication) these plants are actually *H. mnioides*. The synonymizing of *H. andinum* and the discovery of *H. blumii* plants in Ecuador (*Gradstein, Mues & Frahm 6870* [U]) extends the range of this species to include tropical America (Fig. 508).

In its Oceanic range, the plants range in altitude from 400-3550 m.s.m. and often grow on decaying logs and stumps, or occasionally on the trunks of tree ferns, soil, or basic rock in both rainforests and open grasslands with scattered tree ferns (Stephani 1922; Piipo 1984). In China, *H. blumii* occurs on soil, often associated with *Riccardia multifida*, *Pallavicinia subciliata* (Aust.) Steph., and *Callicostella papillata* (Mont.) Mitt. (Gao and Li 1985).

In tropical America, the species ranges from the Andes of Ecuador and Peru, to Guadeloupe, Dominica, and Martinique of the Lesser Antilles (Spruce 1885; Schuster 1967; *Gradstein, Mues & Frahm 6870* [U]). The plants commonly grow on shaded, decaying logs and stumps in humid regions (Spruce 1885; Fulford 1962; Schuster 1967), or over wet rocks in forests (Fulford 1962).

Specimen Examined - EQUADOR: Andes Quitensis, *Spruce s. n.* (lectotype of *H. andinum*, BM); Quito, *Gradstein, Mues & Frahm 6870* (U). PERU: Tavaloso, *Spruce s.n.* (G); s.l., *Spruce s. n.* (NY); s.l., *Spruce s. n.* (BM).

GUADELOUPE: Nuire River, Duss 156 (NY); Pomb-Nouv, Duss 1041 (NY); Sentier du Galion, Allorge s. n. (F, H, G); s.l., Herminier s. n. (G). DOMINICA: s.l., Elliott 2182, Higgins 58a (G); Laudat, Lloyd 98 (NY). MARTINIQUE: Bois du Lorrain, Duss 267 (NY). JAVA: s.l., Blume s. n. (lectotype, STR); Res. Priangan, Verdoorn 61 (F, H, G, L, NY), Verdoorn 60 (F, H, G, L, NY); Tjibodas, Fleischer s. n. (G), Campbell s.n. (NY), Nyman s. n., Rothert 1429, Meijer B3971 (L); Prov. Preanger, Schiffner 398 (G); s.l., Nyman s. n. (G); Huis ten, Wyk 559, Kandang Badek, Wijk 476 (L). SUMATRA: Mt. Korinchi, Meijer 799, 7745, 7621 (L); Brastagi-Tonkkeh, Wijk 1675 (L); Kerinci Jambi, Dransfield 2694 (L); Pajakumbuh, Meijer 6569 (L); Gunung Leuser Nature Reserve, Wilde & Wilde-Duyfjes 14063 (L). PAPUA NEW GUINEA: Alkmaar, Romer 472 (L); West Sepik Prov., Koponen 35418 (F, G), Koponen 35864, 35865, 35568 (G, H); Morobe Prov., Koponen 29332, 30388, 31809, 31813, 31947, 31948, 31949, 31977, Norris 59566, 59814, 60083, 60313, 61595, 61631, 66459 (H), Touw 15228 (L); Bougainville, Ridsdale & Lavarack 31585 (L). PHILIPPINES: Luzon, Mt. Tabayoc, Jacobs B592A, B386 (L); Mindanao, s.l., Elmer 11562 (lectotype of H. giganteum, G). MALAYA: Genting Highland, Furuki 384 (ABSH). CHINA: Guangdong Prov., Hainan Is., Chen et al. X6, 815 (H).

SPECIATION AND PHYLOGENETIC AFFINITIES

As is apparent from the world distribution of *Haplomitrium* (Fig. 508), the region of greatest species diversity for the genus is clearly Australasia. However, the center of origin of the genus is difficult to speculate. Although the region of greatest species diversity is often deemed indicative of the center of origin, this is not always the case. Migration, coupled with an explosion of evolutionary radiation or the occurrence of extinction or emigration at the center of origin may lead to the misinterpretation of a secondarily formed center as the true center of origin (Cain 1944). However, in the case of *Haplomitrium*, the region of greatest species diversity likely does coincide, at least in part, with the putative center of origin of the genus. According to Schuster (1972), *Haplomitrium* is of Gondwanian origin and has been in existence since at least the Carboniferous. Gondwana consisted of Antarctica, South America, Africa, Madagascar, India, Australia, Tasmania, New Guinea, New Caledonia, and New Zealand (Schopf 1970; Schuster 1972). The South American connection explains the occurrence of *Haplomitrium* on this continent.

Schuster (1972) hypothesized that the *Haplomitrium* species which occurred on the Indian bloc (*H. hookeri*, *H. gibbsiae*, and *H. mnioides*) were carried northward during the middle Mesozoic as the Indian bloc migrated. During its movement, India encountered resistance and buckled, thus creating a mountain zone whose northern boundaries were involved in forming the Himalayas when the Indian bloc met Asia. From the Himalayas, *Haplomitrium* migrated throughout the Eurasian bloc. During the millions of years since this migration, selective extinctions which were probably fueled by changing climatic conditions resulted in the present distribution pattern of *Haplomitrium* (Schuster 1972).

According to Cain (1944), the center of origin of a genus is characterized by having the greatest species diversity, the largest number of endemics, and the greatest polymorphy. These traits support Australasia as the center of origin for Haplomitrium. Six of the seven Haplomitrium species occur throughout Australasia, of which three are endemics (H. monoicum, H. intermedium, and H. ovalifolium). The decrease in polymorphy from the center toward the periphery of the range is nicely demonstrated by H. gibbsiae, which is phenotypically most variable throughout New Zealand, while the local populations of South America and India are less variable, particularly in calyptral traits. According to Cain (1944), the most widely distributed species are found near the generic border. Such species may be the oldest of the genus. Haplomitrium hookeri fits this description as it is likely the most primitive element of the genus based on morphogenetic evidence, and it is likewise the most widely distributed Haplomitrium species.

Chandra et al. (1987) contended that the eastern Himalayan region of India constitutes the center of origin of the Haplomitriales based primarily on the great number of species that they believed occupy this region. However, only three species are now recognized from this region since H. indicum, H. dentatum, H. grollei and H. kashyapii are synonymous with H. gibbsiae. In addition, in light of the great age of this genus, it is virtually impossible to explain the dispersal of the species from India to South America and Australasia logically. Thus, the most plausible center of origin, i.e., the area in which speciation has been going on for the longest period of time is Australasia (Gondwanian origin).

The recognition of species characters allow evolutionary progression to be hypothesized within the genus. However, the direction of evolution, i.e., which character states are presumed to be ancestral and which are derived, can often be interpreted in opposing directions. Ultimately, one's decisions regarding the direction of evolution are guided by one's beliefs regarding the fundamentally opposing theories pertaining to the origin of land plants, namely, Bower's antithetic theory (Bower 1908) versus Pringsheim's homologous theory (Pringsheim 1878). I have interpreted the trends within the genus Haplomitrium based upon evolution through reduction. This decision was based on the foundations of the homologous theory and the contention that the selective pressures of the terrestrial environment, paricularly reproductive limitations, restricted the ultimate size of the gametophyte generation. As a direct result, the progressive reduction of the sporophyte was also favored due to the sporophytes dependence on the gametophyte. Schuster (1963, 1966, 1972, 1984b) has proposed a list of presumed derived and primitive hepatic characters and discussed them in depth. Based on these assumptions of derived vs. ancestral character states and on Schuster's (1963, 1966, 1972, 1984b) proposals, the following fundamental phylogenetic progressions from the ancestral to the more derived character states are hypothesized for the genus Haplomitrium.

1) The progressive reduction in the size of one row of leaves (the row which corresponds to the dorsal row in prostrate forms), so that symmetry is altered from the ancestral isophyllous condition to a derived anisophyllous state. Thus, taxa with the most reduced dorsal leaves represent the most highly derived expression of this character.

2) A transition in leaf insertions from one in which all leaves are transversely inserted to one in which the leaves of one row (the dorsal row) remain transverse, while those of the other two rows (the lateral rows) become oblique. So, the ultimate derived condition is one in which the lateral leaves are clearly succubously inserted.

3) An increase in the stability of leaf morphologies coupled with the progression from dissected leaves to entire leaves. Instead of entire, lobed and variously dissected leaves occurring on a single shoot as found in the more primitive elements

of the genus, all leaves of a single shoot become entire (the derived condition).

4) The transition from the production of isolated gametangia, to gametangia being produced acropetally in small numbers (1-3) in leaf axils, to the most derived condition in which large clusters of axillary gametangia are aggregated onto distinct, dilated terminal receptacles.

5) The modification of stem leaves to specialized bracts.

6) A decrease in the potential for apical proliferation of the fertile shoot, i.e., the transition from anacrogyny to functional acrogyny.

7) An increase in the massiveness of the sporophyte protective structure achieved through the increased involvment of stem tissue in its creation, i.e., the size and period of activity of the axial meristematic zone is increased. Therefore, the most primitive state is characterized by a true calyptra, while shoot-calyptrae and perigynia represent the more derived conditions.

8) A reduction in the variability of the number of functional lines of weakness responsible for capsule dehiscence.

9) A tendency toward uniformity in elater morphologies within a single capsule of a taxon. For example, the primitive plastic condition is represented by the occurrence of very long, slender elaters and short, stout elaters in the same capsule. Similarly, some elaters may be unispiralled, while others are bispiralled medianly becoming unispiralled toward the apices, or may be bispiralled throughout. The derived condition is illustrated by the exclusive production of long, slender elaters which are bispiralled throughout.

10) A reduction in spore size.

11) An increase in the amount of fusion of the sporoderm exine elements covering the spores, i.e., tuberculate and verruculate exine patterns are less derived than are ridged patterns in which the ridges anastomose.

As is usually the case with plant taxa, evolutionary advancement is not uniformly expressed within the taxon. Some organisms are very primitive in some respects, yet quite advanced in others. Therefore, creating a linear phylogenetic sequence for the species of *Haplomitrium* is not reasonable. However, based on the recognized evolutionary progressions just discussed, the subgenera and sections can be organized phylogenetically and the relative status of their included species can be likewise hypothesized.

The members of the subgenus *Haplomitrium* section *Haplomitrium* (*H. hookeri*, *H. ovalifolium*, and *H. monoicum*) are the most plesiomorphic group of the genus. These taxa possess numerous ancestral traits whose expressions are often highly plastic. The ancestral characters present in this group include the erect, radial shoots, which in some individuals are isophyllous and bear transversely inserted leaves with margins which may be entire, lobed or variously dissected. The shoots are usually anacrogynous and the gametangia are often free-standing on the axis or scattered acropetally in the axils of unmodified stem leaves, or, they may be in small clusters near the apex of the shoot. The production of a true calyptra is also primitive, as is the variability in capsule dehiscence, the presence of both long and short elaters with varied spiral thickening patterns within a single capsule, and the production of relatively large spores covered by verruculae.

From this complex, *H. hookeri* emerges as the most primitive element of the genus. Its virtual cosmopolitan range and wide array of ancestral traits, plus the great plasticity of these traits suggest that this species is most like the ancestral progenitor of the genus. *Haplomtrium hookeri* exhibits both isophylly and anisophylly, transverse as well as subtransverse leaf insertions, occasional reduction in dorsal leaf size, and the complete generic complement of leaf outlines. In addition, it possesses a range of gametangial loci which are precursory to all others of the genus, namely, gametangia which are free-standing on the stem, scattered in leaf or bract axils, or clustered in distal inflorescences. Thus, both anacryogyny and functional acrogyny are represented in this species. Its sporophytic characters are similarly representative of the presumed ancestral form. The complete range of capsule dehiscence patterns is demonstrated, as are the elater spiral thickening patterns, and the spores, which occasionally are retained in tetrads (like those of H. gibbsiae), are verruculate, but show limited fusion. Thus, the exine patterning may be interpreted as a model of the type from which the tuberculate and ridged exine patterns could have been derived. In addition to this morphogenetic evidence, the distribution of H. hookeri also supports its primitive placement. It is known throughout the entire range of Haplomitrium except in South America, and is the only extant species known from Denmark, the location of the fossil genus Gessella (Haplomitriales).

Haplomitrium monoicum represents the most advanced species of this section as evidenced by its uniformly entire leaves, reduced antheridia, scarcity of slime papillae, and monoicous habit. This species leads to the next, slightly more advanced section, Archibryum, including H. intermedium and H. gibbsiae, in that H. monoicum and H. intermedium both bear entire leaves bordered by enlarged cells, produce a relatively small number of slime cells, and have relatively large median leaf cells of similar size.

The members of the section Archibryum display a range of ancestral vegetative characters which are similar to those exhibited by the members of the section Haplomitrium, but the reproductive features of Archibryum suggest a slightly more derived condition. For example, although axillary gametangia may still be scattered in an acropetal fashion along the shoot, they may also be produced in larger clusters on swollen terminal receptacles where they are surrounded by enlarged bracts, or with apical proliferation, becoming localized at alternating swollen nodes along the stem. The sporophyte protective structures of this group are also more advanced as their ontogenies are not restricted to exclusive archegonial derivation, but additionally may involve stem tissue (*H. intermedium*), or may be exclusively derived from axial meristematic activity (*H. gibbsiae*). In addition, spore sizes are somewhat reduced, and, at least in the case of *H. intermedium*, extensive fusion of the exine has occurred so that the spores are densely covered with ridges.

Haplomitrium intermedium appears to be the least advanced species of this section. Its primitive attributes include frequent terminal branching of the leafy shoot, variability in stolon origin (either from the rhizome or the leafy shoot), variability in gametangial positioning, the frequent occurrence of apical proliferation, and the production of a shoot-calyptra as opposed to the perigynium of *H. gibbsiae*. In contrast, the densely ridged exine which characterizes the spores of *H. intermedium* is more derived than the verrucate sporoderm of *H. gibbsiae*.

In general, *H. gibbsiae* is more derived than *H. intermedium*. It's gametangial positioning is much less variable, with the gametangia usually produced on swollen terminal discs subtended by enlarged bracts and only rarely scattered along the stem. In addition, *H. gibbsiae* produces a perigynium. However, it is plesiomorphic in symmetry, in its variable leaf shapes and leaf insertions, and in the verrucate exine which characterizes its sporoderm.

The advancements in gametangial positioning and sporophyte protective structures figure prominently in the role played by the section *Archibryum* in bridging the gap between the subgenus *Haplomitrium* and the subgenus *Calobryum* (which includes *H. mnioides* and *H. blumii*). The vegetative, reproductive, and sporophytic characteristics of the subgenus *Calobryum* point to it as being the most derived of the genus. The shoots are always anisophyllous with the reduced row of leaves being transversely inserted while the lateral leaves are clearly succubous, and, with the exception of the bracts, the leaves are almost exclusively entire. The gametangia are always produced in large numbers on a terminal, dilated disc surrounded by enlarged bracts, in some cases apparently without the potential for apical proliferation. In addition, stem tissue is always involved in the creation of the sporophyte protective structure, capsule dehiscence is usually restricted to a single suture, and the elaters are long, slender, and uniformly bispiralled throughout.

The species of this subgenus are very closely related in that they differ only in a few species specific character states. Although the vegetative shoots of H. mnioides are prostrate (a derived condition), it is probably more ancestral than H. blumii in that it retains the potential for apical proliferation while H. blumii does not, and its spores are larger and covered by unfused tuberculae as opposed to the small spores of H. blumii which are covered by frequently anastomosing ridges.

Haplomitrium has traditionally been regarded as one of the most, if not the most, ancestral element of the hepatics and has been hypothesized to be the taxon which joined the leafy and thalloid hepatics (Wettstein 1903-1908; Verdoorn 1932; Evans 1939; Smith 1955; Schuster 1966, 1967, 1971, 1972, 1984b). In contrast, a few regarded Haplomitrium as a derived group (Cavers 1911; Grolle 1969). The interpretation of its phylogenetic status hinges on one's assumptions regarding the characteristics of the presumed hepatic ancestors, i.e., whether the hepatic prototype was composed of a dorsiventral thallus or a cylindrical axis. Those who favor the former view regard Haplomitrium as being specialized since it is erect and radial, while those who accept the latter assumption view Haplomitrium as being ancestral. E. O. Campbell (1971) and Schuster (1966, 1979, 1984b) suggested that the ancestral prototypes were basically composed of robust, parenchymatous, erect, terete, leafless axes. Schuster (1966) compared the basic structure of these organisms to that of Horneophyton

Barghoorn & Darrah or Rhynia Kidston & Lang. From this idealized concept, Schuster (1966, 1979, 1984b) hypothesized the ancestral hepatics as being anchored by a leafless rhizomatous system, from which erect, radial shoots arose. These shoots grew via tetrahedral apical cells which were responsible for the production of three ranks of isophyllous, transversely inserted leaves. Copious mucilage was produced from abundant slime papillae which were born on the shoots. In general, genetic control over the expression of traits was not highly fixed, and the anatomy of the axes was relatively simple, but, a conducting strand was probably present. The gametophytes were unisexual with large, isolated gametangia, while the terminal sporophytes were chlorophyllose, massive, and complex with a capsule wall composed of at least 8-12 layers of cells.

Crandall-Stotler (1981) also based her concepts regarding the primal hepatics on a fundamental cylindrical form. However, she proposed that thalloid and leafy hepatic habits are equally primitive adaptations since both types of structural modifications could have arisen independantly from cylindrical axes. The dorsiventrally flattened, broadened, thalloid habit may have arisen from creeping axes in which dorsiventral gradients altered the "...cell division and expansion potentials in lateral planes...", while erect, leafy forms may have come from erect axes on which leaflike enations arose from epidermal initials in response to pressures for apical protection and the need for increased photosynthetic surface area (Crandall-Stotler 1981). The reduction or elaboration of characters within each form could then be induced as a result of different ecological niche requirements (Crandall-Stotler 1981).

The supporters of the antithetic theory and of dorsiventral prototypes view evolution differently. Grolle (1969) defends his hypothesis of ancestral dorsiventrality by observing that the gametophytes of higher plants are always dorsiventral. However, as pointed out by Schuster (1971), this is not uniformly true, since radial gametophytes do occur in the more primitive lycopods and ferns such as the Stromatopteridaceae, Gleicheniaceae, and Schizaeaceae, while flat prothalli occur in more derived ferns such as the Marattiales.

More powerful evidence supporting both the dorsiventral ancestral thallus and the antithetic theory comes from the work of Graham (1985, 1989). The flat gametophyte thallus of the charalian alga Coleochaete Breb. (a haplobiontic alga) produces zygotes which are retained and nourished on the haploid parental thallus. This feature, along with its other numerous land plant characters are used to suggest that Coleochaete may serve as a model of the ancestral land plants and illustates how a delay in meiosis could give rise to the first embryophytes.

Both Grolle's and Graham's hypotheses assume that this dorsiventral ancestor gave rise to both the hepatics and the tracheophytes. However, this is not necessarily the case. It is only logical to assume that numerous groups of plants were involved in the colonization of terrestrial environments and that these groups "solved" the problems of terrestrial life in their specific niches in different ways. Therefore, it is conceivable that in some groups, certain selective pressures favored the increasing complexity of the diploid generation and the reduction of the haploid generation (as is exemplified in modern day tracheophytes), while in other groups, the sporophyte generation was reduced while the gametophyte generation became dominant (as is exemplified in extant bryophyte groups). In other words, organisms similar to Coleochaete may well have given rise to the tracheophytes as per the antithetic theory, while a different group of progenitors may have yielded the hepatics as per the homologous theory.

Haplomitrium displays numerous characters which attest to its primitive status. The most convincing feature is its shared pattern of antheridial and archegonial ontogeny, which attests to the homology of primal gametangia. In addition, the homologous production of slime papillae on the rhizome and the scales and leaves on the aerial shoots of Haplomitrium demonstrates the presumed evolution of leaves from slime papillae (Schuster 1984a). Other evidence includes the great plasticity exhibited by Haplomitrium and the formation of numerous sporophytes per shoot as a result of imperfect genetic control, the production of massive gametangia, copious slime production, the lack of rhizoids, and the presumed primitive construction of its spermatozoid (at least in the two Haplomitrium species studied by Carothers and Duckett [1978, 1979] and Carothers and Rushing [1988]). The extreme primitive status of Haplomitrium and the fact that it is erect and radial suggests that it has evolved from an ancestor whose morphology was based on an erect, cylindrical axis. In fact, Haplomitrium displays a wide array of traits which also characterize the hypothesized erect, radial hepatic ancestor as described by Schuster (1966, 1979, 1984b). These include a habit typified by a leafless, prostrate rhizomatous system from which radial shoots bearing 3-ranked leaves arise, with each type of shoot growing via a tetrahedral apical cell. In addition, numerous slime papillae exist on all plant parts and the shoot anatomy is fairly generalized, but a well-defined conducting strand is present. Branching is primitive in that branches may potentially arise from any or all three merophyte rows to produce a "bush-like" plant. The gametophytes are unisexual (except for those of H. monoicum) and large gametangia are produced from any or all three merophyte rows, with similar ontogenetic patterns displayed by the antheridia and archegonia at least in the early stages of development. The sporophyte seta is massive and the derived feature of asexual reproduction by special propagules is extremely rare. This large assemblage of characters illustrates the primitive nature of the genus. However, some characters displayed by the genus are quite advanced. These include the unistratose capsule wall with its unique longitudinal, annular thickening pattern, the production by most species of shoot-calyptrae or perigynia, the evolution of functional acrogyny, and the relatively small spore sizes, as well as the presence of at least some degree of anisophyllous symmetry in members of each species, the reduction of the dorsal row of leaves, and the slight to distinctly succubous insertion of the lateral leaves. This assemblage of derived traits plus the unique Haplomitrium-type spermatozoid ultrastructure, which suggests the isolation of Haplomitrium from all other bryophytes, and its unique pattern of sporeling development, implies that Haplomitrium should not be considered as the ancestor to either the Metzgeriales or the Jungermanniales. This fact was also recognized by Schuster (1967, 1971, 1984b) who suggested that it may, however, serve as a model from which the Metzgeriales has evolved. Additional evidence which negates Haplomitrium as being an ancestor of the simple thalloid or leafy hepatics is the evolutionary status of the the Metzgerialian group with which Haplomitrium shows its greatest affinities. Of all of the diverse taxa grouped within the Metzgeriales, Haplomitium is most like the Pallaviciniaceae, a family that is regarded as one of the more derived rather than primitive elements of the order (Renzaglia 1982; Schuster 1984b). Members of the Pallaviciniaceae, as characterized by Renzaglia (1982), are the only Metzgerialian taxa which are rhizomatous and possess central conducting strands of pitted hydroids. The rhizomes of Hattorianthus Schust. & Inoue are even prostrate and subterannean (Schuster and Inoue 1975; Renzaglia 1982), but all species of the Pallaviciniaceae possess rhizoids. Although the gametophytes of the Pallaviciniaceae are not leafy, some members do possess aerial thalli which may be variously lobed, e.g., Jensenia Lindb. and Symphyogyna (Evans 1925; Hässel de Menéndez 1961; Renzaglia 1982). The two groups are also similar in branching ontogeny in that the branch primordium undergoes a period of dormancy before the branch apical cell is formed.

the primordia may not complete their development into mature branches (Renzaglia 1982), thus creating a situation reminiscent of the basal "bumps" and spurs found in Haplomitrium species. Like Haplomitrium (except H. monoicum), fertile members of the Pallaviciniaceae are dioicous and heteromorphic, and genera such as Symphyogyna and Xenothallus Schust. form massive shoot-calyptrae after fertilization (Renzaglia 1982). These two genera also share a few sporophytic features with Haplomitrium. Both have cylindrical capsules, elaters which are long, slender, and attenuate, (Hässel de Menéndez 1961; Schuster 1984a), and a shared pattern of spore wall ontogeny characterized by callosic exine precursors (Brown and Lemmon 1986, 1987). Similarly, some species of Jensenia, Symphyogyna, and Podomitrium possess sporoderms which range from tuberculate, to variously ridged, to densely vermiculose (Hässel de Menéndez 1961; Taylor et al. 1974).

Therefore, although Haplomitrium is a very ancient taxon which retains a wide array of ancestral characters it is not the common ancestor of leafy and/or the simple thalloid hepatics. Instead, it is a unique and plastic taxon which probably shares a recent common ancestor with the Metzgeriales. This common ancestry supports Crandall-Stotler's (1981) hypothesis regarding the primitive equivalance of the thalloid and leafy hepatic forms and the fundamental cylindrical morphology of the primal entities. Further evidence for this concept may be found in the sporeling ontogenies of the two groups. In the Metzgeriales, sporeling development is typified by the production of flattened thalli or by the initial production of prostrate cylinders which, through an alteration in apical cell geometries or segmentation, are converted to flattened thalli (Bartholomew 1985, 1986). In contrast, sporeling ontogeny in Haplomitrium is characterized by the early production of numerous erect, leafless, cylindrical axes from a massive protonema.

In summary, then, *Haplomitrium* is a very old taxon, probably of Gondwanian origin, which has retained many of the primitive characters and modalities of its ancestors. It comprises seven extant plastic species, each defined by relatively broad species limits, with *H. hookeri* being the most primitive entity of the genus and representing a model from which all other *Haplomitrium* species may have been derived. *Haplomitrium*'s closest affinities lie with another order of great age, the Metzgeriales. The previously hypothesized close phylogenetic links between *Haplomitrium* and the postulated primitive elements *Takakia* and *Andreaeobryum* do not exist. Furthermore, *Haplomitrium*'s wide array of plesiomorphic traits is complimented by an assemblage of clearly apomorphic characters which isolate *Haplomitrium* and negate its role as the ancestral element which gave rise to either the Metzgeriales or the Jungermanniales. Therefore, *Haplomitrium* is best classified as the sole extant genus comprising the Haplomitriales, coordinate with the Metzgeriales, with close affinities in the Pallaviciniaceae. This supports the alignment of these two orders in a subclass (Metzgeriidae) of the Jungermanniopsida. This arrangement best illustrates the derivation of the Haplomitriales and Metzgeriales from a recent common ancestor (while separating it from the Jungermanniales), and does not imply the direct descent of any bryophyte taxon from *Haplomitrium*. The common ancestry of the Haplomitriales and Metzgeriales further suggests that the primal hepatic entities were fundamentally cylindrical in form as opposed to being composed of a flattened, broadened thallus, and that from this cylindrical morphology, both dorsiventral and erect, radial gametophytes could arise independently as a response to ecologically controlled selective pressures.

REFERENCES

Allison, K. W. & J. Child. 1975. Liverworts of New Zealand. 300 pp. University of Otaga Press, Dunedin.

Andreas, J. 1899. Ueber den Bau der Wand und die Oeffnungsweise des Lebermoos-sporogons. Flora 86: 161-213.

Asakawa, Y., S. Hattori, M. Mizutani, N. Tokunaga & T. Takemoto. 1979. Chemosystematics of bryophytes. III. Terpenoids of the primitive Hepaticae, *Takakia* and *Haplomitrium*. J. Hattori Bot. Lab. 46: 77-90.

__________, N. Tokunaga, M. Toyota, T. Takemoto, S. Hattori, M. Mizutani & C. Suire. 1979. Chemosystematics of bryophytes. II. The distribution of terpenoids in Hepaticae and Anthocerotae. J. Hattori Bot. Lab. 46:67-76.

Barlow, P. W. 1983. Nucleolus-associated bodies (karyosomes) in dividing and differentiating plant cells. Protoplasma 115: 1-10.

Bartholomew, S. E. 1985. A survey of sporeling development in selected representatives of the Metzgeriales (Hepatophyta). 115 pp. M. S. thesis, East Tennessee State University, Johnson City.

__________. 1986. Sporeling development in select representatives of the Pallaviciniaceae: *Pallavicinia subciliata*, *Symphyogyna aspera*, and *Podomitrium phyllanthus*. Lindbergia 12: 111-118.

__________. 1987 [1988]. Apical organization and leaf ontogeny in *Haplomitrium mnioides* (Lindb.) Schust. Bryologist 90: 383-389.

Basile, D. V. 1967. The influence of hydroxy-L-proline on ontogeny and morphogenesis of the liverwort, _Scapania nemorosa_. Amer. J. Bot. 54: 977-983.

__________. 1968. The influence of L-proline and hydroxy-L-proline on ontogeny of _Scapania nemorosa_. Bull. Torrey Bot. Club 95: 127-134.

__________. 1973. Hydroxy-L-proline and 2, 2'-dipyridyl- induced phenovariation in the liverwort _Jungermannia lanceolata_. Bull. Torrey Bot. Club 100: 350-352.

__________. 1974. A possible role for hydroxyproline- proteins in leafy liverwort phylogeny. J. Hattori Bot. Lab. 38: 91-98.

__________. 1975. A comparison of some macronutrient media used to culture bryophytes. Bryologist 78: 403-413.

__________. 1979. Hydroxyproline-induced changes in form, apical development, and cell wall protein in the liverwort _Plagiochila arctica_. Amer. J. Bot. 66: 776-783.

Beneke, W. 1903. Über die Keimung der Brütknospen von _Lunularia cruciata_. Bot. Zeitung. 61: 19-46.

Berrie, G. K. 1959. The cytology of _Haplomitrium hookeri_. Bryologist 62: 1-5.

__________. 1962. Australian liverworts. I. _Haplomitrium intermedium_, sp. nov. (Calobryales). Proc. Linn. Soc. New South Wales 87: 191-195.

__________ & P. M. Webster. 1982. Ultrastructure of plastids and mitochondria in gemmae of _Marchantia polymorpha_ L. Ann. Bot. (London) 50: 199-206.

Bissing, D. 1974. Haupt's gelatin adhesive mixed with formalin for affixing paraffin to slides. Stain Tech. 49: 117.

Böcher, T. 1963. Phytogeography of Middle West Greenland. Meddel. Gronland 148: 1-289 + 10 pl.

Bold, H. 1949. The morphology of Chlamydomonas chlamydogama, sp. nov. Bull. Torrey Bot. Club 76: 101-108.

__________, C. J. Alexopoulos & T. Delevoryas. 1980. Morphology of plants and fungi, ed. 4. x + 819 pp. Harper & Row, New York.

Bopp, M. 1983. Developmental physiology of bryophytes. In R. M. Schuster (ed.), New Manual of Bryology. Vol. 1. pp. 276-324. Hattori Bot. Lab., Nichinan.

Bower, F. O. 1908. The origin of a land flora. 727 pp. London.

Brown, R. C. & B. E. Lemmon. 1986. Spore wall development in the liverwort, Haplomitrium hookeri. Canad. J. Bot. 64: 1174-1182.

__________ & __________. 1987. Involvement of callose in determination of exine patterning in three hepatics of the subclass Jungermanniidae. Mem. New York Bot. Gard. 45: 111-121.

__________, __________ & K. S. Renzaglia. 1986. Sporocytic control of spore wall pattern in liverworts. Amer. J. Bot. 73: 593-596.

Buch, H. 1930. Über die Entstehung der verschiedenen Blattflachenstellungen bei den Lebermoosen. Ann. Bryol. 3: 25-40.

________. 1932. Experimentelle Morphologie. In F. Verdoorn (ed.), Manual of Bryology. pp. 73-88. Martinus Nijhoff, The Hague.

________. 1936. Suomen Maksasammalet. 116 pp., 35 figs. Helsinki.

Burr, R. J., B. G. Butterfield & C. Hébant. 1974. A correlated scanning and transmission electron microscope study of the water-conducting elements in the gametophytes of Haplomitrium gibbsiae and Hymenophyton flabellatum. Bryologist 77: 612-17.

Cain, S. A. 1944. Foundations of plant geography. 556 pp + xiv. Hafner Publ. Co., New York.

Campbell, D. H. 1920. Studies in some East Indian Hepaticae. Calobryum blumei, N. ab E. Ann. Bot. (London) 34: 1-12, 25 fig.

________. 1940. The evolution of land plants (Embryophyta). vii + 732 pp. Stanford Univ. Press, Stanford, California.

________ & F. Williams. 1914. A morphological study of some members of the genus Pallavicinia. Leland Stanford Jr. Univ. Publ. 1914: 1-44, figs. 1-23.

Campbell, E. O. 1954. The structure and development of Monoclea forsteri Hook. Trans. Proc. Roy. Soc. New Zealand 82: 237-248.

________. 1959. The structure and development of Calobryum gibbsiae Steph. Trans. Proc. Roy. Soc. New Zealand 87: 243-254.

________. 1971. Problems in the origin and classification of bryophytes with particular reference to liverworts. New Zealand J. Bot. 9: 678-688.

__________. 1987. Steereomitrium minutum gen. et sp. nov. (Calobryales). Mem. New York Bot. Gard. 45: 569-574.

Carothers, Z. B. & J. G. Duckett. 1978. A comparative study of the multilayered structure in developing bryophyte spermatozoids. In C. Suire (ed.), Congres International de Bryologie. pp. 95-112. Bryophyt. Biblioth. 13.

__________ & __________. 1979. Spermatogenesis in the systematics and phylogeny of the Hepaticae and Antherocotae. In G. C. S. Clarke & J. G. Duckett (eds.), Bryophyte Systematics. Systemat. Assc. Spec. Vol. 14. pp. 425-445. Academic Press, Inc., London.

__________ & A. E. Rushing. 1988. Comparative morphology of the bryophyte blepharoplast. In N. G. Miller (ed.), Advances in Bryology. Vol. 3. Bryophyte Ultrastructure. pp. 95-134. J. Cramer, Berlin.

__________ & __________. 1989. Blepharoplast structure in Treubia tasmanica. Amer. J. Bot. Suppl. 76(6): 5 (Abstr. #11).

Carrington, B. 1874. British Hepaticae. Part I. 88 pp. 16 pl. London.

Cavers, F. 1911. The interrelationships of the bryophytes. New Phytol. Reprint 4: 1-203, 72 figs.

Chandra, V., D. Kumar & S. C. Srivastava. 1987. Calobryales: distribution and phytogeographical discussion. Geophytology 17: 227-232.

Clarke, G. C. S. 1979. Spore morphology and bryophyte systematics. In G. C. S. Clarke & J. G. Duckett (eds.), Bryophyte Systematics. Systemat.

Assc. Spec. Vol. 14. pp. 231-250. Academic Press, Inc., London.

Clayton-Green, K. A., T. J. G. A. Green, & B. Staples. 1977. Studies of Dawsonia superba L. antherozoid dispersal. Bryologist 80: 439-444.

Cooke, M. C. 1894. Handbook of British Hepaticae. v-vii + 310 pp., 7 figs. W. H. Allen & Co., Ltd., London.

Crandall, B. J. 1969. Morphology and development of branches in the leafy Hepaticae. vii + 270 pp. Beih. Nova Hedwigia 30. J. Cramer, Vaduz.

Crandall-Stotler, B. 1974. Oil bodies of liverworts. Abstr. Ill. State Acad. Sci. 1974: 1.

__________. 1976. The apical cell and early development of Pleurozia purpurea Lindb. Lindbergia 3: 197-208.

__________. 1981. Morphology/anatomy of hepatics and anthocerotes. In W. Schultze-Motel (ed.), Advances in Bryology. Vol. 1. pp. 315-398. J. Cramer, Vaduz.

__________. 1984. Musci, hepatics and anthocerotes - an essay on analogues. In R. M. Schuster (ed.), New Manual of Bryology, Vol. 2. pp. 1093-1129. Hattori Botanical Laboratory, Nichinan.

__________. 1986. Morphogenesis, developmental anatomy and bryophyte phylogenetics: Contraindications of monophyly. J. Bryol. 14: 1-23.

__________ & J. J. Bozzola. 1988. Fine structure of the meristematic cells of Takakia lepidozioides Hatt. et H. Inoue (Takakiophyta). J. Hattori Bot. Lab. 64: 197-218.

__________ & __________. 1989. Apical cells and meristem dynamics in Takakia. Amer. J. Bot. Suppl. 76(6): 6 (Abstract #13).

Diers, L. 1970. Origin of plastids: cytological results and interpretations including some genetical aspects. In P. Miller (ed.), Control of Organelle Development. Soc. Exp. Biol. Symp. 24: 129-145. University Press, Cambridge.

Douin, C. 1908. Le pèdicelle de la capsule des hépatiques. Bull. Soc. Bot. France 55: 194-202, 270-276, 360-366, 368-376.

Downey, J. A. & D. V. Basile. 1989. New evidence for lignin in bryophytes: UV, IR and proton NMR analysis of a thioglycolic acid extraction from sporophytes of Pellia epiphylla. Amer. J. Bot. Suppl. 76(6): 191 (Abstract #504).

Duckett, J. G. 1986. Ultrastructure in bryophyte systematics and evolution: an evaluation. J. Bryol. 14: 25-42.

__________, Z. B. Carothers & C. C. J. Miller. 1982. Comparative spermatology and bryophyte phylogeny. J. Hattori Bot. Lab. 53: 107-125.

__________, __________ & __________. 1983. Gametogenesis. In R. M. Schuster (ed.), New Manual of Bryology. Vol. 1. pp. 232-275. Hattori Bot. Lab., Nichinan.

__________ & K. S. Renzaglia. 1987. Transmission electron microscopy of oil bodies in hepatics: games with systematics and cytochemistry. Bull. Brit. Bryol. Soc. 49: 17-19.

__________ & __________. 1988a. Cell and molecular biology of bryophytes: ultimate limits to the resolution of phylogenetic problems. Bot. J. Linn. Soc. 98: 225-246.

__________ & __________. 1988b. Ultrastructure and development of plastids in the bryophytes. In N. G. Miller (ed.), Advances in Bryology. Vol. 3. Bryophyte Ultrastructure. pp. 33-94. J. Cramer, Berlin.

__________, __________ & E. A. Smith. 1988. Preparative techniques for transmission electron microscopy of bryophytes. In J. M. Glime (ed.), Methods in Bryology. pp. 181-192. Hattori Bot. Lab., Nichinan.

Engel, J. J. 1981. Haplomitrium monoicum, a remarkable new species of Calobryales (Hepaticae) from New Caledonia, together with a reclassification of subg. Haplomitrium. Ann. Missouri Bot. Gard. 68: 668-676.

Esau, K. 1977. Anatomy of seed plants. ed. 2. xx + 550 pp. John Wiley & Sons, Inc., New York.

Evans, A. W. 1912. Branching in the leafy Hepaticae. Ann. Bot. (London) 26: 1-280.

__________. 1917. Notes on New England Hepaticae, XIV. Rhodora 19: 263-272.

__________. 1925. The lobate species of Symphyogyna. Trans. Connecticut Acad. Arts 27: 1-50.

__________. 1939. The classification of the Hepaticae. Bot. Rev. 5: 49-96.

Evert, R. F., R. D. Warmbrodt, & S. E. Eichhorn. 1989. Sieve-pore development in some leptosporangiate ferns. Amer. J. Bot. 76: 1404-1413.

Frey, W. 1971. Blattenwicklung bei Laubmoose. Dissertation Tubingen. pp. 463-556. J. Cramer, Lehre.

Frye, T. C. & L. Clark. 1943. Hepaticae of North America, Part II. Univ. Wash. Publ. Biol. 6: 166-336.

Fulford, M. 1956 [1957]. The young stages of the leafy Hepaticae: a résumé. Phytomorphology 6: 199-235.

__________. 1962 [1963]. Manual of the leafy Hepaticae of Latin America, Part I. Mem. New York Bot. Gard. 11: 1-172.

__________. 1965. Evolutionary trends and convergence in the Hepaticae. Bryologist 68: 1-31.

__________ & V. M. Diller. 1956. Studies on the growth of Haplomitrium. I. Organic media. Rev. Bryol. Lichenol. 25: 239-246.

__________, J. Taylor & R. Hatcher. 1958. The "calyptra" of Calobryum blumii Nees. Phytomorphology 8: 298-302.

Furuki, T. 1986. Gemmae of Haplomitrium hookeri (Smith) Nees. Hikobia 9: 495-496.

Galatis, B. & P. Apostolakos. 1977. On the fine structure of differentiating mucilage papillae of Marchantia. Canad. J. Bot. 55: 772-795.

__________, __________ & C. Katsaros. 1978a. Ultrastructural studies on the oil bodies of Marchantia paleacea Bert. I. Early stages of oil-body cell differentiation: origination of the oil body. Canad. J. Bot. 56: 2252-2267.

__________, __________ & __________. 1978b. Histochemical studies on the oil-bodies of Marchantia paleacea Bert. Protoplasma 97: 13-29.

__________, C. Katsaros & P. Apostolakos. 1978. Ultrastructural studies on the oil bodies of Marchantia paleacea Bert. II. Advanced stages of oil-body cell differentiation: synthesis of lipophilic material. Canad. J. Bot. 56: 2268-2285.

Gams, H. 1957. Kleine Kryptogamenflora. IV. Die Moos- und Farnpflanzen. Verlag, Stuttgart.

Gao, C.-H. & D.-K. Li. 1985. A preliminary study of the Chinese Haplomitriaceae (Class Hepaticae). Wuyi Science Journal 5: 231-234.

Gibbs, L. S. 1911. The hepatics of New Zealand. J. Bot. 49: 261-266.

Goebel, K. v. 1891. Morphologische und biologische Studien. IV. Ueber javanische Lebermoose. 2. Calobryum blumii Nees. Ann. Jard. Bot. Buitenzorg 9: 11-25 + pl. II-III.

__________. 1902. Über Homologien in der Entwicklung mannlicher und weiblicher Geschlechtorgane. Flora 90: 279-305.

__________. 1930. Organographie der Pflanzen. Bryophyten-Pteridophyten. i-x + pp. 643-1378. Zweiter Teil. Dritte Auflage. G. Fisher, Jena.

Gottsche, C. M. 1843. Anatomische-physiologische Untersuchungen über Haplomitrium Hookeri N. v. E. mit Vergleichung anderer Lebermoose. Nova Actorum Acad. Caes. Leop.-Carol. Nat. Cur. 20: 267-389 + pl. XIII-XX.

__________, J. B. G. Lindenberg & C. G. Nees ad Esenbeck. 1844-1847. Synopsis Hepaticarum. vii + 834 pp. Meissner, Hamburg.

Gradstein, S. R. 1974. Report on the first I. A. B. meeting at Boulder, 1973. Taxon 23: 211-213.

Graham, L. E. 1985. The origin of the life cycle of land plants. Amer. Sci. 73: 178-186.

__________. 1989. Lignin-like compounds and sporopollenin in Coleochaete, an algal model for land plant ancestry. Science 245: 399-401.

Gray, P. & F. M. Pickle. 1956. Iron mordanted safranin and celestine blue for staining skeletal elements in plant tissues. Phytomorphology 6: 196-198.

Gray, S. F. 1821. A natural arrangement of British plants. Vol. I. 852 pp., 21 pl. Paternoster-Row, London.

Grolle, R. 1963. Über ein Calobryum von den Philippinen. J. Hattori Bot. Lab. 26: 5-9.

__________. 1964. Miscellanea hepaticologica 1-10. Oester. Bot. Z. 111: 185-192.

__________. 1969. Review of "The Hepaticae and Anthocerotae of North America. Vol I." Nova Hedwigia 16: 539-542.

__________. 1983. Nomina generica Hepaticarum; references, types and synonymies. Acta Bot. Fenn. 121: 1-62.

Grubb, P. J. 1970. Observations on the structure and biology of Haplomitrium and Takakia, hepatics with roots. New Phytol. 69: 303-326.

Harvey-Gibson, R. J. & D. Miller-Brown. 1927. Fertilization of Bryophyta. (Preliminary note). Ann. Bot. (London) 41: 190-191.

Hässel de Menéndez, G. G. 1961. Las especies Sudamericans del género Pallavicinia. Bol. Soc. Argent. Bot. 9: 261-282.

__________. 1970. El género Haplomitrium (Hepaticae) en el extremo Austral del Sud. Revista Mus. Argentino Cienc. Nat. "Bernardino Rivadavia" 3: 235-238.

__________. 1975 [1976]. Calobryales. In S. A. Guarrera, I. Gamundi de Arnos & D. Rabinovich de Halperin (eds.), Flora Criptogamica de Tierra del Fuego. 15. pp. 15-20. Buenos Aires.

Hatcher, R. E. 1965. Towards the establishment of a pure culture collection of Hepaticae. Bryologist 68: 227-231.

Hattori, S. 1951. Oil bodies of Japanese Hepaticae. I. J. Hattori Bot. Lab. 5: 69-97 + 7 pl.

__________. 1953. Oil bodies of Japanese Hepaticae. II. J. Hattori Bot. Lab. 10: 63-78 + 3 pl.

__________ & M. Mizutani. 1958. What is Takakia lepidozioides? J. Hattori Bot. Lab. 20: 295-303 + 3 fig.

Hébant, C. 1964. Signification et évolution des tissus conducteurs chez les bryophytes. Naturalia Monspel., Sér. Bot. 16: 79-86.

__________. 1973a. Acid phosphomonoesterase activities (B-glycerophosphatase and naphtol AS-MX phosphatase) in conducting tissues of bryophytes. Protoplasma 77: 231-241.

__________. 1973b. Diversity of structure of the water-conducting elements in liverworts and mosses. J. Hattori Bot. Lab. 37: 229-234.

__________. 1975. On the occurrence of lysosomal acid phosphatase activity in the differentiating water-conducting strand of Takakia and its evolutionary significance. Phytomorphology 25: 279-282.

__________. 1977. The conducting tissue of bryophytes. Bryopht. Biblioth. 10: 1-157 + 80 pl.

__________. 1978. Development of pores in water-conducting cells of the liverwort Hymenophyton flabellatum (Metzgeriales, Bryophytes). Protoplasma 96: 205-208.

Heckman, C. A. 1970. Spore wall structure in the Jungermanniales. Grana 10: 109-119.

Herzog, T. 1925. Anatomie der Lebermoose. In K. Linsbauer (ed.), Handbuch der Pflanzenanatomie. Abt. II. Pt. 2, Bryophyten. 112 pp. Gebruder Borntraeger, Berlin.

Hewson, H. J. 1973. Sexual dimorphism in *Haplomitrium intermedium*. Proc. Linn. New South Wales 98: 29-34.

Hölzl, J. 1969. Elektronmikroskopische Beobachtungen an den Blattzellen des Lebermooses *Mylia taylori*. Oesterr. Bot. Z. 117: 293-304.

Horikawa, Y. 1929. Studies on the Hepaticae of Japan. II. Sci. Rep. Tohoku Imp. Univ., Ser. 4, Biol. 2: 395-429 + pl. 16-18.

Hooker, W. J. [1812-] 1816. British Jungermanniae. xxvii + 84 pp., 20 pl. London.

Hurey-Py, G. 1948. Note préliminaire sur l'action de l'acid naphtylacétique sur la germination et la croissance des tiges feuillées de *Funaria hygrometrica*. C. R. Hebd. Seances Acad. Sci., Ser. D., Sci. Nat. 227: 1256-1258.

Husnot, T. 1922. Flore analytique et descriptive des Hepatiques de France et des Contrees Voisines. Hepaticologica Gallica, ed. 2. 163 pp. + 23 pls. Paris.

Inoue, H. 1960. Studies in spore germination and the earlier stages of gametophyte development in the Marchantiales. J. Hattori Bot. Lab. 23: 148-191.

_________. 1964. Regeneration of leaf of *Calobryum rotundifolium*. J. Jap. Bot. 39: 91-93.

_________. 1971. *Haplomitrium hookeri* (Smith) Nees in Japan. J. Jap. Bot. 46: 353-357.

_________. 1974. Illustrations of Japanese Hepaticae. Vol 1. ix + 189 pp. Tsukiji Shokan Publishing Co., Ltd., Tokyo.

__________ & R. Hibino. 1984. Studies on spore morphology of hepatics (1). Metzgeriales. Bull. Natl. Sci. Mus. Ser. B (Bot.) 10: 177-189.

Johansen, D. A. 1935. Dehydration and infiltration. Science 82: 253-254.

__________. 1940. Plant microtechnique. vii + 523 pp. McGraw-Hill Book Co., New York.

Kitagawa, N. 1984. Morphological modifications induced by the long-period, aseptic cultivation of four species of liverworts. Acta Phytotax. Geobot. 35: 103-112.

Knapp, E. 1930. Untersuchungen über die Hüllorgane um Archegonien und Sporogonien der akrogynen Jungermaniacen. Bot. Abh. 16: 1-168.

Kumar, D. 1987. A remarkable *Calobryum* from India. Geophytology 17: 110-113.

__________ & R. Udar. 1976. *Calobryum dentatum* Kumar et Udar sp. nov.: a new species of *Calobryum* from India. J. Indian Bot. Soc. 55: 23-30.

__________ & __________. 1977. A new species of *Haplomitrium*, *H. grollei* Kumar et Udar, from Darjeeling, Eastern Himalayas, India. Geophytology 7: 260-263.

Lampa, E. 1902. Untersuchungen an einigen Lebermoosen. Sitzungsber. Kaiserl. Akad. Wiss., Math.-Naturwiss. Cl. 111: 477-489.

Lamport, D. T. A. 1965. The protein component of primary cell walls. Advances Bot. Res. 2: 151-218.

Landwehr, J. 1980. Atlas nederlandse Levermossen. 287 pp. Koninklijke Nederlandse Naturhistorische Vereniging.

Ledbetter, M. C. & K. R. Porter. 1970. Introduction to the fine structure of plant cells. x + 188 pp. Springer-Verlag, Berlin.

Leitgeb, H. 1868. Beitrage zur Entwicklungsgeschichte der Pflanzenorgane. I. Wachsthum des Stammches von Fontinalis antipyretica. Sitzb. Akad. Wiss. Wien 57: 308-342.

__________. 1875. Untersuchungen über die Lebermoose. II. Die foliosen Jungermannieen. iv + 95 pp., 12 pl. H. Dabis, Jena.

Ligrone, R. & R. Gambardella. 1988. The sporophyte-gametophyte junction in bryophytes. In N. G. Miller (ed.), Advances in Bryology. Vol 3. pp. 225-274. J. Cramer, Berlin.

Lilienfeld, F. 1911 [1912]. Beiträge zur Kenntnis der Art Haplomitrium Hookeri Nees. Bull. Int. Acad. Sci. Cracovie, Cl. Sci. Math., Ser. B, Sci. Nat. 1911: 315-399 + tab. XII.

Lindberg, S. O. 1874. Manipulus muscorum secundus. F. Fl. Fenn. 13: 353-417.

__________. 1875. Hepaticae in Hibernia mense Julii 1873 lectae. Helsingfors.

__________. 1885. Scalia Hookeri et Fossombroniae scandinavicae vivae descriptae. Rev. Bryol. Lichenol. 12: 33-48.

Lorbeer, G. 1934. Die Zytologie der Lebermoose mit besonderer Berucksichtigung allgemeiner Chromosomenfragen. I. Teil. Jahrb. Wiss. Bot. 80: 565-818 + 110 fig.

Macvicar, S. M. 1926. The student's handbook of British hepatics. 490 pp. V. V. Sumfield, Eastbourne.

Marchant, H. J. 1976. Plasmodesmata in algae and fungi. In B. E. S. Gunning & A. W. Robards (eds.), Intercellular communication in plants: Studies on plasmodesmata. pp. 59-80. Springer-Verlag, Berlin.

McCormick, F. A. 1914. A study of Symphyogyna aspera. Bot. Gaz. (London) 58: 401-418, pl. 30-32.

McFarland, K. D., D. K. Smith, & P. G. Davison. 1989. Observations on antheridial and apical organization in Takakia ceratophylla. Amer. J. Bot. Suppl. 76(6): 11 (Abstract #27).

Mehra, P. N. & D. Kumar. 1980. Cytology of two Indian Calobryales. J. Hattori Bot. Lab. 47: 57-61.

__________ & __________ . 1990. Some observations on the embryology of Calobryum indicum. J. Hattori Bot. Lab. 68: 239-253.

Meylan, C. H. 1924. Les hepatiques de la Suisse. 318 pp. Fretz Freres, S. A., Zurich.

Mitten, W. 1891. An enumeration of all species of Musci and Hepaticae recorded from Japan. Trans. Linn. Soc. London, Bot., Ser. 2, 3: 153-206 + 51 pl.

Miyoshi, N. 1967. Electron-microscopic observations of spores in the Hepaticae. 1. Calobryum rotundifolium (Mitt.) Schiffn. Bull. Okayama Coll. Sci. 3: 35-40.

Müller, K. 1906-1911. Die Lebermoose Deutschlands, Oesterreichs und der Schweiz. In Rabenhorst's Kryptogamen-Flora, Zweite Auflage. pp. 309-313, 396-400. Eduard Kummer, Leipzig.

__________. 1939. Untersuchungen über die Ölkörper der Lebermoose. Ber. Deutsch. Bot. Ges. 57: 326-370.

__________. 1948. Morphologische und anatomische Untersuchungen an Antheridien beblatterten Jungermannien. Bot. Not. 1948(4): 71-80, fig. 1.

__________. 1951. Die Lebermoose Europas. In Rabenhorst's Kryptogamen-Flora. ed. 3, part 1. pp. 1-160. Geest & Portig, Leipzig.

Murray, B. M. 1988. Systematics of the Andreaeopsida (Bryophyta): Two orders with links to Takakia. Beih. Nova Hedwigia 90: 289-336.

Nees von Esenbeck, C. G. 1830. Hepaticae Javanicae. 86 pp. Poland.

__________. 1833. Naturgeschichte der Europaischen Lebermoose, Vol. 1. 348 pp., 6 fig. Berlin.

__________. 1836. Naturgeschichte der Europaischen Lebermoose, Vol. 2. 500 pp. Berlin.

__________. 1838. Naturgeschichte der Europaischen Lebermoose, Vol. 3. 594 pp. Breslau.

Nehira, K. 1961. The germination of spores in Hepaticae 1. Calobryum rotundifolium (Mitt.) Schiffn., Bazzania albicans Steph. and Heteroscyphus planus (Mitt.) Schiffn. Hikobia 2: 185-189.

__________. 1963. The germination of spores in Hepaticae 5. Megaceros tosanus Steph. and Dendroceros japonicus Steph. Hikobia 3: 184-190.

__________. 1966. Sporelings in the Jungermanniales. J. Sci. Hiroshima Univ. Ser. B., 2 Bot. 11: 1-49.

__________. 1983. Spore germination, protonema development and sporeling development. In R. M. Schuster (ed.), New Manual of Bryology. Vol. 1. pp. 343-385. Hattori Bot. Lab., Nichinan.

Niedhart, H. V. 1979. Comparative studies of sporogenesis in bryophytes. In G. C. S. Clarke & J. G. Duckett (eds.), Bryophyte systematics. pp. 251-280. Bryophyte Systematics Assoc. Special Vol. 14. Academic Press, London

Nishida, Y. 1978. Studies on the sporeling types in mosses. J. Hattori Bot. Lab. 44: 371-454.

Parihar, N. S. 1957. An introduction to Embryophyta. I. Bryophyta. vi + 308 pp. Allahabad.

Paton, J. & M. F. V. Corley. 1969. Haplomitrium hookeri (Sm.) Nees in alpine habitats in Scotland. Trans. Brit. Bryol. Soc. 5: 827-828.

Pearson, W. H. 1902. The Hepaticae of the British Isles. Vols. 1 & 2. 526 pp. + 228 pl. Lovell Reeve & Co., Ltd., London.

Pfeffer, W. 1874. Die Ölkörper der Lebermoose. Flora 57: 2-6, 17-27, 33-43 + 1 pl.

Pickett-Heaps, J. D. 1975. Green algae: Structure, reproduction and evolution in selected genera. viii + 606 pp. Sinauer Assoc., Inc., Sunderland, Massachusetts.

Pihakaski, K. 1966. An electron microscopy study on the oil bodies of two hepatic species. Protoplasma 62: 393-399.

__________. 1968. A study of the ultrastructure of the shoot apex and leaf cells in two liverworts, with special reference to the oil bodies. Protoplasma 66: 79-103.

__________. 1972a. Histochemical studies on the oil bodies of the liverworts *Pellia epiphylla* and *Bazzania trilobata*. Ann. Bot. Fenn. 9: 65-76.

__________. 1972b. Studies on the hepatic oil bodies, with special reference to the vacuole-like oil bodies of *Pellia epiphylla*. Publ. Univ. Turku 170: 1-25.

Piippo, S. 1984. Bryophyte flora of the Huron Peninsula, Papua New Guinea. III. Haplomitriaceae, Lepicoleaceae, Herbertaceae, Pseudolepicoleaceae, Trichocoleaceae, Schistochilaceae, Balantiopsaceae, Pleuroziaceae & Porellaceae (Hepaticae). Ann. Bot. Fenn. 21:21-48.

Poulsen, C. 1974. Further contributions to the knowledge of the Paleozoic of Slagelse no. 1, western Sealand. Dan. Geol. Unders. II. Raekke 101: 28-33.

Pringsheim, N. 1878. Über Sprossung der Moosfrüchte und den Generationswechsel der Thallophyten. Jahrb. Wiss. Bot. 11: 1-46.

Proskauer, J. 1951. Notes on Hepaticae. II. 4. On *Monoclea*. Bryologist 50: 243-266.

__________. 1962. On *Takakia*, especially its mucilage hairs. J. Hattori Bot. Lab. 25: 217-223 + 11 fig.

Randolph, L. R. 1935. A new fixing fluid and revised schedule for the paraffin method in plant cytology. Stain Tech. 10: 95-96.

Remy, W. 1982. Lower Devonian gametophytes: relation to phylogeny of land plants. Science 215: 1625-1627.

__________ & R. Remy. 1980a. Devonian gametophytes with anatomically preserved gametangia. Science 208: 295-296.

__________ & __________. 1980b. Lyonophyton rhyniensis nov. gen. et nov. sp., ein Gametophyt aus dem Chert von Rhynie (Schottland). Argument. Paleobot. 6: 37-72.

__________, __________, H. Hass, S. C. Schultka, & F. Franzmeyer. 1980. Sciadophyton Steinmann--ein Gametophyt aus dem Siegen. Argument. Paleobot. 6: 73-94.

Renzaglia, K. S. 1978. A comparative morphology and developmental anatomy of the Anthocerotophyta. J. Hattori Bot. Lab. 44: 31-90.

__________. 1982. A comparative developmental investigation of the gametophyte generation in the Metzgeriales (Hepatophyta). x + 254 pp. J. Cramer, Vaduz. Bryophyt. Biblioth. 24.

__________ & S. E. Bartholomew. 1985. Sporeling development in Fossombronia cristula Aust. with special reference to the apical organization and growth. Bryologist 88: 337-343.

Reynolds, D. N. 1980. Gametic dispersal in Mnium ciliare. Bryologist 83: 73-77.

Robards, A. W. 1976. Plasmodesmata in higher plants. In B. E. S. Gunning & A. W. Robards (eds.), Intercellular communication in plants: Studies on plasmodesmata. pp. 15-57. Springer-Verlag, Berlin.

Saito, K. 1975. A monograph of Japanese Pottiaceae. J. Hattori Bot. Lab. 39: 373-537.

Schiffner, V. 1893. Ueber exotische Hepaticae. Deutsch. Akad. Naturf. 40:219-316.

__________. 1899. Ueber einige Hepaticae aus Japan. Oesterr. Bot. Z. 49: 385-390.

__________. 1900. Die hepaticae der Flora von Buitenzorg. 220 pp. Leiden.

__________. 1941. Kritische Bemerkungen über die europaischen Lebermoose. Part 27. 27 pp. Horn, Austria.

Schljakov, R. N. 1972. On the higher taxa of liverworts (Hepaticae s. str.). Bot. Zh. (Leningr.) 57: 496-508.

Schofield, W. B. 1968. Bryophytes of British Columbia II. Hepatics of particular interest. J. Hattori Bot. Lab. 31: 265-282.

__________ & C. Hébant. 1984. The morphology and anatomy of the moss gametophore. In R. M. Schuster (ed.), New Manual of Bryology. Vol. 2. pp. 627-657. Hattori Bot. Lab., Nichinan.

Schopf, J. M. 1970. Relation of floras of the Southern Hemisphere to continental drift. Taxon 19: 657-674.

Schostakowitsch, W. 1894. Ueber die Reproductions- und Regenerationserscheinungen bei den Lebermoosen. Flora 79: 350-384.

Schuster, R. M. 1953. Boreal Hepaticae, a manual of the liverworts of Minnesota and adjacent regions. Amer. Midl. Naturalist 49: 257-684.

__________. 1963. Studies on Antipodal Hepaticae. I. Annotated keys to the genera of antipodal Hepaticae with special reference to New Zealand and Tasmania. J. Hattori Bot. Lab. 26: 185-309.

__________. 1966. The Hepaticae and Anthocerotae of North America east of the hundredth meridian. Vol. 1. xvii + 802 pp. Columbia University Press, New York.

__________. 1967. Studies on Hepaticae XVI. Calobryales. Nova Hedwigia 13: 1-63 + 12 pl.

__________. 1971. Two new antipodal species of <u>Haplomitrium</u> (Calobryales). Bryologist 74: 131-143.

__________. 1972. Continental movements, "Wallace's Line" and Indomalayan-Australasian dispersal of land plants: some eclectic concepts. Bot. Rev. (Lancaster) 38: 3-85.

__________. 1979. The phylogeny of the Hepaticae. <u>In</u> G. C. S. Clarke & J. G. Duckett (eds.), Bryophyte Systematics. Systematics Assoc. Vol. 14: 41-82. Academic Press, London.

__________. 1982. Exogenous branching and its phylogenetic significance. J. Hattori Bot. Lab. 51: 1-50.

__________. 1984a. Evolution, phylogeny and classification of the Hepaticae. <u>In</u> R. M. Schuster (ed.), New Manual of Bryology, Vol. 2. pp. 892-1070. Hattori Botanical Laboratory, Nichinan.

__________. 1984b. Comparative anatomy and morphology of the Hepaticae. In R. M. Schuster (ed.), New Manual of Bryology. Vol. 2. pp. 760-891. Hattori Bot. Lab., Nichinan.

__________ & J. Engel. 1977. Austral Hepaticae, V. The Schistochilaceae of South America. J. Hattori Bot. Lab. 42: 273-423.

__________ & H. Inoue. 1975. Studies on Pallaviciniaceae and Allisoniaceae (Metzgeriales) in Japan. Bull. Natl. Sci. Mus. 1: 101-107.

Sharma, K. K., V. M. Diller, & M. Fulford. 1960. Studies on the growth of Haplomitrium. II. Media containing amino acids. Bryologist 63: 203-212.

Schweitzer, H.-J. 1981. Der Generationswechsel Rheinischer Psilophyten. Bonner Palaobot. Mitteil. 8: 1-19.

Siessegger, B. 1966. Die Elateren der Lebermoose, ihr Bau und ihre Taxonomische Verwendbarkeit. Ph. D. dissertation, Eberhard-Karls-Universitat, Tubingen. 87 pp.

Slavik, B. 1950. Specific bodies in the cells of hepatics (Hepaticae). Stud. Bot. Cechoslov. 11: 226-244.

Sloover, J. d. 1959. Considération sur la valeur spécifique de Moerckia flotoviana (Nees) Schiffn., Dilaenacée nouvelle pour la flore belge. Bull. Jard. Bot. Etat 29: 157-181.

Smith, D. K. 1990. Sporophyte of Takakia discovered. Bryol. Times 57/58: 1, 4.

Smith, G. M. 1955. Cryptogamic botany: Bryophytes and Pteridophytes. Vol. 2. ed. 2. 399 pp., 254 fig. McGraw-Hill, New York.

Smith, J. E. (J. Sowerby, proprietor). 1813. English Botany, Vol. 36. J. Davis, London.

Smith, J. L. 1964. Water-conducting system of _Symphyogyna_. Nature 202: 617.

__________. 1966. The liverworts _Pallavicinia_ and _Symphyogyna_ and their conducting system. Univ. Calif. Publ. Bot. 39: 1-83, pl. 1-18.

Spruce, R. 1885. Hepticae of the Amazon and the Andes of Peru and Ecuador. Trans. Bot. Soc. Edinburgh 15: 309-588.

Steinkamp, M. P. & W. T. Doyle. 1984. Spore wall ultrastructure in the liverwort _Fossombronia longiseta_. Canad. J. Bot. 62: 1871-1879.

Stephani, F. 1898-1924. Species Hepaticarum. Vol. I-VI. Georg, Geneva. I: 1-113, 1898-1900; II: 1-615, 1901-5; III: 1-693, 1906-9; IV: 1-822, 1909-12; V: 1-1044, 1912-17; VI: 1-763, 1917-24.

Steward, F. C., J. K. Pollard, A. A. Patchett, & B. Witkop. 1958. The effects of selected nitrogen compounds on the growth of plant tissue cultures. Biochim. Biophys. Acta 28: 308-317.

Stewart, G. H. 1978. Oil bodies of the New Zealand leafy Hepaticae (Jungermanniales). New Zealand J. Bot. 16: 185-205.

Stewart, W. N. 1983. Paleobotany and the evolution of plants. x + 406 pp. Cambridge University Press, New York.

Stotler, R. & B. Crandall-Stotler. 1977. A checklist of the liverworts and hornworts of North America. Bryologist 80: 405-428.

Suire, C. 1970. Recherches cytologiques sur deux Hepatiques: Pellia epiphylla (L) Corda (Metzgeriale), et Radula complanata (L) Dum. (Jungermannile) ergastome, sporogenese et spermatogenese. Botaniste 53: 125-392.

Tatuno, S. 1941. Über die Chromosomen der Lebermoose von Japan. Proc. Imp. Acad. Tokyo 17: 396-399.

Taylor, J., P. J. Hollingsworth & W. C. Bigelow. 1974. Scanning electron microscopy of liverwort spores and elaters. Bryologist 77: 281-327.

Thomas, R. J. 1980. Cell elongation in hepatics: the seta system. Bull. Torrey Bot. Club 107: 339-345.

Udar, R. & S. Chandra. 1961. Calobryum Nees - a genus new to Indian flora. Curr. Sci. 30: 22-23.

__________ & __________. 1965. A new species of Calobryum Nees, C. indicum Udar et Chandra, from Darjeeling, Eastern Himalayas, India. Rev. Bryol. Lichenol. 33: 555-559.

__________ & V. Chandra. 1965. Discovery of Haplomitrium hookeri (Smith) Nees in Asiatic flora. Curr. Sci. 38: 618.

__________ & D. Kumar. 1977. Haplomitrium hookeri in Western Himalayas, India. Bryologist 80: 340-342.

__________ & __________. 1982. A new Haplomitrium from India. J. Indian Bot. Soc. 61: 73-79.

__________, V. Chandra & S. Chandra. 1968. Calobryum blumii Nees - A taxon new to Indian flora. Curr. Sci. 37: 265.

__________ & S. C. Srivastava. 1983. Reproductive biology of some Indian liverworts. Phytomorphology 33: 37-46.

__________ & __________. 1984. Scanning electron microscopy of spores of some Indian liverworts. J. Hattori Bot. Lab. 56: 97-103.

Verdier, J. L. 1965. Contribution a l'etude au-microscope electronique des propagules de Lunularia cruciata (L) Dum. Rev. Cytol. Biol. Veg. 28: 279-301.

Verdoorn, F. 1932. Classification of hepatics. In F. Verdoorn (ed.), Manual of Bryology. pp. 413-432. M. Nijhoff, The Hague.

Watson, E. V. 1971. The structure and life of bryophytes. ed. 3. 212 pp. Hutchinson & Lo, Ltd., London.

Wettstein, R. v. 1903-1908. Handbuck der systematischen Botanik. II. Band. 577 pp. Leipzig.

Winkler, S. 1969. Anatomische Untersuchungen über Leitstrange der Gametophyten bei den Lebermoosen. Oesterr. Bot. Z. 117: 348-364.

Worley, I. A. 1969. Haplomitrium hookeri from Western North America. Bryologist 72: 225-232.

Yang, B.-Y. 1966. The geographical distribution and growth habits of Haplomitrium. Taiwania 12: 9-19 + 5 pl.

__________. 1967. Spore germination and leafy gametophyte of Haplomitrium rotundifolium developed in culture. Taiwania 13: 153-167 + 6 pl.

__________. 1968. Induced aberrant gametophytes from *Haplomitrium rotundifolium*. Taiwania 14: 81-88.

__________, F. M. Hsu & S. M. Lee. 1968. Spore germination and leafy gametophyte of *Haplomitrium blumii* developed in antiseptic culture. Taiwania 14: 73-79.

FIGURES

FIGURES 1-8. Vegetative gametophytic features of *Haplomitrium hookeri* (Smith) Nees. -1. Light micrograph of a leafy shoot arising from the rhizomatous system. -2-8. Scanning electron micrographs. 2. Leafy shoot with distant, patent, transversely inserted leaves. 3. Leafy shoot with clustered, imbricate leaves. 4. Nearly orbicular leaf. 5. Leaf with three blunt lobes. 6. Ovate-triangular leaf with undulate margin and terminal slime papilla. 7. Branched rhizomatous system with descending stolons. Note the numerous slime papillae. 8. Apical view of a rhizome apex bearing slime cells and scales.

Key: ls, leafy shoot; r, rhizome; sa, scale; sn, stolon; sp, slime papilla. (Fig. 1: *Lindberg s.n.* [MANCH]; 2, 4, 5, 8: *Lorbeer H56* [PR]; 3: *Schofield & Spence 83691* [F]; 6 *Worley & Rawson 12542* [H]; 9: *Loitlesberger 1348* [AD].)

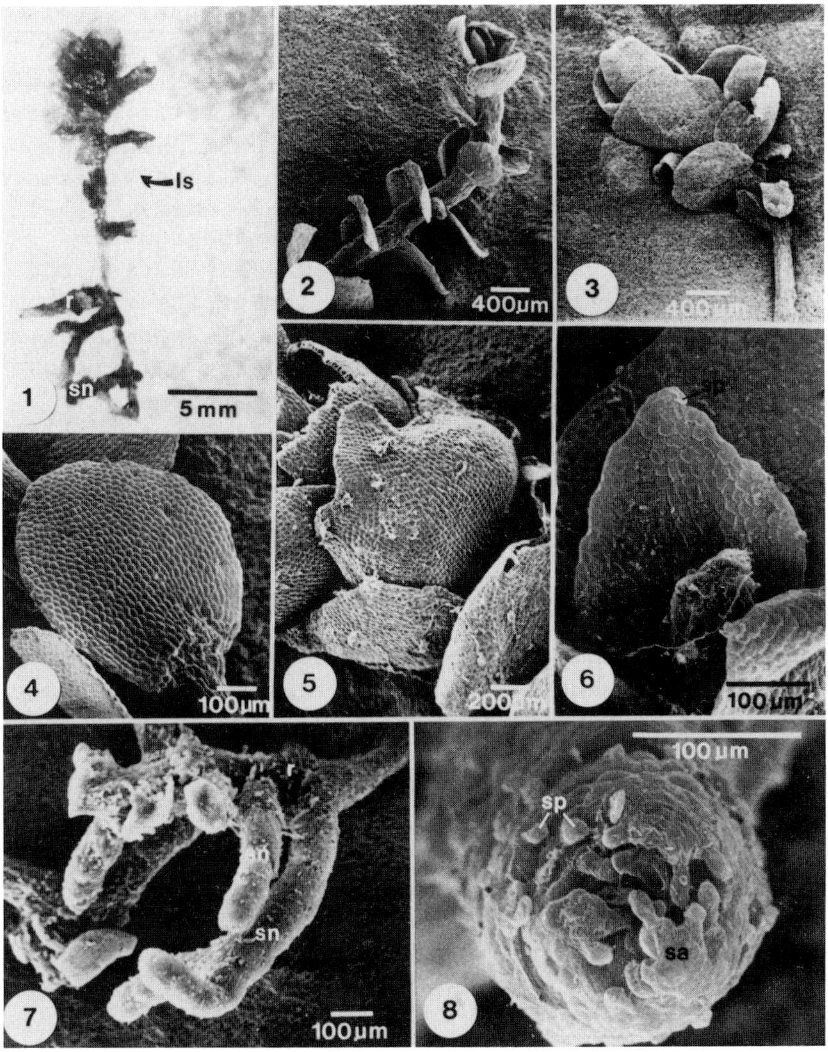

ls
r
sn
1
5 mm
2
400µm
3
400µm
4
100µm
5
200µm
6
sp
100µm
7
r
sn
100µm
8
100 µm
sp
sa

FIGURES 9-20. Vegetative leaves of *Haplomitrium hookeri* (Smith) Nees, *H. monoicum* Engel, and *H. intermedium* Berrie. -9-16. *H. hookeri*. 9a-c. Leaves taken from the midregion of one shoot. 9a-b. Reduced dorsal leaves. 9c. Lateral leaf. 10a-d. Leaves removed from a single shoot. 10a-b. Slightly reduced dorsal leaves. 10c-d. Larger lateral leaves. 11-13. Leaves of different shapes from the midregion of shoots. 14a-b. Leaves of different shapes from the basal portion of the same shoot. 15a-b. Leaves from the apical region of one shoot. 16a-b. Cross sections through a leaf. 16a. Basal half showing unistratose condition. 16b. Extreme base near the point of insertion revealing a small bistratose region. -17-18. *H. monoicum*. 17a-c. Leaves from a single shoot. 17a. Slightly reduced dorsal leaf. 17b-c. Lateral leaves. 18a-b. Transections through a lateral leaf. 18a. Midregion showing unistratose composition. 18b. Near point of insertion revealing a small, median bistratose region. -19-20. *H. intermedium*. 19a-d. Leaves taken from the same shoot. 19a-b. Reduced dorsal leaves. 19c-d. Lateral leaves. Note the undulate margin of (d). 20a-b. Transections through a leaf. 20a. Basal quarter showing the unistratose composition. 20b. Near point of leaf insertion showing a small, bistratose region.

Key: sp, slime papilla. (Figs. 9-16: *Jensen 1347* [AD]; 19, 20: *Selkirk s.n.* [ABSH]; 17, 18: *Crosby 14165* [F].)

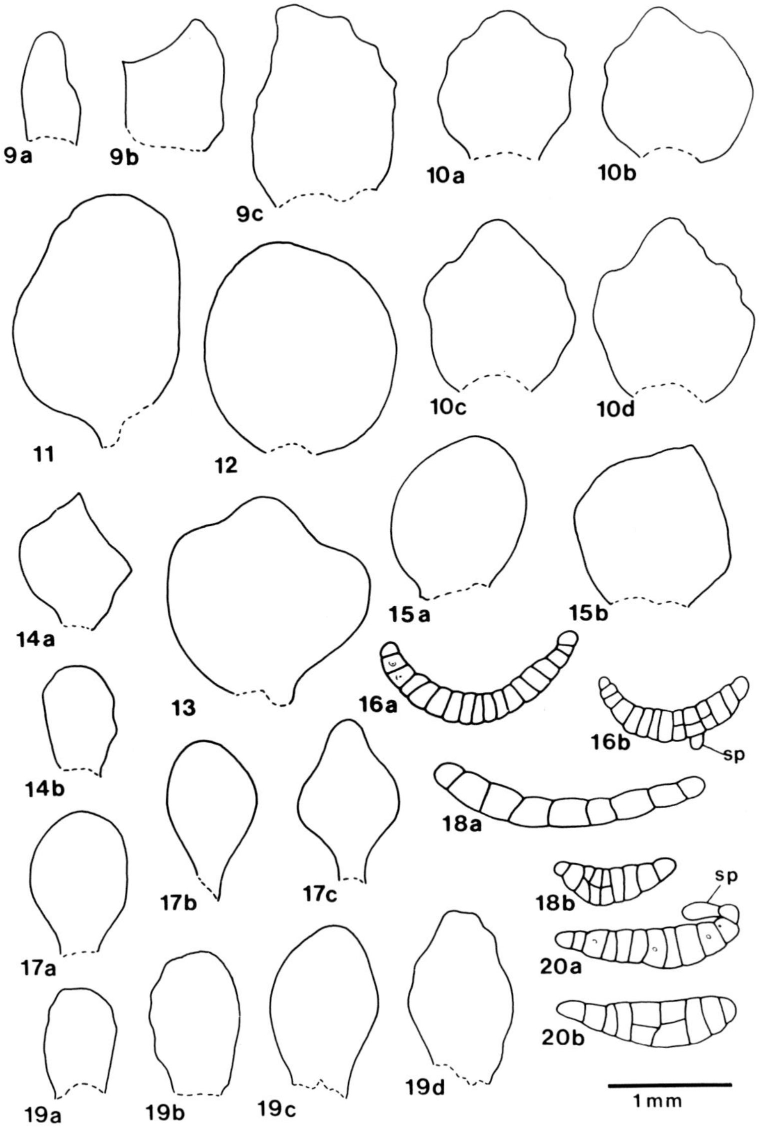
9a
9b
9c
10a
10b
11
12
10c
10d
14a
13
15a
15b
16a
16b
sp
14b
18a
17b
17c
18b
sp
17a
20a
20b
19a
19b
19c
19d
1 mm

FIGURES 21-25. Vegetative gametophytic features of *Haplomitrium monoicum* Engel. -21. Light micrograph of a sterile plant. Note the rhizome. -22-25. Scanning electron micrographs. 22. Distal portion of a leafy shoot. Note the entire leaves with 1-celled borders of enlarged cells. 23. Apical portion of a leafy axis. Note the decurrent leaves (arrow). 24. One-celled leaf border composed of enlarged cells. 25a-b. Margin of leaf with tiny unicellular slime papilla.

Key: lb, leaf border; ls, leafy shoot; r, rhizome; sn, stolon; sp, slime papilla. (Figs. 21-25: *Crosby 14165* [F].)

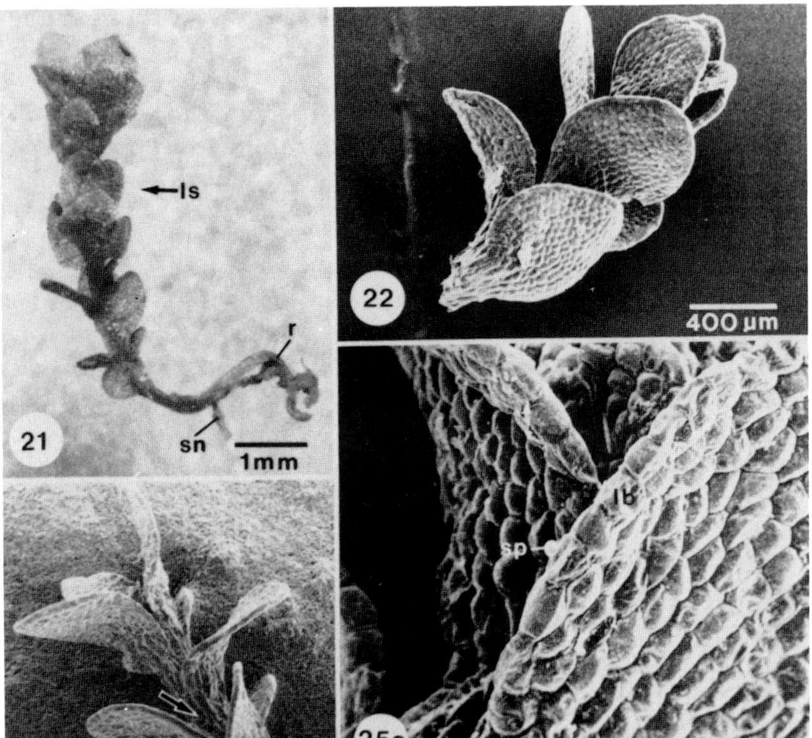

21
ls
r
sn
1mm
22
400 µm
23
400 µm
24
lb
lb
100 µm
25a
lb
sp
100 µm
25b
sp
20 µm

FIGURES 26-34. Vegetative gametophytic features of *Haplomitrium intermedium* Berrie. -26-27. Light micrographs. 26. Sterile plant. 27. Short rhizome between two leafy shoots. Note the descending stolons. -28-34. Scanning electron micrographs. 28. Distal portion of a vegetative leafy shoot. Note the long internodes between leaves, and the absence of slime cells. 29. Leaf with a basal "tooth". 30. Transversely inserted leaf with 1-celled border and marginal slime cell. 31. Apical portion of a leafy shoot showing axillary terminal branching. 32. Axillary origin of a stolon. 33. Young branch forming from a stolon. 34a-b. Stolons. 34a. Branched stolon. 34b. Apical portions. Note the infrequent occurrence of slime papillae.

Key: asn, axillary stolon; b, branch; lb, leaf border; ls, leafy shoot; r, rhizome; sn, stolon; sp, slime papilla. (Figs. 26, 27, 29: *Berrie 3/62* [SYD]; 28, 30-32, 34: *Selkirk s.n.* [ABSH]; 33: *Berrie 2/62/1* [SYD].)

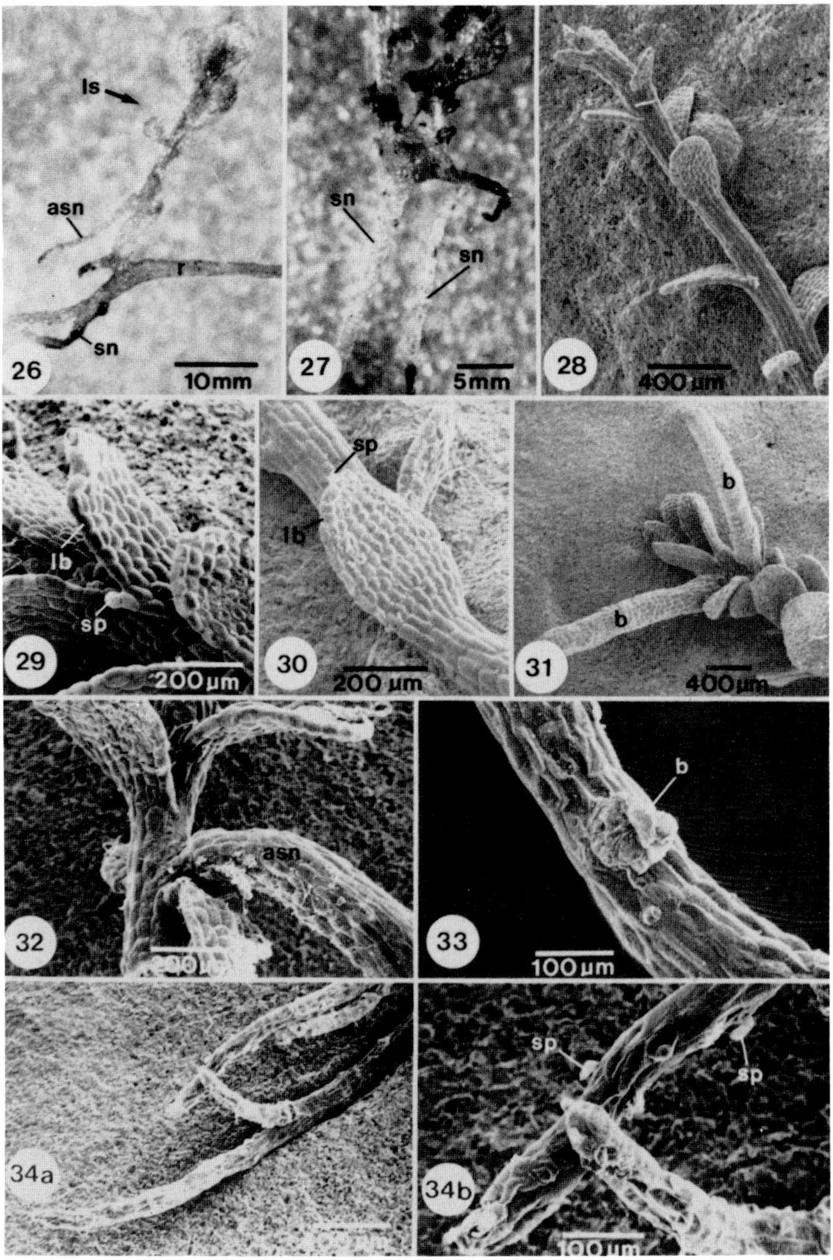

ls
asn
r
sn
26
10mm
sn
sn
27
5mm
28
400 µm
lb
sp
29
200 µm
sp
lb
30
200 µm
b
b
31
400µm
asn
32
b
33
100 µm
34a
sp
sp
34b
100 µm

FIGURES 35-41. Vegetative gametophytic features of *Haplomitrium gibbsiae* (Steph.) Schust. and *H. kashyapii* Udar & Kumar. -35-39. *H. gibbsiae*. 35-36. Light micrographs. 35. Plants with numerous leafy shoots arising from the freely branched rhizomatous systems. Note the unelongated spurs. 36. Distal portion of an anisophyllous shoot which is leafy throughout. Dorsal view. 37-39. Scanning electron micrographs. 37. Distal portion of an anisophyllous shoot with succubously inserted lateral leaves. Ventral view. 38. Orbicular leaf with entire margin. 39. Apical portion of a leaf bearing marginal slime papillae. -40-41. Light micrographs of *H. kashyapii* shoots. 40. Shoot with leaves clustered distally. 41. Ventral view of the distal portion of an anisophyllous shoot which is leafy throughout.

Key: dl, dorsal leaf; ll, lateral leaf; ls, leafy shoot; r, rhizome; sn, stolon; sp, slime cell; su, lateral spur. (Figs. 35, 37-39: *Sneddon s.n.* [ABSH]; 36: *Allison H1443* [CHR]; 40-41: *Udar & Kumar C-73/1* [LWU].)

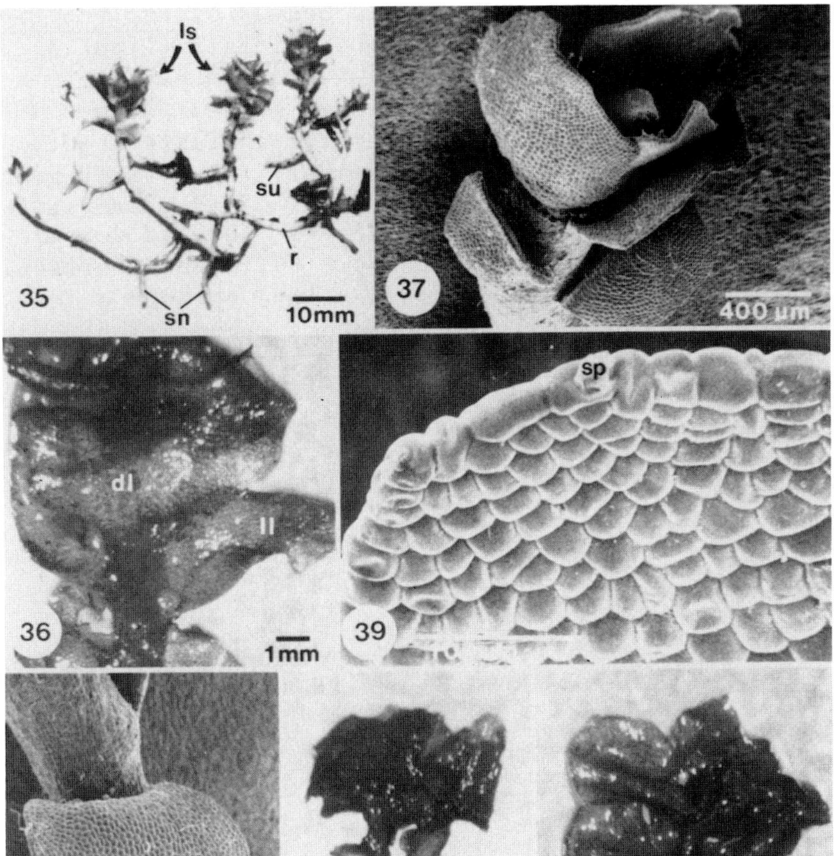
ls
su
r
sn
35
10mm
37
400 µm
dl
ll
36
1mm
sp
39
38
40
1mm
41
1mm

FIGURES 42-47. Vegetative leaves of *Haplomitrium gibbsiae* (Steph.) Schust. and *H. mnioides* (Lindb.) Schust. -42-44. *H. gibbsiae*. 42a-c. Leaves of various shapes taken from a single shoot. 43a-c. Leaves of various shapes taken from a single shoot. 44a-b. Transections through a leaf. 44a. Midregion. 44b. Base. -45-47. *H. mnioides*. 45a-d. Leaves removed from the same stem. 45a-b. Reduced dorsal leaves. 45c-d. Lateral leaves. 46. Nearly orbicular lateral leaf. 47a-b. Cross sections through a leaf. 47a. Midregion showing unistratose composition. 47b. Base showing polystratose field.

Key: sp, slime papilla. (Figs. 42-44: *Fife 8453* [ABSH]; 45: *Iwatsuki 142* [H]; 46, 47: *Furuki 7408* [ABSH].)

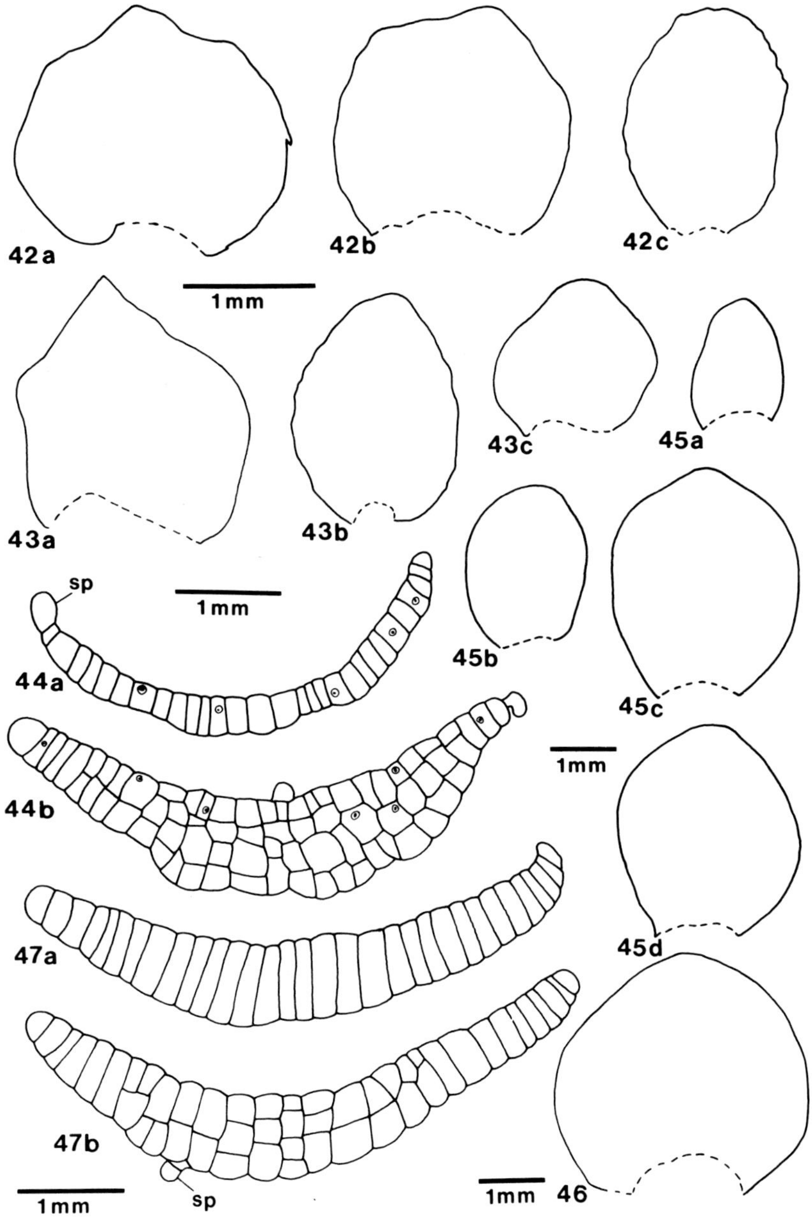
42a
42b
42c
1mm
43a
43b
43c
45a
1mm
sp
44a
44b
45b
45c
1mm
47a
45d
47b
sp
1mm
1mm
46

FIGURES 48-55. Vegetative gametophytic features of *Haplomitrium mnioides* (Lindb.) Schust. -48. Light micrograph showing habit of plant. -49-55. Scanning electron micrographs. 49a-b. Distal portion of a prostrate leafy shoot. 49a. Dorsal view. 49b. Ventral view. 50. Basal region of a shoot bearing reduced leaves. 51. Single-stalked slime cells on leaf lamina. 52. Young, unelongated branches (arrows) scattered at the base of a leafy shoot. 53. Apex of a rhizome which is tilting dorsally and becoming leafy. 54. Young, unelongated branches (arrows) along the terminal portion of a rhizome. 55. Stolon apex. Note the numerous slime papillae.

Key: dl, dorsal leaf; le, young leaf; ll, lateral leaf; ls, leafy shoot; r, rhizome; sn, stolon; sp, slime papilla. (Figs. 48, 53: *Doei s.n.* [ABSH]; 49, 50: *Inoue & Gradstein 506* [F]; 51, 52, 54, 55: *Mizutani s.n.* [ABSH].)

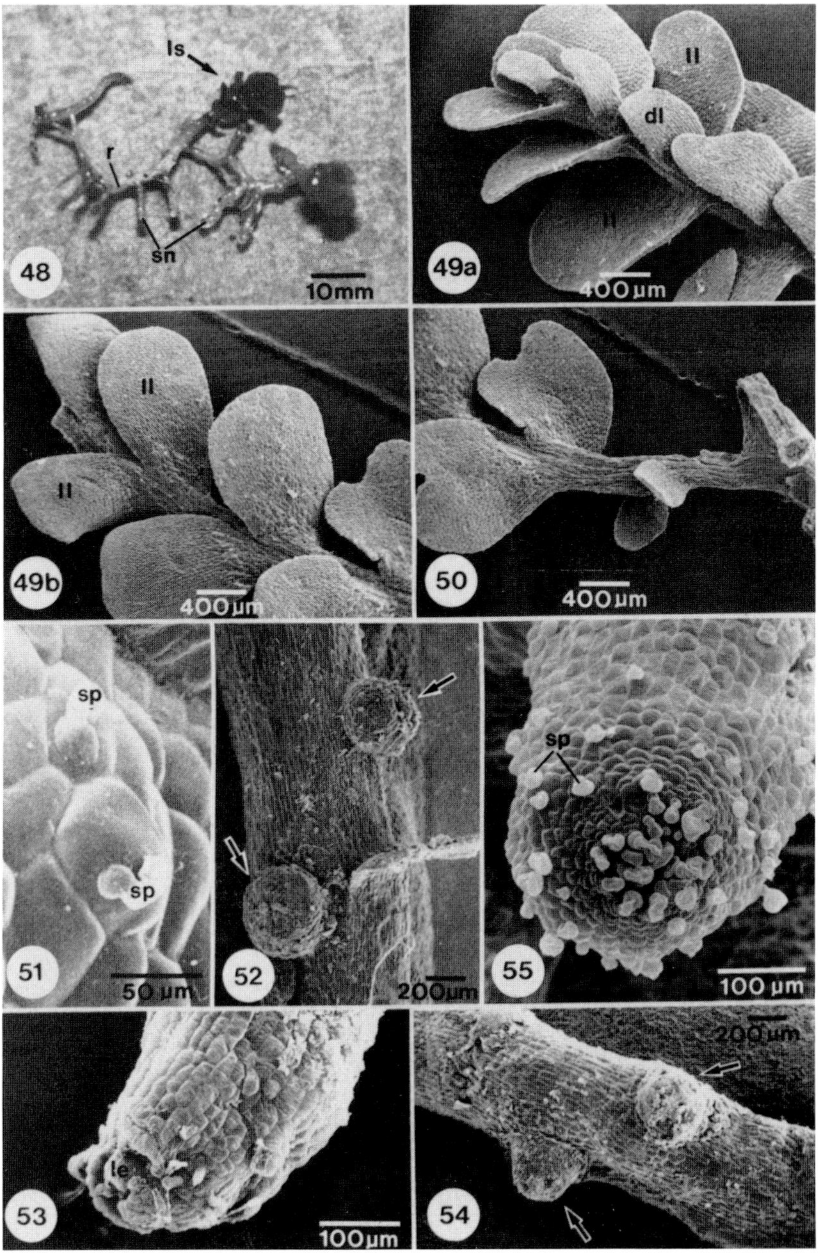

48
ls
r
sn
10mm
49a
ll
dl
400 µm
49b
400 µm
50
400 µm
51
sp
50 µm
52
200µm
55
sp
100 µm
53
te
100µm
54
200 µm

FIGURES 56-60. Vegetative gametophytic features of *Haplomitrium blumii* (Nees) Schust. -56-57. Light micrographs. 56. Habit. 57. Young, unelongated branch (arrow) at the base of a leafy shoot. -58-60. Scanning electron micrographs. 58. Reduced dorsal leaf. 59. Rhizomatous system of a plant. Note the descending stolons. 60. Distal portion of a stolon covered with numerous slime cells.

Key: dl, dorsal leaf; ll, lateral leaf; ls, leafy shoot; r, rhizome; sn, stolon; sp, slime papilla. (Fig. 56: *Wijk 1675* [L]; 57: *Furuki 384* [ABSH]; 58: *Norris 61631* [H]; 59, 60: *Koponen 35418* [G].)

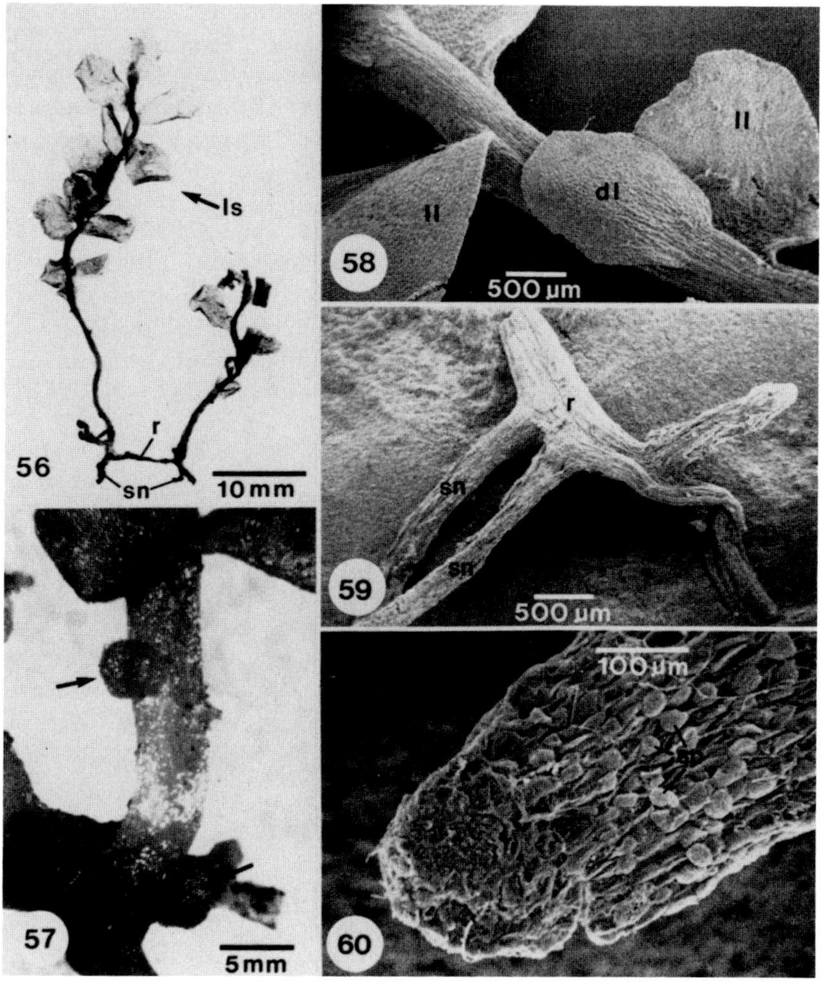
56
ls
r
sn
10 mm
57
5 mm
58
ll
dl
ll
500 µm
59
r
sn
sn
500 µm
60
100 µm

FIGURES 61-65. Vegetative leaves of *Haplomitrium blumii* (Nees) Schust. and *H. andinum* (Spruce) Schust. -61-62. *H. blumii*. 61a-d. Leaves removed from a single shoot. 61a-b. Lateral leaves. 61c. Dorsal leaf from the fourth series. 61d. Apical dorsal leaf. 62a-b. Cross sections through a leaf. 62a. Midportion showing unistratose composition. 62b. Basal portion showing the extensive polystratose field. -63-65. *H. andinum*. 63a-c. Leaves from the same shoot. 63a. Lateral leaf. 63b. Dorsal leaf from the third series. 63c. Dorsal leaf from the fourth series. 64a-b. Lateral leaves of various shapes. 65a-b. Cross section through a leaf. 65a. Midregion. 65b. Base.

Key: sp, slime papilla. (Fig. 61: *Koponen 35568* [G]; 62: *Furuki 384* [ABSH]; 63-65: *Allorge s.n.* [F].)

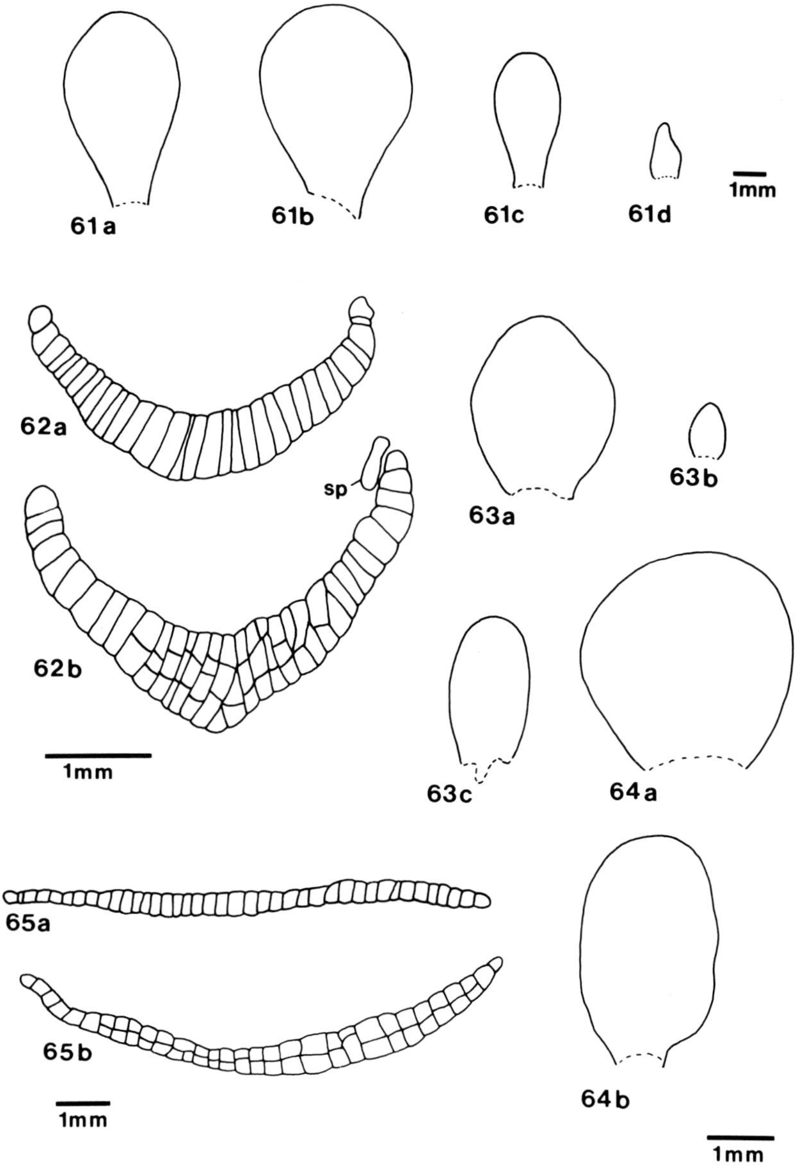
61a
61b
61c
61d
1mm
62a
sp
62b
1mm
63a
63b
63c
64a
65a
65b
1mm
64b
1mm

FIGURES 66-67. Vegetative gametophytic features of *Haplomitrium andinum* (Spruce) Schust. -66. Light micrograph. Dorsal view of a plant. -67a-b. Scanning electron micrographs. Dorsal view of leafy shoot. 67a. First two apical rows of extremely reduced dorsal leaves. 67b. Dorsal leaf on midportion of the shoot.

Key: dl, dorsal leaf; ll, lateral leaf; ls, leafy shoot; r, rhizomatous system. (Figs. 66, 67: *Allorge s. n.* [F].)

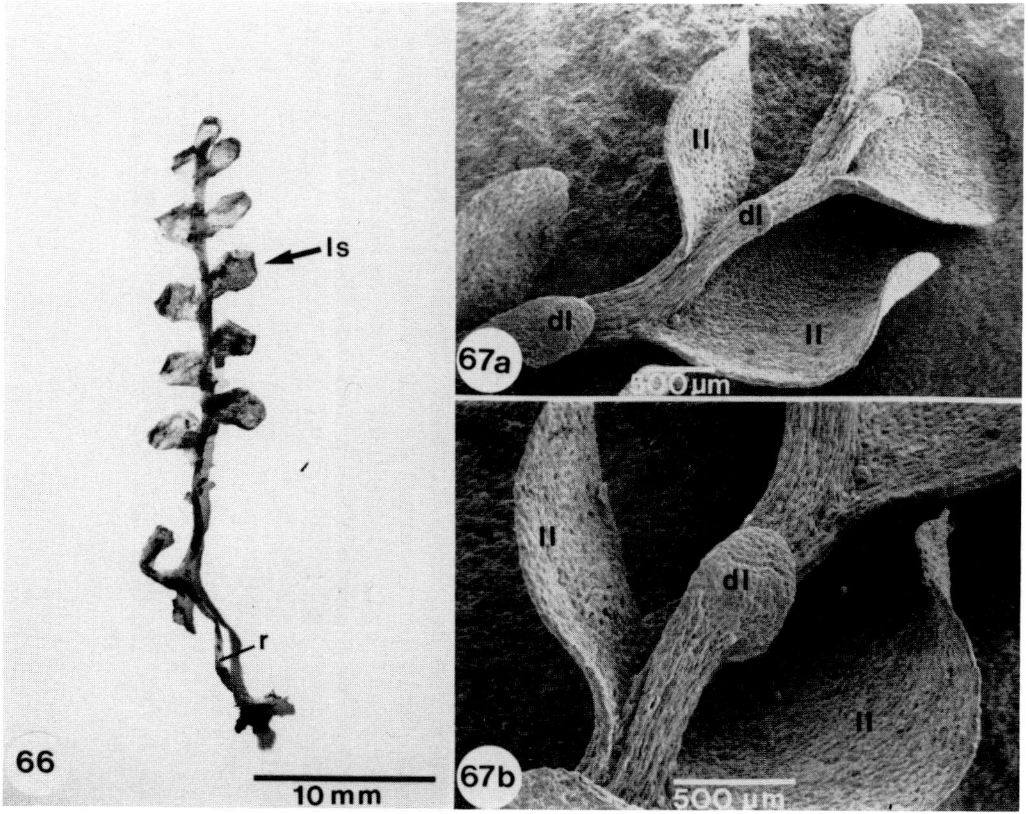
66
ls
r
10 mm
67a
ll
dl
ll
dl
ll
00 µm
67b
ll
dl
ll
500 µm

FIGURES 68-72. Apical cells of *Haplomitrium* Nees and their derivatives. -68. Diagrammatic representation of a tetrahedral apical cell. -69-70. Sections taken through the leafy apices of *H. hookeri* (Smith) Nees. 69a-b. Serial transections made at 10 µm intervals showing apical cells with equal cutting faces. Derivatives are formed in a dextrorse spiral. The apical tilt is evident in that merophyte X is visible in 69b, but not in 69a. 70. Longitudinal section. The dorsal tilt of the apex is evident in that the ventral height of the apical cell is greater than the dorsal height. -71a-b. Serial transection taken through a leafy apex of *H. gibbsiae* (Steph.) Schust. showing the apical cell as an isosceles triangle. Note that the first division of the young derivative (wall 1-1) is near the kathodic end and the second division (wall 2-2) occurs in the broader anodic portion. 71b is 10 µm below 71a. -72a-b. Serial transverse sections through the leafy apex of *H. blumii* made at 10 µm intervals. The dorsal cutting face of the apical cell is reduced. The apical tilt is evident in that merophyte VII is not visible in 72a, but is apparent in 72b. Note the sinistrorse segmentation spiral.

Key: df, dorsal cutting face; vn, ventral surface; x, apical cell. (Figs. 69, 70: *Lorbeer H56* [PR]; 71: *Sneddon s.n.* [ABSH]; 72: *Furuki 384* [ABSH].)

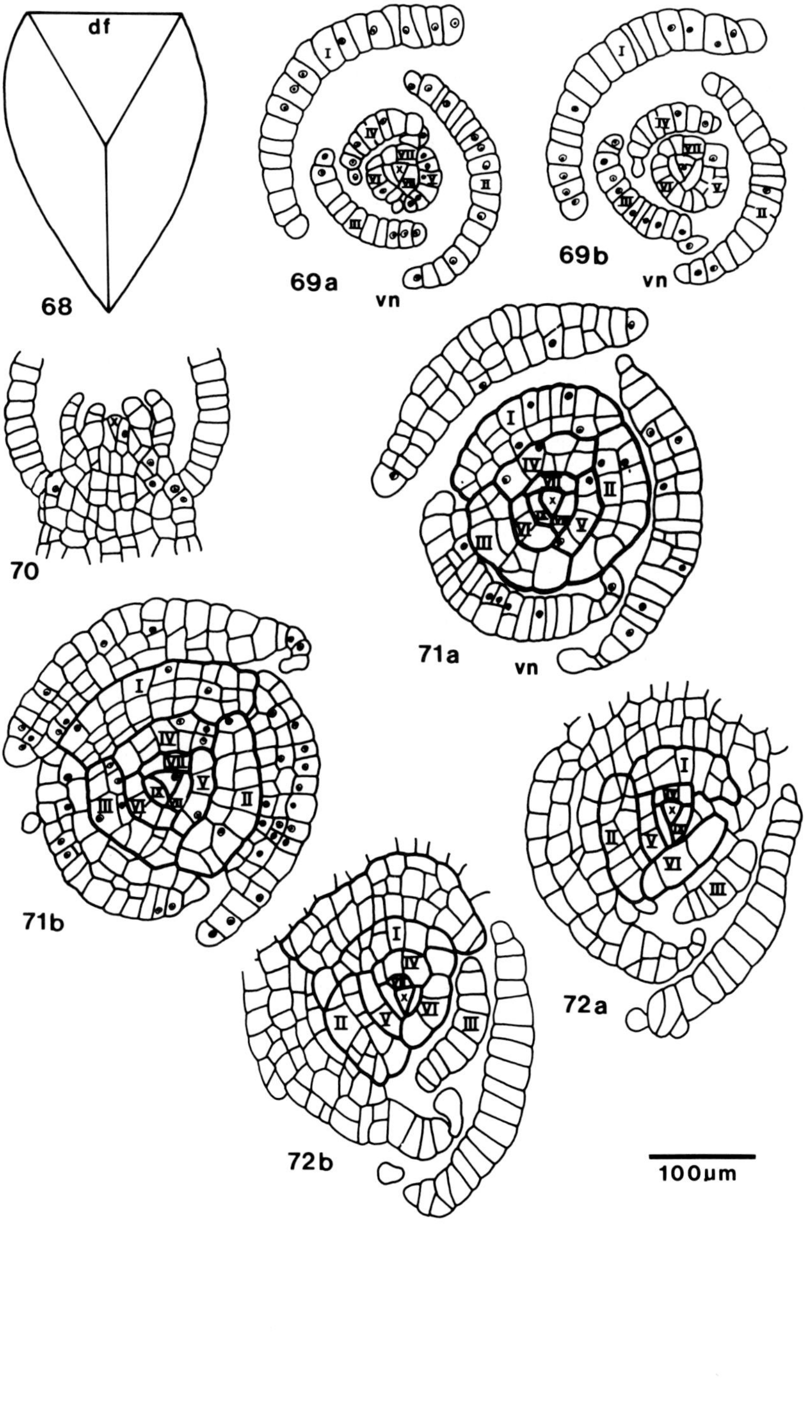
df
68
69a
vn
69b
vn
70
71a
vn
71b
72a
72b
100µm

FIGURES 73-77. Apical cells and derivatives of *Haplomitrium* Nees. -73a-b. Serial longitudinal sections through a leafy apex of *H. mnioides* (Lindb.) Schust. Walls 3-3, 3^1-3^1, 4-4, and 4^1-4^1 delimit the stem tissue initials (epidermal initial, cortex initial, and hydroid initial). Walls 5-5 and 5^1-5^1 denote further development in the epidermal tissue. 73a. Median section through a dorsally tilted apical cell showing its bulging midregion. Note that one leaf is produced per merophyte. 73b. Section taken 20 um below 73a showing the sequence of merophyte development. Walls 1-1 and 2-2 delimit a central wedge-shaped cell, 3a-3a and 4a-4a mirror walls 1-1 and 2-2, while 5a-5a creates a wedge-shaped leaf apical cell. -74-75. Stolon apical cells and their derivatives. 74. *H. gibbsiae* (Steph.) Schust. Transection. Note that the cutting face preceding the youngest segment is longer than the other two due to the oblique wall forming the youngest derivative. 75. *H. intermedium* Berrie. Longitudinal section. Note that one slime papilla is produced per merophyte. -76-77. Diagrammatic illustrations of merophyte development. 76. Three-celled derivative with triangular acroscopic and basiscopic surfaces and rectangular anodic, kathodic and free surfaces. Walls 1-1 and 2-2 denote the first two, opposite, oblique, anticlinal divisions of the merophyte. 77. Older, 9-celled merophyte with epidermal, cortex, and hydroid initials delimited.

Key: ac, acroscopic; a, anodic; ba, basiscopic; c, cortex initial; cc, central cell; ep, epidermal initial; h, hydroid initial; ka, kathodic; li, leaf apical cell; st, stem initial; sp, slime papilla; x, apical cell. (Fig. 73: *Doei s.n.* [ABSH]; 74: *Sneddon s.n.* [ABSH]; 75: *Lorbeer H56* [PR].)

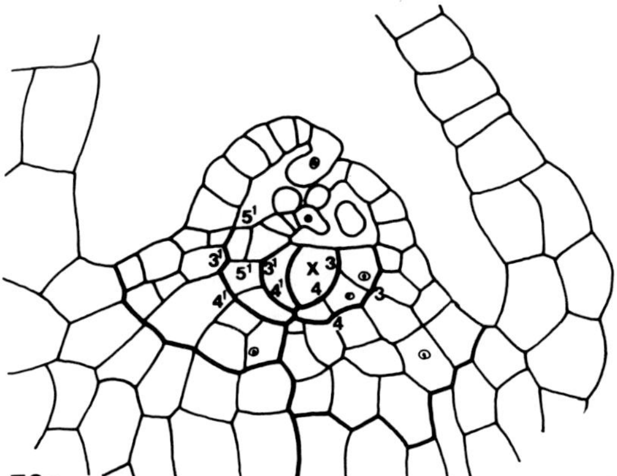

73a

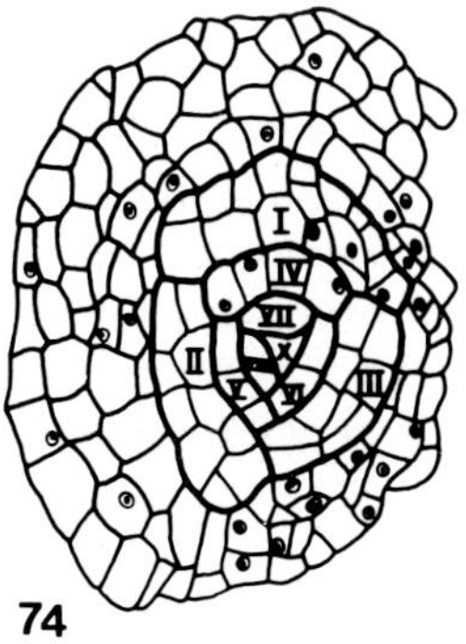

74

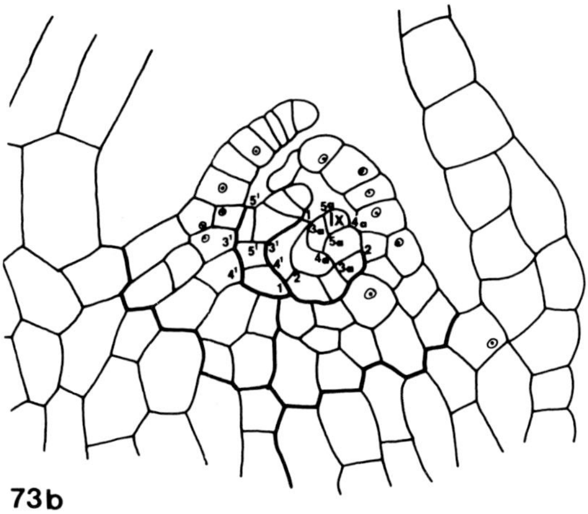

73b

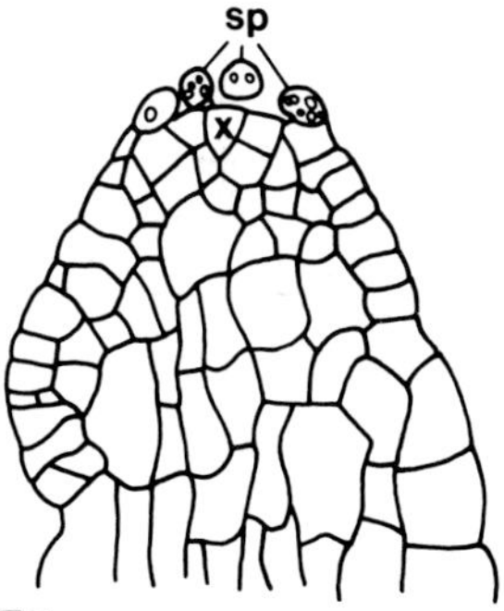

75

100 µm

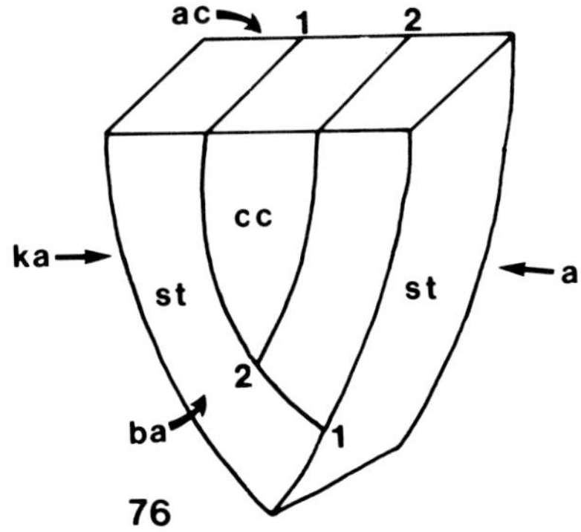

76

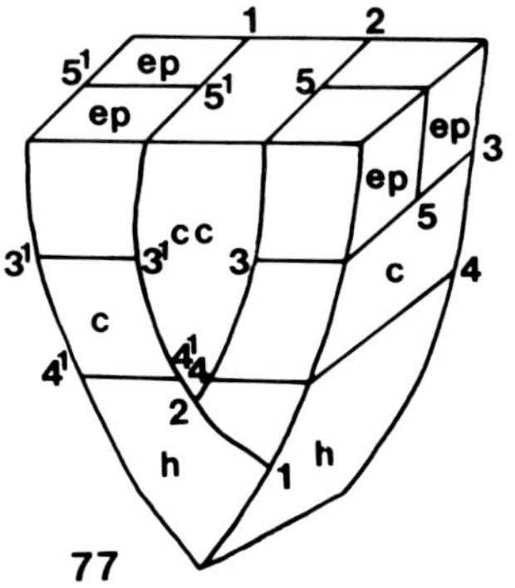

77

FIGURES 78-80. Fine structure of the leafy shoot apical cells and derivatives of *Haplomitrium hookeri* (Smith) Nees and *H. gibbsiae* (Steph.) Schust. -78-79. *H. hookeri*. 78a-b. Transection. 78a. Note the dextrorse spiral of derivatives (with the merophytes numbered consecutively--I-IV), the broad anodic portions of the derivatives, the kathodic positioning of the first divisions of the derivatives, and the nuclear positions. 78b. Note the curve of the developing wall at the kathodic end of the youngest segment (arrow) and the elongate wall preceding the cutting face. 79. Longitudinal section. -80. *H. gibbsiae*. Transection through an apical cell and a portion of its youngest derivatives.

Key: g, dictyosome; l, lipid; m, mitochondrion; n, nucleus; nu, nucleolus; ob, undifferentiated oil body; p, plastid; v, vacuole; x, apical cell. (Figs. 78-79: *Lorbeer H56* [PR]; 80: *Fife 8453* [ABSH].)

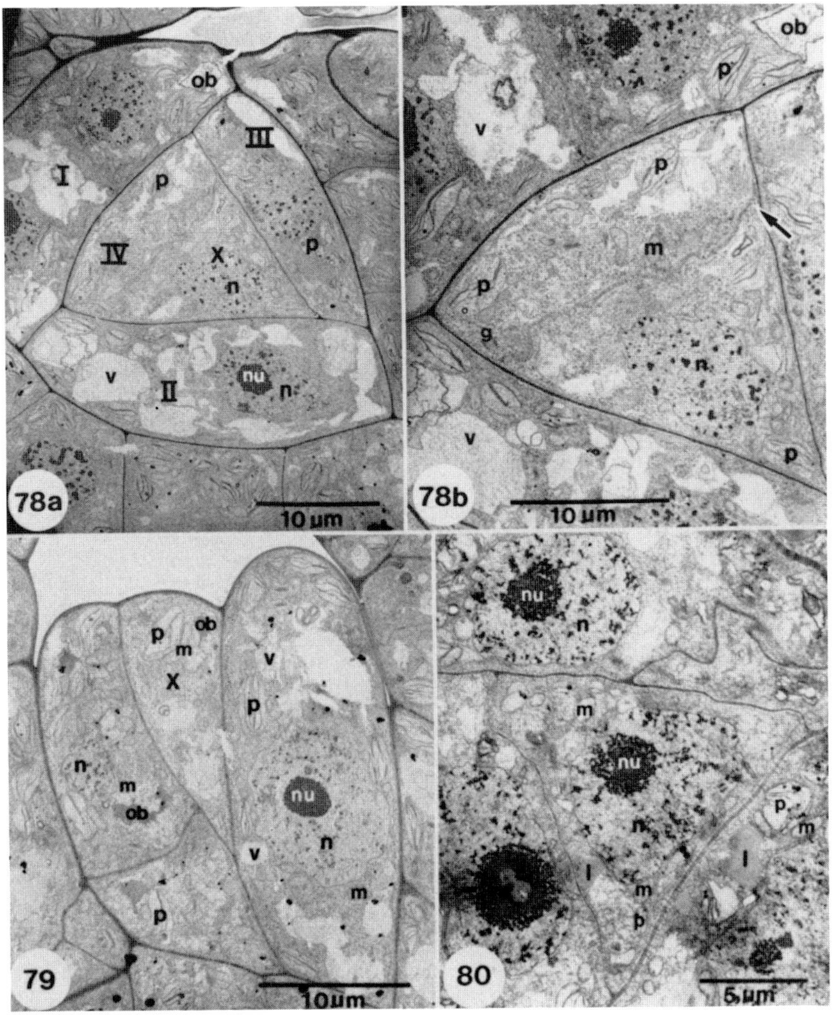
78a
ob
I
III
IV
II
p
X
n
v
nu
10 µm
78b
ob
p
v
m
g
n
10 µm
79
p
m
ob
X
v
n
nu
10µm
80
nu
n
m
l
p
5 µm

FIGURES 81-85. Fine structure of the apical cell and derivatives of <u>Haplomitrium hookeri</u> (Smith) Nees. -81a-b. Longitudinal section. 81a. Distal portion of an apical cell showing the thickened free surface wall. Note the young plastid, elongate mitochondria, "precursory" oil bodies, and the ER which almost parallels the free surface and the cutting faces of the cell. 81b. Groups of large vesicles positioned between the plasmalemma and free surface cell wall. Also note the rough ER. -82. Young wall forming in a derivative. Note that portions of the ER have become trapped in the growing wall (arrow). -83. Apical cell cutting face which is traversed by plasmodesmata with desmotubules (arrows). -84. Portion of a derivative possessing young plastids with organized thylakoids, starch grains, and oil droplets. -85. Portion of a merophyte possessing a plastid (arrow a) and a mitochondrion dividing by constriction (arrow b).

Key: d, oil droplet; er, endoplasmic reticulum; fr, free surface; g, dictyosome; m, mitochondrion; n, nucleus; ob, oil body; p, plastid; pd, plasmodesma; pl, plasmalemma; sg, starch grain; t, thylakoid; vs, vesicle; w, cell wall. (Figs. 81-85: <u>Lorbeer H56</u> [PR].)

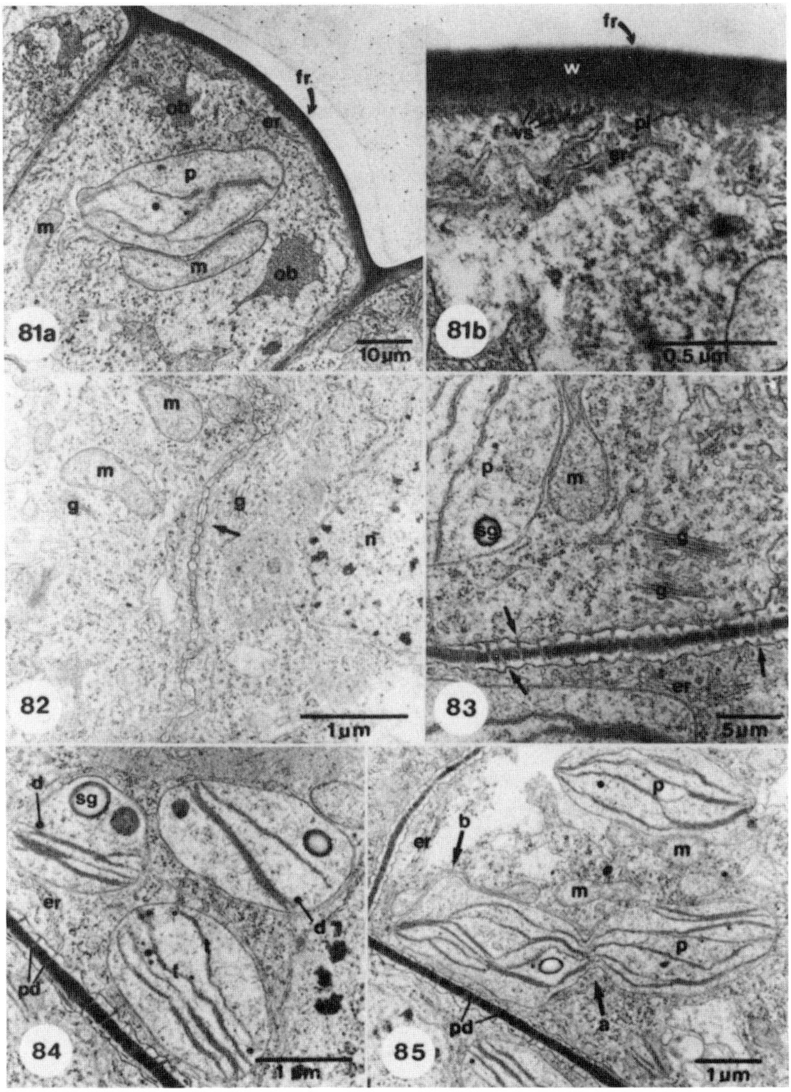
81a
fr.
ob
er
p
m
m
ob
10µm
81b
fr
w
vs
pl
er
0.5 µm
82
m
m
g
g
n
1µm
83
p
m
sg
g
g
er
5µm
84
d
sg
er
d
f
f
pd
1 µm
85
p
er
b
m
m
p
a
pd
1µm

FIGURES 86-90. Fine structure of cells in the distal portions of Haplomitrium Nees shoots. -86-88. Progressively older merophytes of H. hookeri (Smith) Nees. Note the increasing organization of the lamellae of the young plastids into thylakoids with grana, the larger, more abundant starch grains in the chloroplasts, and the increased vacuolar volume and oil body abundance in the older merophytes. 86. Merophyte located ca. 18 µm posterior to the basal point of the apical cell. 87. Merophyte located ca. 32 µm posterior to the apical cell. 88. Merophyte located ca. 43 µm posterior to the apical cell. -89. Mature vegetative cell of H. blumii (Nees) Schust. -90. Mature chloroplasts of a shoot cell of H. blumii. Note the well organized thylakoids, grana stacks, and the large starch reserves.

Key: d, oil droplet; gr, granum; k, karyosome; m, mitochondrion; n, nucleus; nu, nucleolus; ob, oil body; p, plastid; sg, starch grain; v, vacuole. (Figs. 86-88: Lorbeer H56 [PR]; 89-90: Furuki 384 [ABSH].)

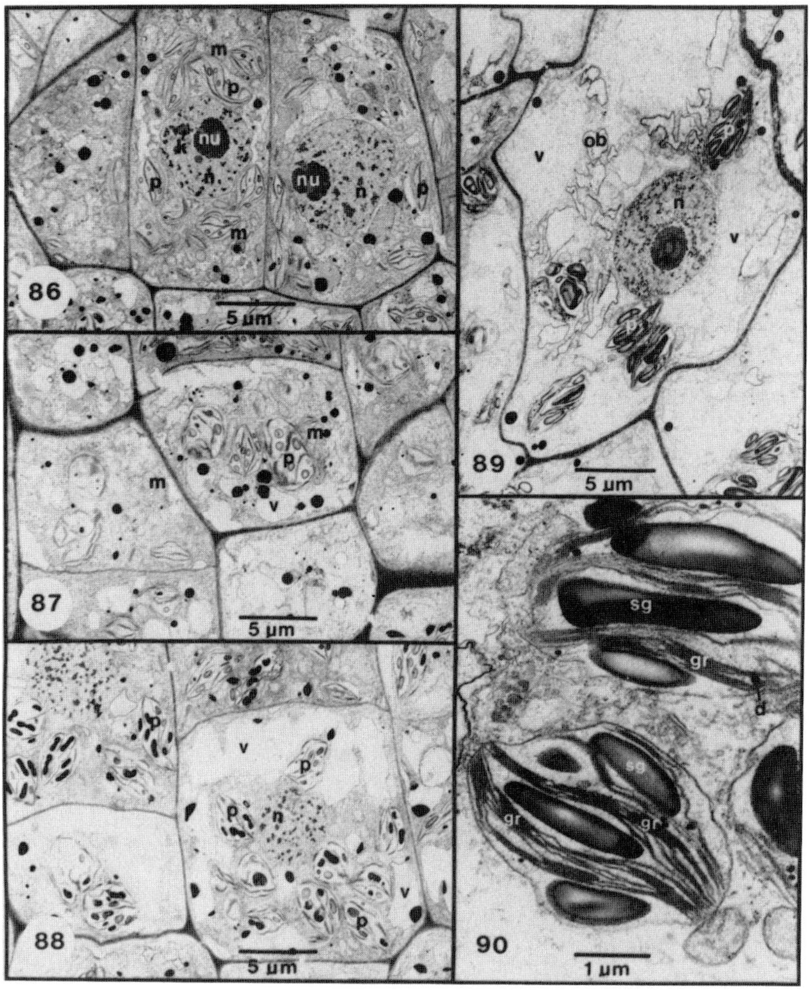
86
m
p
nu
n
nu
n
p
p
m
5 µm
87
m
p
v
m
5 µm
88
p
v
p
p
n
v
p
5 µm
89
v
ob
n
v
5 µm
90
sg
gr
d
sg
gr
gr
1 µm

FIGURES 91-95. Scanning electron micrographs of *Haplomitrium* Nees shoots. -91-94. *H. mnioides* (Lindb.) Nees. 91. Apical view of *a* very young shoot showing scales formed from irregular divisions in the basal cells of the slime papillae. 92. Scales near the upturned region of the rhizome. 93. Apical view of an unelongated branch. Note the apical scale. 94. Lateral view of a young shoot which has branched from the rhizome and which bears a single leaf. -95. Stolon of *H. blumii* (Nees) Schust. bearing light-induced branches and leaves.

Key: b, branch; le, leaf; r, rhizome; sa, scale; sn, stolon; sp, slime papilla. (Figs. 91-94: *Doei s.n.* [ABSH]; 95: *Furuki 384* [ABSH].)

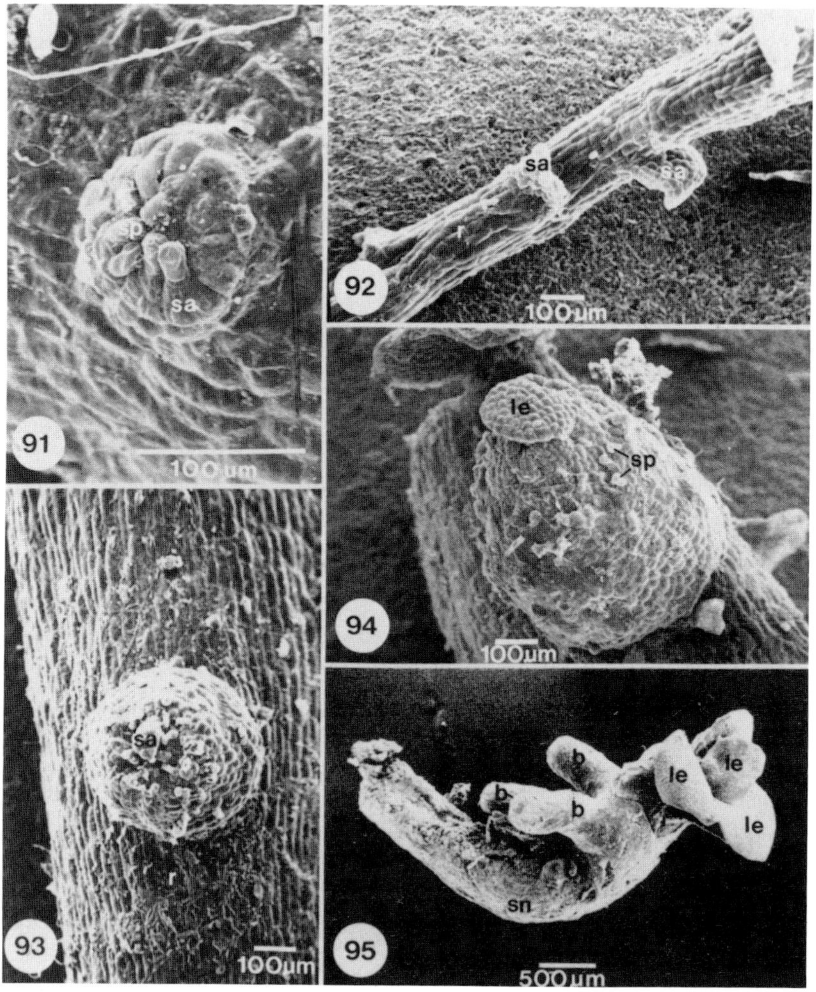
sp
sa
91
100 µm
sa
sa
92
100 µm
93
sa
r
100µm
le
sp
94
100µm
b
b
b
le
le
le
sn
95
500 µm

FIGURES 96-102. Light micrographs of sections made through the midregions of *Haplomitrium* Nees shoots showing the central conducting strands. -96-97. *H. intermedium* Berrie. 96. Leafy shoot transection. 97. Stolon transection. -98. Leafy shoot transection of *H. monoicum* Engel. -99-100. Leafy shoot transections of *H. chilensis* showing the variations in the diameters of the central strands. 99. Female plant. 100. Male plant. -101-102. *H. mnioides* (Lindb.) Schust. leafy shoots. 101. Transection. 102. Longitudinal section.

Key: c, cortex; ep, epidermis; h, central strand of hydroids; le, leaf. (Figs. 96-97: *McKay 3951* [MUCV]; 98: *Crosby 14165* [F]; 99-100: *Engel 11513B* [F]; 101-102: *Doei s.n.* [ABSH].)

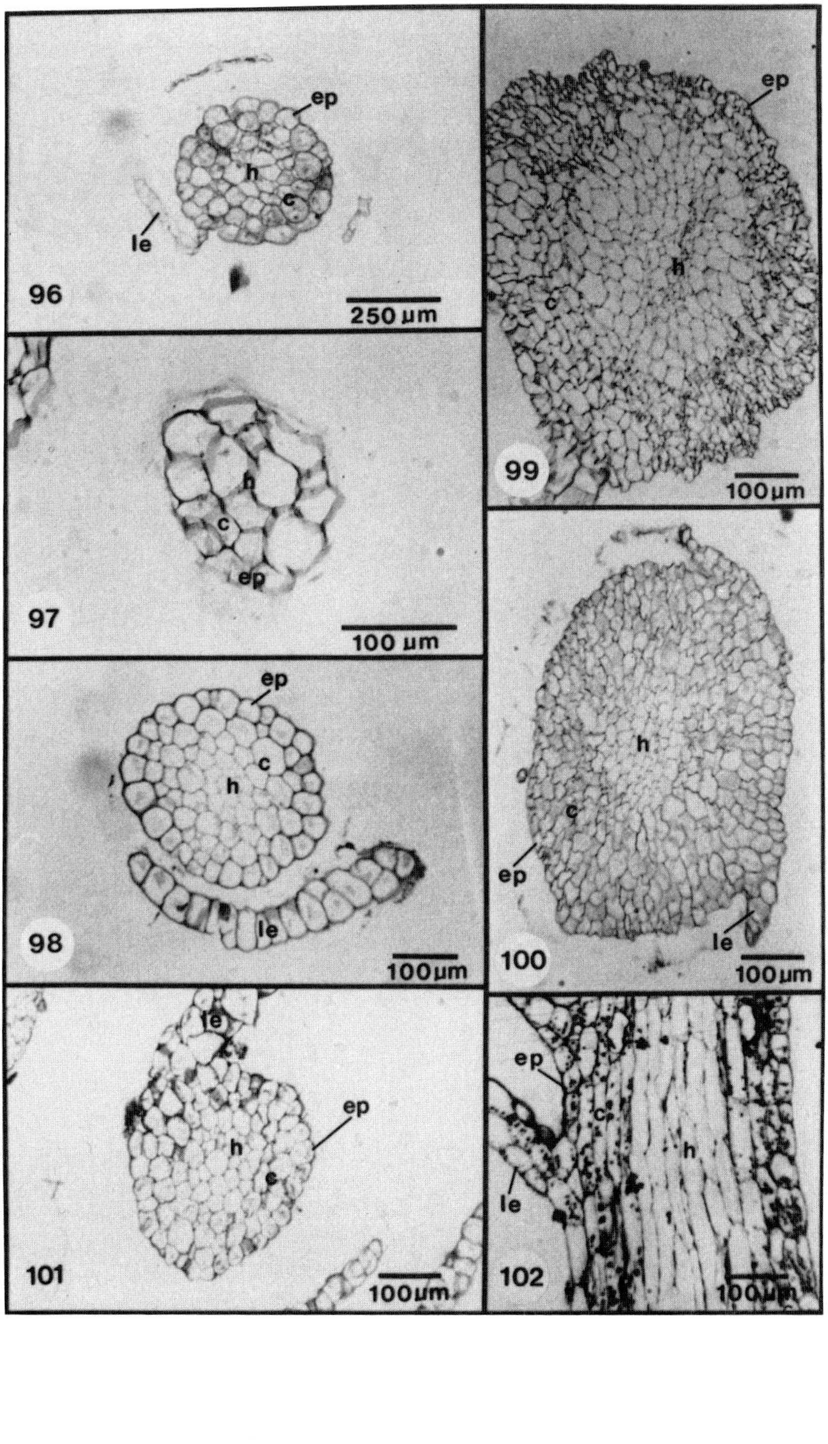
96
ep
h
c
le
250 µm
97
h
c
ep
100 µm
98
ep
c
h
le
100µm
99
ep
h
c
100µm
100
h
c
ep
le
100µm
101
le
ep
h
c
100µm
102
ep
c
h
le
100µm

FIGURES 103-106. *Haplomitrium* Nees shoots. -103-104. Scanning electron micrographs of *H. hookeri* (Smith) Nees shoots in transection. 103. Leafy shoot. 104. Stolon. -105-106. Transmission electron micrographs. 105. Leafy shoot epidermal cells of *H. hookeri*. 106. Stolon epidermal cell of *H. mnioides* (Lindb.) Schust.

Key: c, cortex; ep, epidermis; er, endoplasmic reticulum; h, central strand of hydroids; k, karyosome; m, mitochondrion; n, nucleus; nu, nucleolus; ob, oil body; p, plastid; v, vacuole. (Figs. 103-105: *Lorbeer H56* [PR]; 106: *Doei s.n.* [ABSH].)

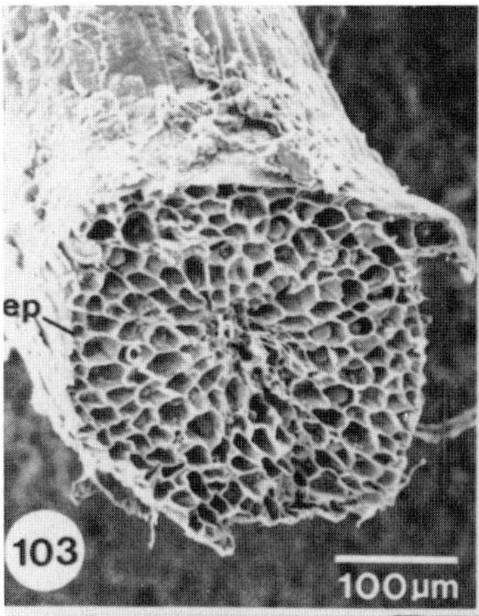
ep
103
100 µm

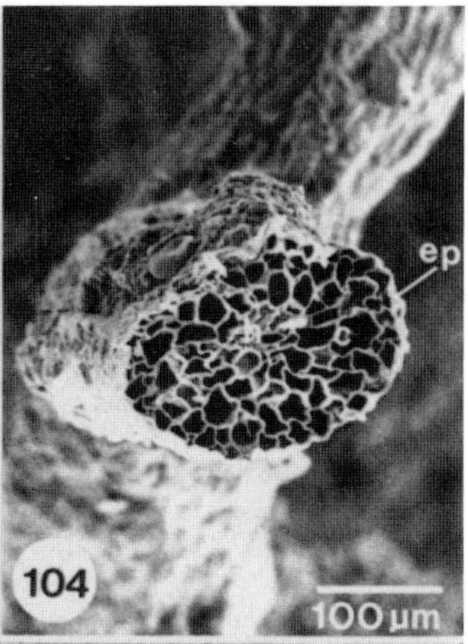
ep
104
100 µm

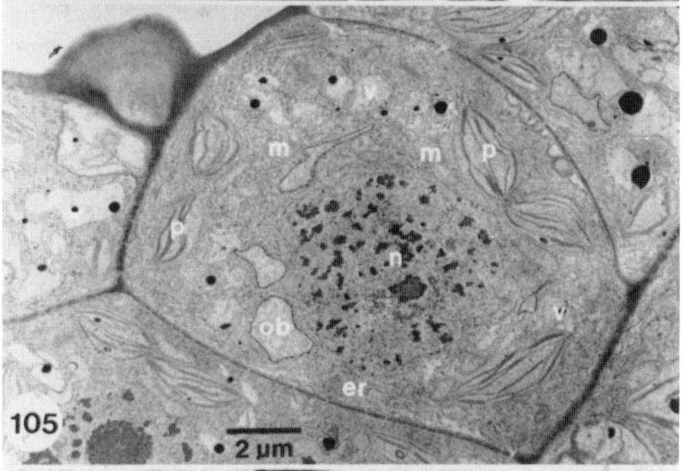
v
m
m
p
p
n
ob
v
er
105
2 µm

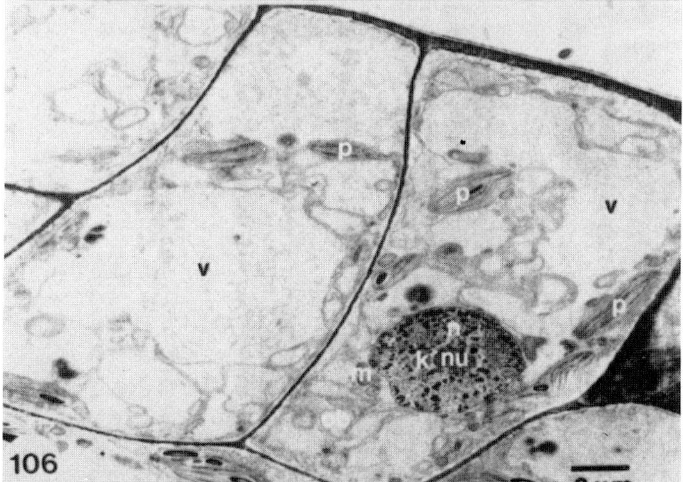
p
p
v
v
p
n
k nu
m
106
2 µm

FIGURES 107-110. Fine structure of shoot slime papillae and cortex cells of *Haplomitrium* Nees. -107-108. Slime cells. 107. Mucilage cell of a leafy shoot of *H. hookeri* (Smith) Nees. 108a-b. Slime cell of *H. mnioides* (Lindb.) Nees stolon. 108a. Near longitudinal section. Note the nucleus, mitochondria, ER, and degenerating organelles. 108b. Distal portion of a cell. Note the homogeneous composition of the cell wall, large vacuoles, tightly packed ER lamellae, and degenerating organelles. -109-110. Cortex cells of a mature leafy shoot. 109. *H. hookeri*. 110. *H. mnioides*.

Key: do, degenerating organelle; er, endoplasmic reticulum; g, dictyosome; l, lipid; m, mitochondrion; n, nucleus; nu, nucleolus; ob, oil body; mu, mucilage; sg, starch grain; sp, slime papilla; p, chloroplast; v, vacuole; w, cell wall. (Figs. 107, 109: *Lorbeer H56* [PR]; 108, 110: *Doei s.n.* [ABSH].)

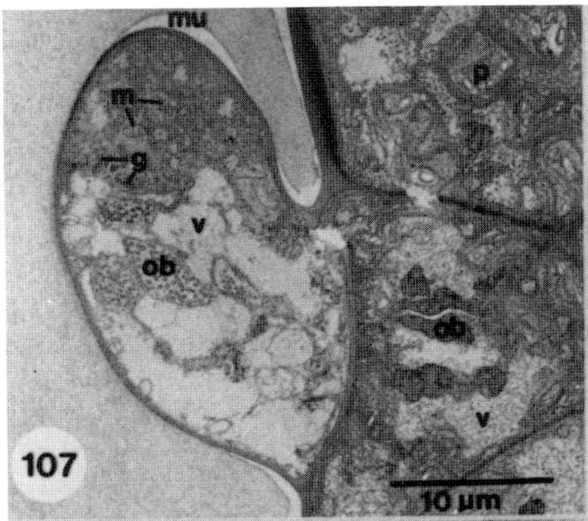
107
mu
p
m
g
v
ob
ob
v
10 µm

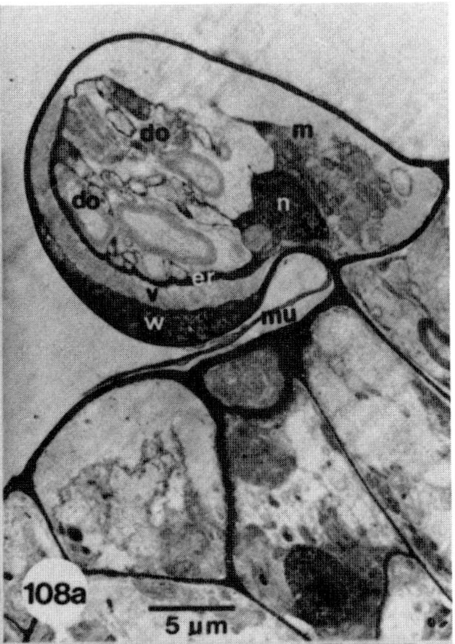
108a
do
m
do
n
er
v
w
mu
5 µm

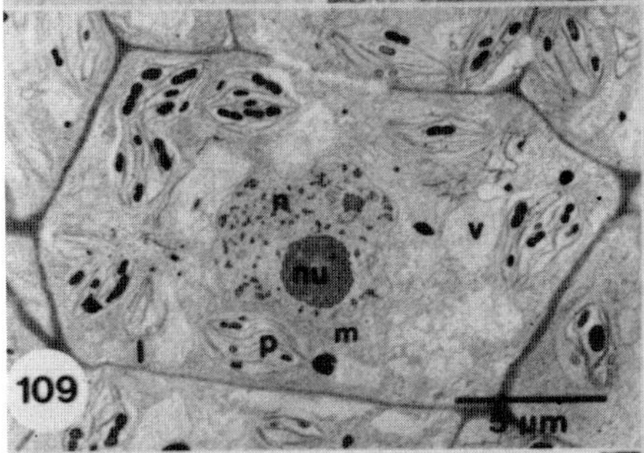
109
nu
v
p
m
l

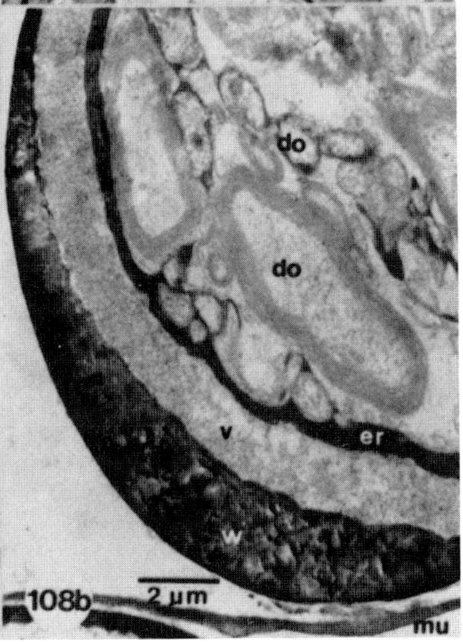
108b
do
do
v
er
w
2 µm
mu

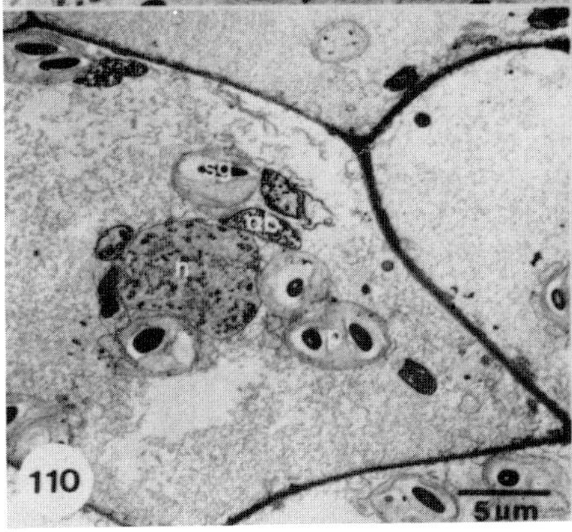
110
5 µm

FIGURES 111-114. Transmission electron micrographs of the early stages of hydroid differentiation. -111-112. Longitudinal sections through cells of the central strand of *H. gibbsiae* (Steph.) Schust. 111. Young elongate cell which still contains cytoplasm with centrally positioned organelles. 112. Centrally clustered organelles of a young cell. Note the nucleus, oil bodies, and degenerating plastids. -113-114. Longitudinal sections through the cells of the central strand of *H. mnioides* (Lindb.) Schust. 113. Cell with a large central vacuole and displaced organelles. 114. Organelles between the tonoplast of the large central vacuole and the plasmalemma.

Key: k, karyosome; l, lipid droplet; m, mitochondrion; n, nucleus; nu, nucleolus; ob, oil body; p, plastid; sg, starch grain; v, vacuole. (Figs. 111-112: *Fife 8453* [ABSH]; 113-114: *Doei s.n.* [ABSH].)

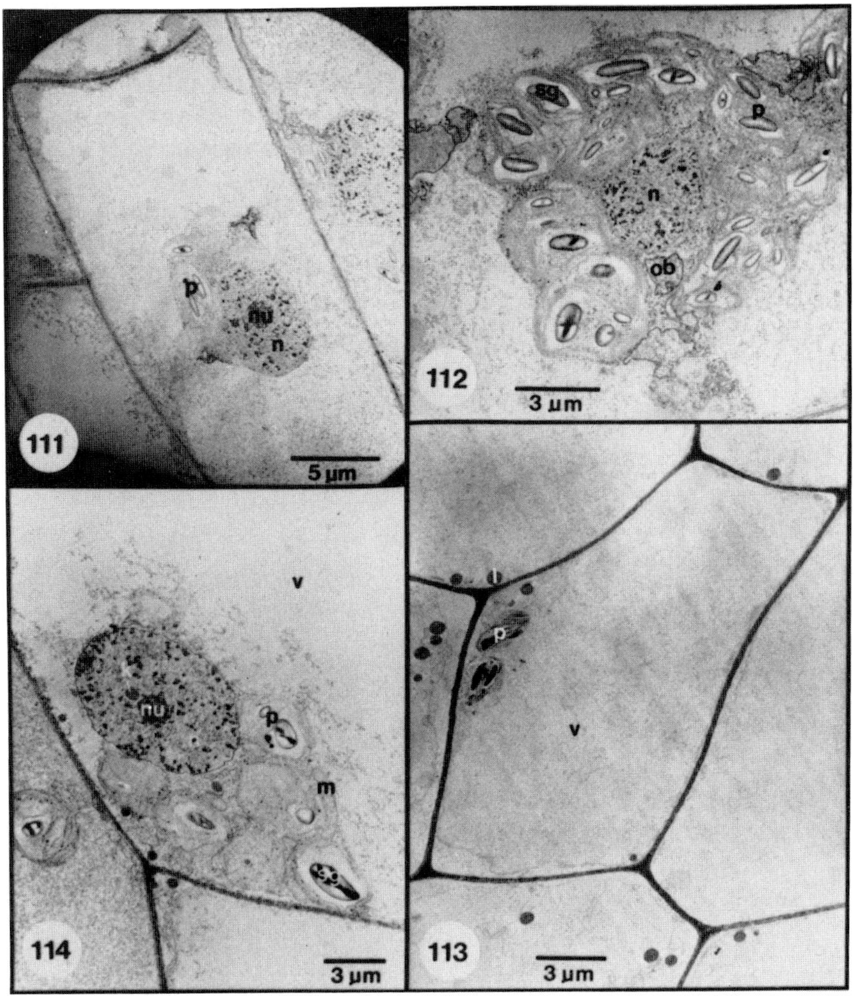

111
p
nu
n
5 µm
112
sg
p
n
ob
3 µm
114
v
nu
p
m
3 µm
113
p
v
3 µm

FIGURES 115-119. Transmission electron micrographs showing hydroid ontogeny in _Haplomitrium mnioides_ (Lindb.) Schust. Longitudinal sections. -115. End wall of a young hydroid with electron dense regions (*) surrounding each plasmodesma which is being hydrolyzed. Note the associated ER. -116. End wall with depressions (arrows) formed from the removal of wall material from around the plasmodesmata. -117. Plasmodesma of an end wall from which the wall material has been removed to expose the desmotubule. -118. Portion of an end wall showing one plasmodesma with an exposed, isolated desmotubule. -119. Portion of an end wall of an older hydroid with widened, empty, plasmodesmatal channels.

Key: dt, desmotubule; er, endoplasmic reticulum; pc, empty plasmodesmatal channel; vs, vesicle. (Figs. 115-119: _Furuki 7408_ [ABSH].)

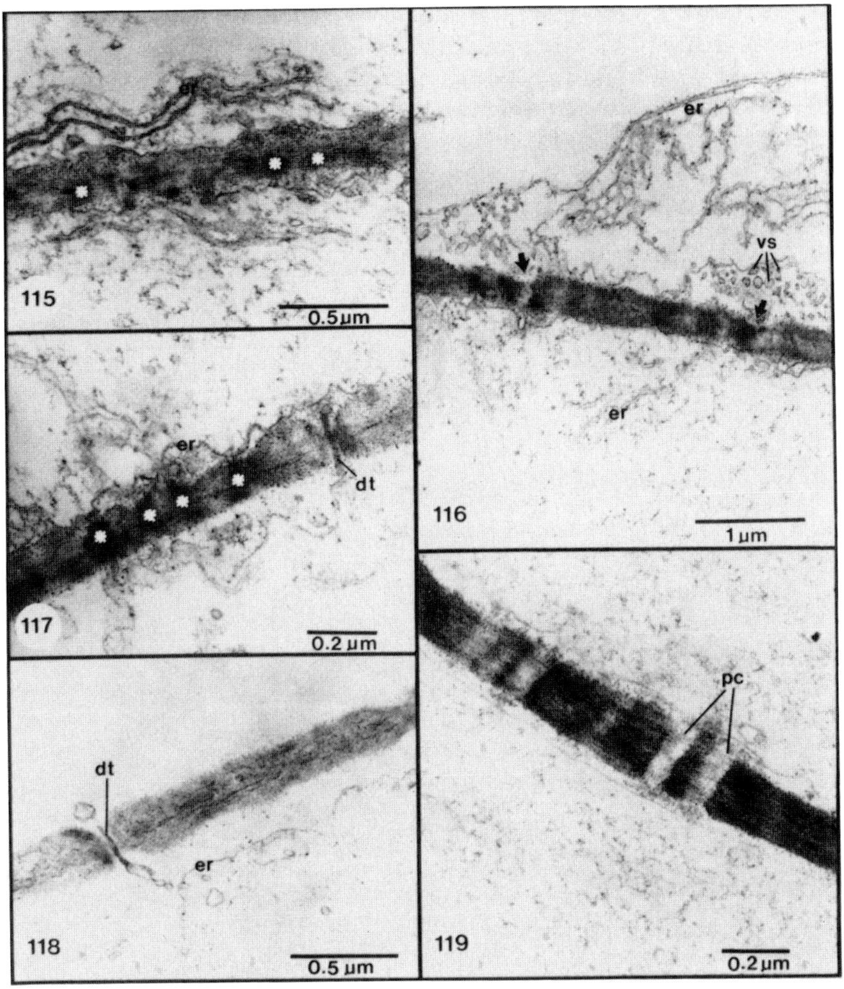

er
115
0.5 µm
er
vs
er
116
1 µm
er
dt
117
0.2 µm
dt
er
118
0.5 µm
pc
119
0.2 µm

FIGURES 120-126. Mature hydroids of *Haplomitrium* Nees. -120-123. Transmission electron micrographs of *H. mnioides* (Lindb.) Nees. Longitudinal sections. 120. Portion of an end wall with pits. 121-122. Pitted lateral walls. 123. Invaginated pitted end wall. -124-126. Scanning electron micrographs. 124. Pitted lateral hydroid wall of *H. intermedium* Berrie. 125-126. Hydroid pits of *H. blumii* (Nees) Schust. 125. Lateral wall. 126. End wall.

Key: pt, pit. (Figs. 120-123: *Furuki 7408* [ABSH]; 124: *Berrie 2/62/1* [SYD]; 125-126: *Campbell s.n.* [NY].)

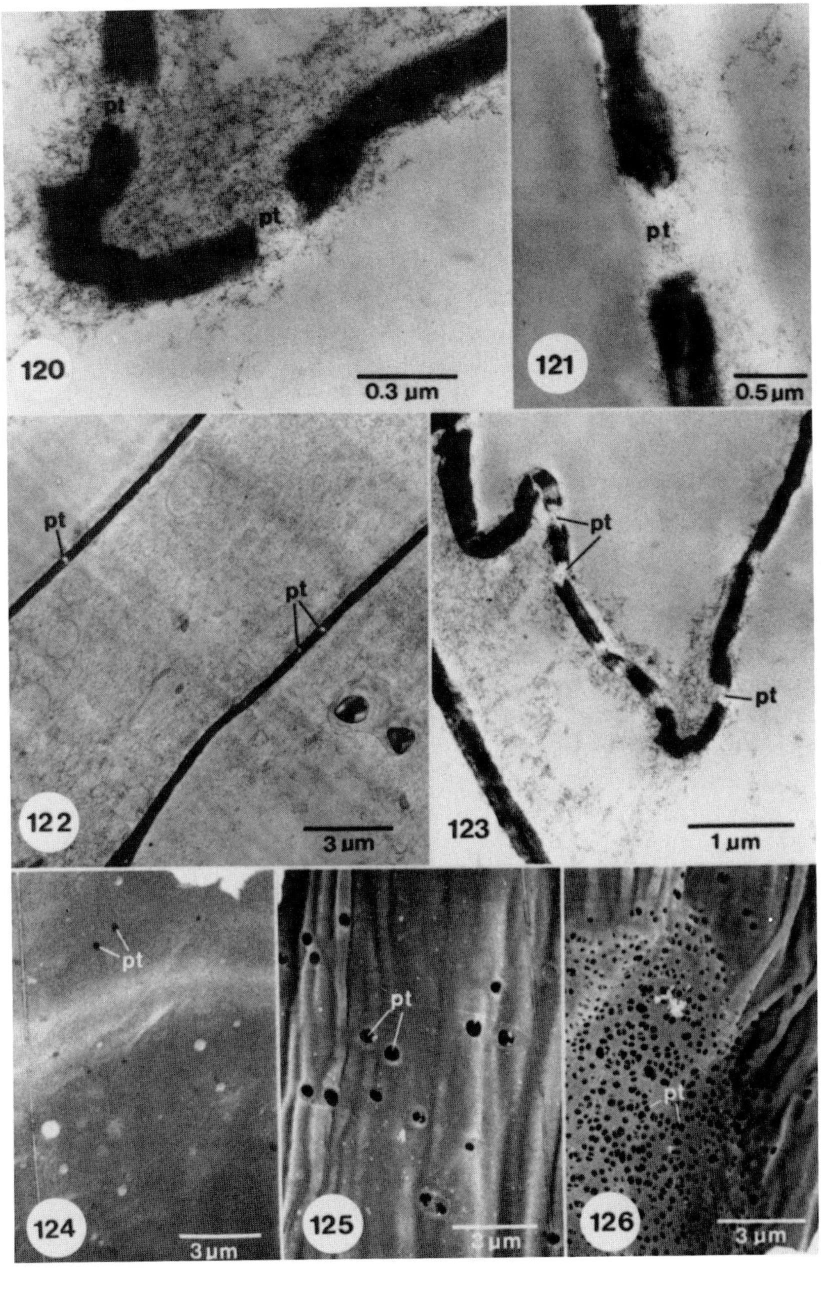

120
pt
pt
0.3 µm
121
pt
0.5 µm
122
pt
pt
3 µm
123
pt
pt
1 µm
124
pt
3 µm
125
pt
3 µm
126
pt
3 µm

FIGURES 127-134. Scanning electron micrographs showing the stages of leaf ontogeny in Haplomitrium mnioides (Lindb.) Schust. -127. Triangular leaf formed by segmentation from a depressed-ovoid apical cell and longitudinal divisions in the newly formed segments. -128. Habit of a leafy apex showing a mature entire leaf. Note the apical cell with two cutting faces at the tip of arrow a, the field of early apical growth (arrows b_1 and b_2), and the region of basal growth composed of straight rows of cells. -129. Surface view of a reduced dorsal leaf showing the pattern of growth. -130. Male bract with a pluristratose basal field (indicated by the arrows). -131. Apex of a leaf from vermiculite culture with each portion of the divided apical cell beginning to segment. -132. Lateral view of a bifid leaf showing three segmentations. -133. Segmentation from a newly activated apical cell on the margin of a leaf forming a lateral projection. -134. Female bracts with lateral projections (indicated by the arrows). Note that the straight rows of cells resulting from the regular pattern of divisions in the basal cells are undisturbed by the apical activity of each lateral outgrowth (depicted by arrow).

Key: dx, daughter apical cell; li, leaf apical cell; px, apical cell of lateral projection of incised leaf; sp, slime papilla. (Figs. 127-134: Doei s.n. [ABSH].) (Photos used with permission from The Bryologist.)

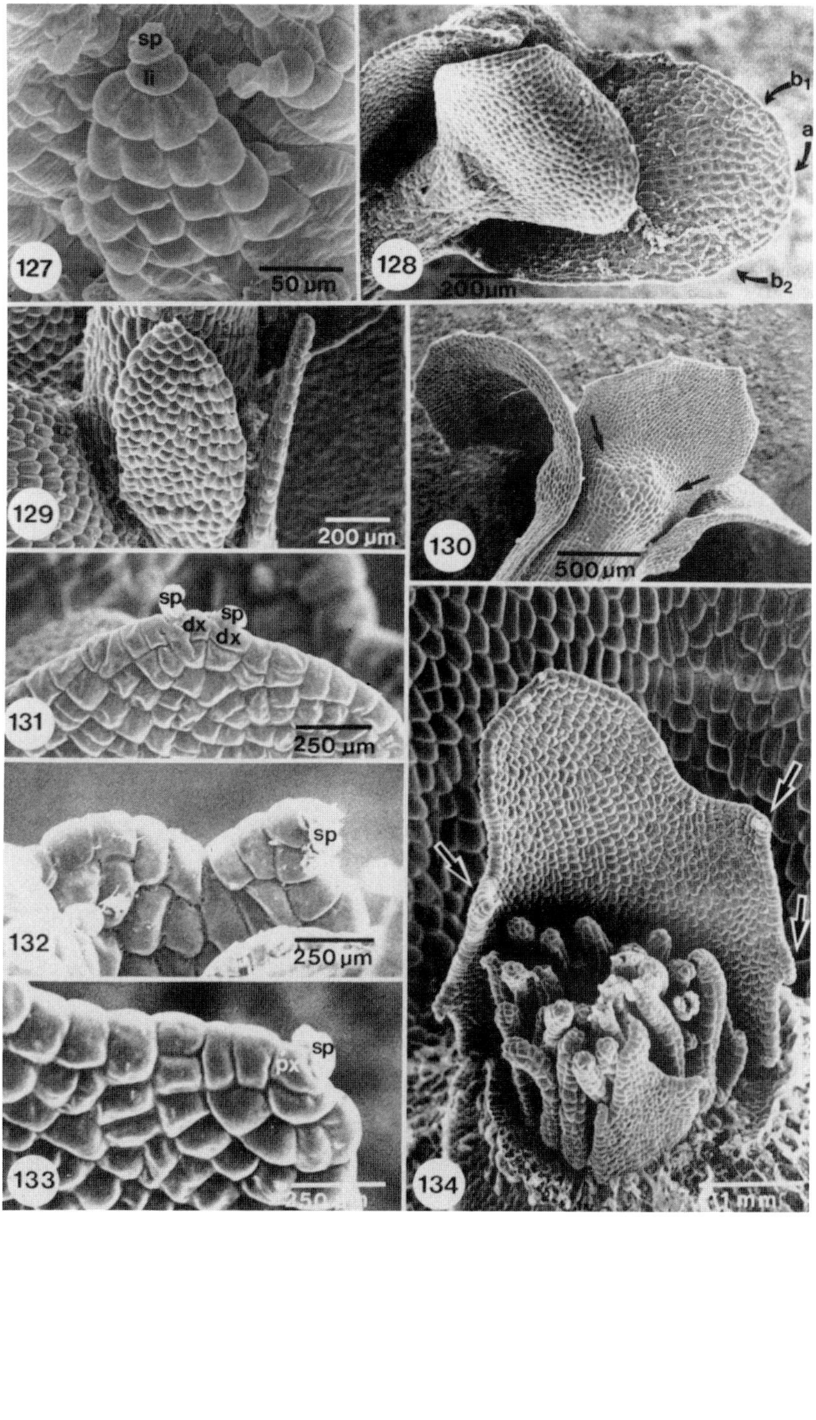
127
sp
li
50 µm
128
b1
a
b2
200µm
129
200 µm
130
500 µm
131
sp
dx
sp
dx
250 µm
132
sp
250 µm
133
sp
px
134
1 mm

FIGURES 135-138. Fine structure of *Haplomitrium* Nees leaves. -135. Longitudinal section made through a young triangular-shaped leaf of *H. hookeri* (Smith) Nees. -136. Longitudinal section through a slightly older *H. mnioides* (Lindb.) Schust. leaf. -137. Mature leaf cells of *H. gibbsiae* (Steph.) Schust. -138. Chloroplasts of a mature *H. hookeri* leaf cell. Note the distinct thylakoids, grana, and large starch grains.

Key: gr, granum; l, lipid; m, mitochondrion; n, nucleus; nu, nucleolus; ob, oil body; p, plastid; sg, starch grain; t, thylakoid; v, vacuole. (Figs. 135, 138: *Lorbeer H56* [PR]; 136: *Doei s.n.* [ABSH]; 137: *Fife 8453* [ABSH].)

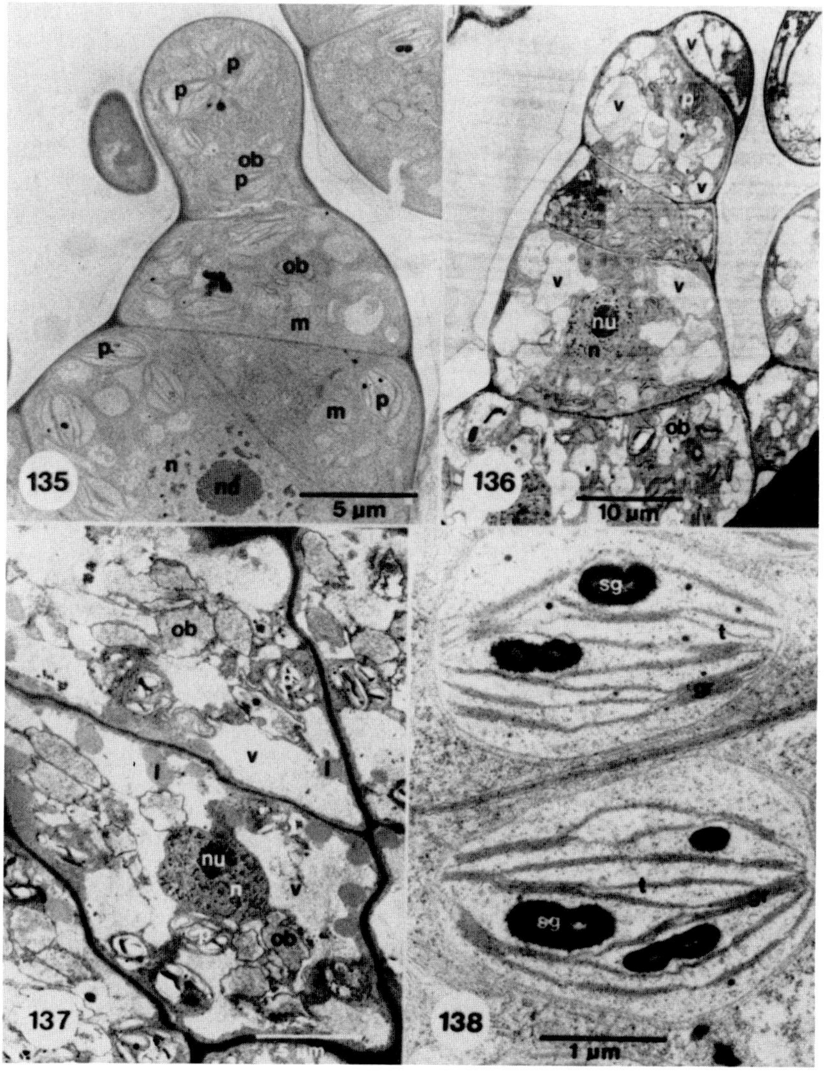
135
p
p
p
ob
p
ob
m
p
m
p
n
nd
5 µm
136
v
v
p
v
v
v
nu
n
ob
10 µm
137
ob
v
l
l
nu
n
v
ob
138
sg
t
t
sg
1 µm

FIGURES 139-147. Light micrographs showing the oil bodies in the cells of *Haplomitrium* Nees and *Steereomitrium* E. O. Campb. -139-140. *H. gibbsiae* (Steph.) Schust. 139. Leafy shoot. 140. Stolon. -141. Slime cell of *H. hookeri* (Smith) Nees. -142. Immature seta of *H. mnioides* (Lindb.) Schust. -143-144. *H. intermedium* Berrie. 143. Young leaf. 144. Older leaf. Note the spherules contained in the oil bodies. -145. Young leaf of *H. hookeri*. -146. Young leaf of *S. minutum* E. O. Campb. -147. Young leaf of *H. gibbsiae*.

Key: ob, oil body; sl, spherule. (Figs. 139, 140: *Fife 8453* [ABSH]; 141, 145: *Lorbeer H56* [PR]; 142: *Doei s.n.* [ABSH]; 143-144: *Selkirk s.n.* [ABSH]; 146: *Campbell s.n.* [ABSH].)

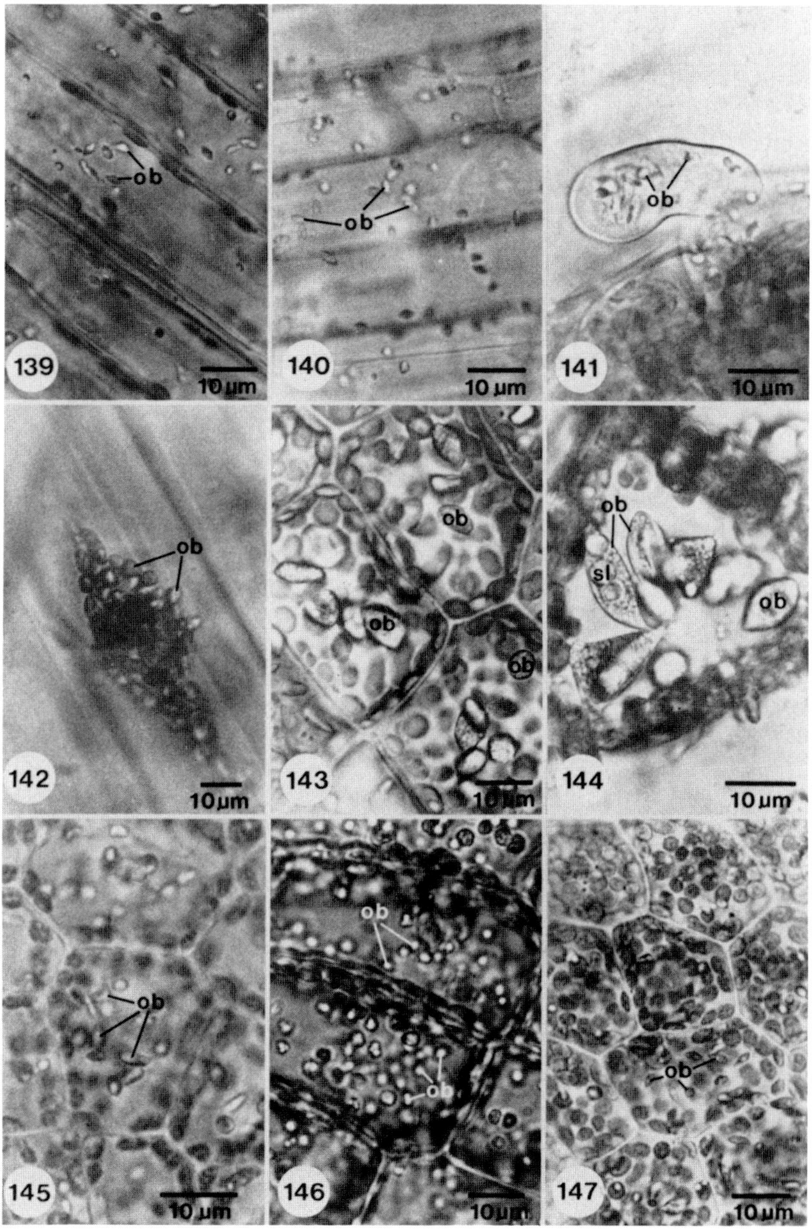
ob
139
10 µm
ob
140
10 µm
ob
141
10 µm
ob
142
10 µm
ob
ob
ob
143
ob
sl
ob
144
10 µm
ob
145
10 µm
ob
ob
146
ob
147
10 µm

FIGURES 148-154. Oil bodies from the leaf cells of *Haplomitrium* Nees. -148-151. Light micrographs. 148-149. *H. gibbsiae* (Steph.) Schust. 148. Mature cells. 149. Old, dying cells. Note the botryoidal texture of the oil bodies. 150. Young leaf of *H. mnioides* (Lindb.) Schust. 151. Young leaf of *H. blumii* (Nees) Schust. -152-154. Transmission electron micrographs. 152-153. *H. gibbsiae*. 152. Leaf cell. Note the electron dense oil body membranes. 153. Portion of an oil body showing the inner oil droplets, stroma, membrane, and peripheral ER. 154. Portion of *H. blumii* oil body membrane with vesicles lining the inner surface of the membrane.

Key: er, endoplasmic reticulum; me, oil body membrane; ob, oil body; og, inner oil globule; p, plastid; sm, stroma; vs, vesicle. (Figs. 148, 149, 152, 153: *Fife 8453* [ABSH]; 150: *Doei s.n.* [ABSH]; 151, 154: *Furuki 384* [ABSH].)

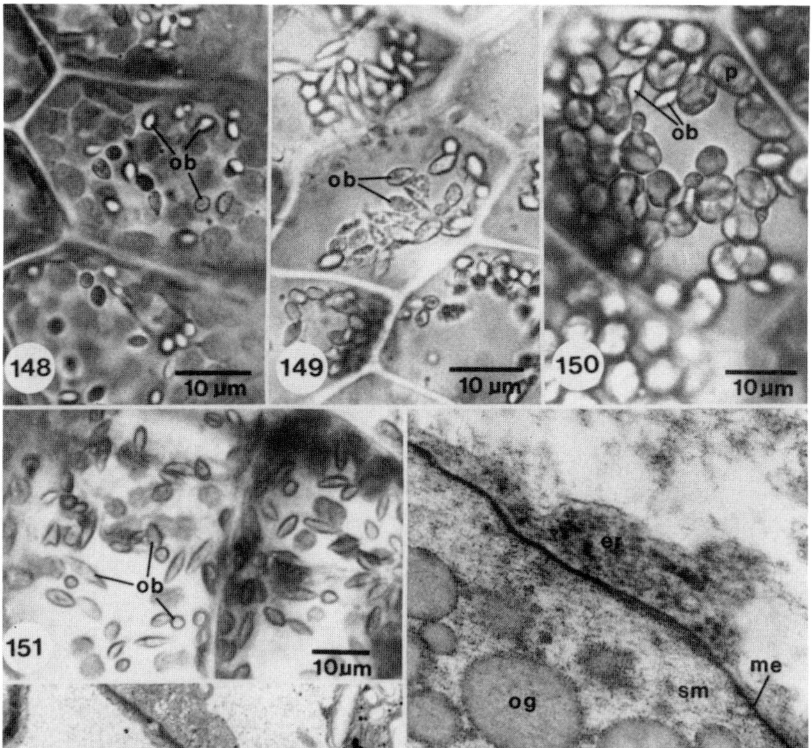
148
ob
10 µm
149
ob
10 µm
150
p
ob
10 µm
151
ob
10µm
152
p
ob
p
ob
me
1µm
153
er
me
og
sm
0.1µm
154
vs
vs
vs
me
sm
0.1µm

FIGURES 155-159. Transmission electron micrographs of the oil bodies of _Haplomitrium_ Nees. -155. _H. intermedium_ Berrie. -156. _H. hookeri_ (Smith) Nees. -157. _H. gibbsiae_ (Steph.) Schust. -158. _H. mnioides_ (Lindb.) Schust. -159. _H. blumii_ (Nees) Schust.

Key: er, endoplasmic reticulum; me, membrane; og, inner oil globule; p, chloroplast; sm, stroma. (Fig. 155: _Selkirk s.n._ [ABSH]; 156: _Lorbeer H56_ [PR]; 157: _Fife 8453_ [ABSH]; 158: _Doei s.n._ [ABSH]; 159: _Furuki 384_ [ABSH].)

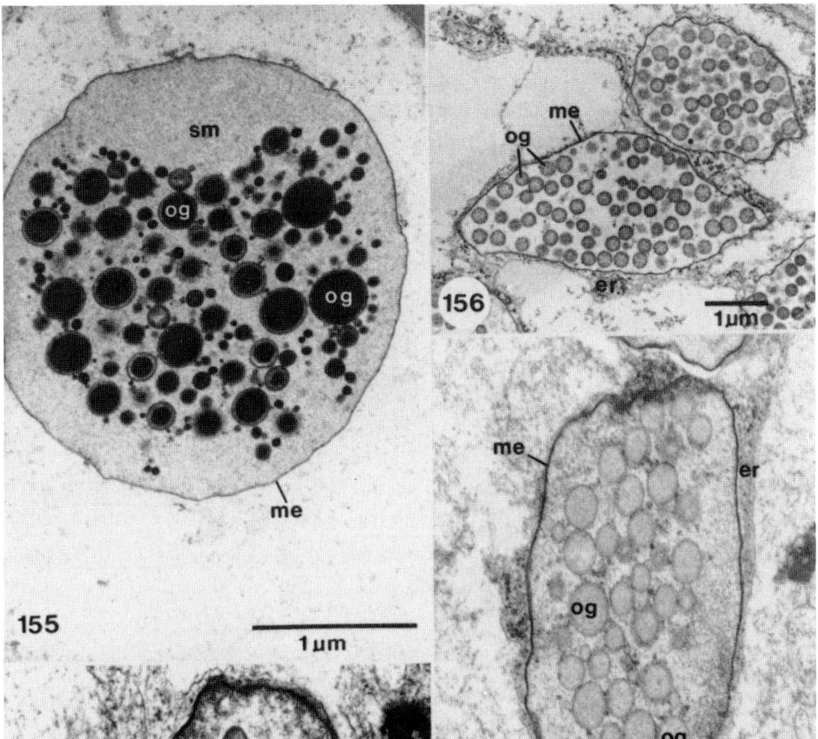
155
sm
og
og
me
1 µm
156
me
og
er
1µm
157
me
er
og
og
1µm
158
og
er
og
me
0.5 µm
159
me
p
og
sm
er
0.5µm

FIGURES 160-165. Scanning electron micrographs of branching in *Haplomitrium* Nees. -160. Branched rhizomatous system of *H. hookeri* (Smith) Nees. -161-162. *H. mnioides* (Lindb.) Schust. 161. Young, supra-axillary branch ("bud") formed at a 90° angle from the main shoot. 162. Axillary branch. -163. Young branches formed at 90° angles from a leafy shoot of *H. blumii* (Nees) Schust. Note branch b_1 is axillary while branch b_2 is not directly associated with a leaf. -164a-b. Leafy shoot of *H. intermedium* Berrie with terminal, axillary branches. 164a. Distal portion of a shoot. 164b. Axillary origin of a branch. -165. Elongated, leafy, terminal branch of *H. hookeri* which lends a dichotomous appearance to the plant.

Key: b, branch; le, leaf. (Fig. 160: *Worley & Rawson 12542* [H]; 161: *Mizutani s.n.* [ABSH]; 162: *Doei s.n.* [ABSH]; 163: *Furuki 384* [ABSH]; 164: *Selkirk s.n.* [ABSH]; 165: *Lorbeer H56* [PR].)

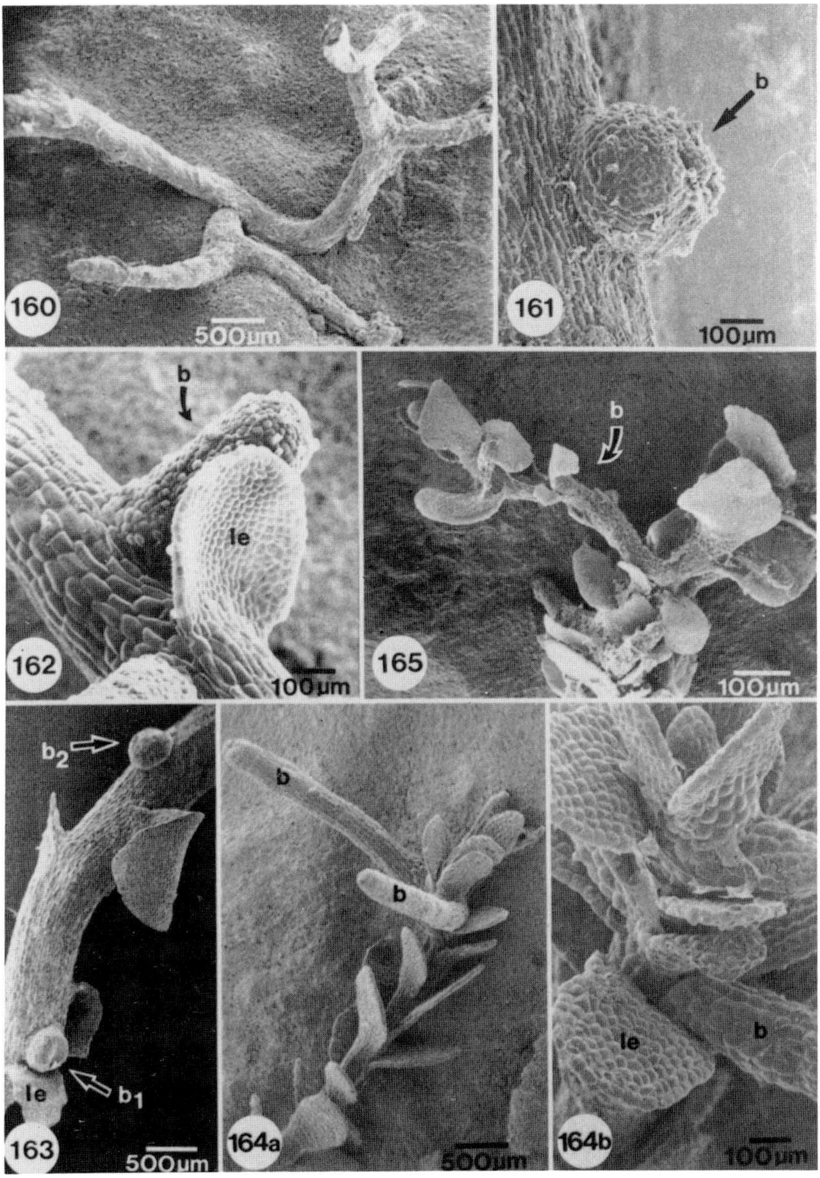
160
500µm
161
b
100µm
162
b
le
100µm
165
b
100µm
163
b_2
b_1
le
500µm
164a
b
b
500µm
164b
le
b
100µm

FIGURES 166-170. Illustrations of branching in *Haplomitrium* Nees. -166. Habit drawing of a portion of a *H. hookeri* (Smith) Nees plant showing a leafy branch emerging at a 45° angle from the main shoot. -167. Longitudinal section through the apical portion of a leafy *H. hookeri* shoot showing a branch initial and a dormant tetrahedral branch apical cell. Note that they are located in the second epidermal cell basiscopic to the leaf above. -168. Transection through the apex of a leafy *H. mnioides* (Lindb.) Schust. shoot showing the branch apical cell in close proximity to the tetrahedral apical cell of the main shoot. The shoot is homodromous. -169. Longitudinal section through a leafy shoot of *H. gibbsiae* (Steph.) Schust. showing a branch apical cell which has been displaced into the axil of the leaf below. -170. Longitudinal section through a leafy shoot of *H. hookeri* showing a branch apical cell located midway between the leaves.

Key: b, branch; bi, branch initial; bx, branch apical cell; le, leaf; sp, slime papilla; x, apical cell. (Fig. 166: *Schofield & Spence 83691* [F]; 167, 170: *Lorbeer H56* [PR]; 168: *Doei s.n.* [ABSH]; 169: *Fife 8453* [ABSH].)

166

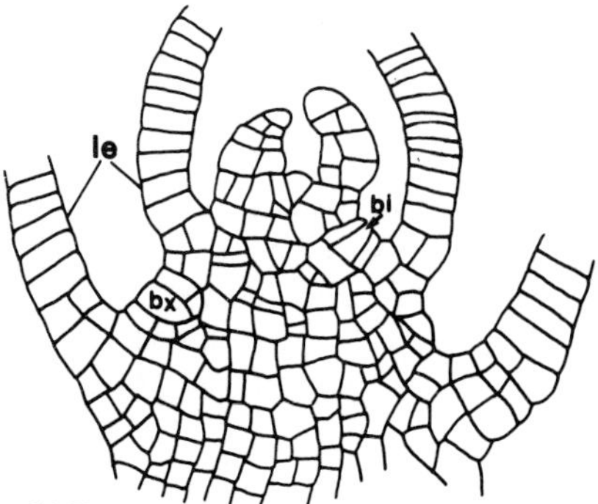

167

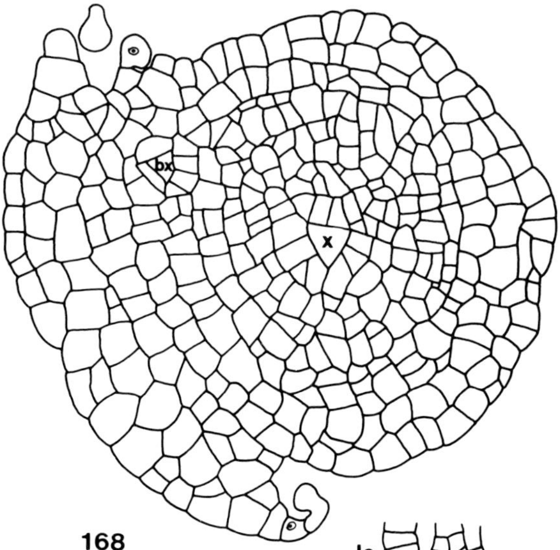

168

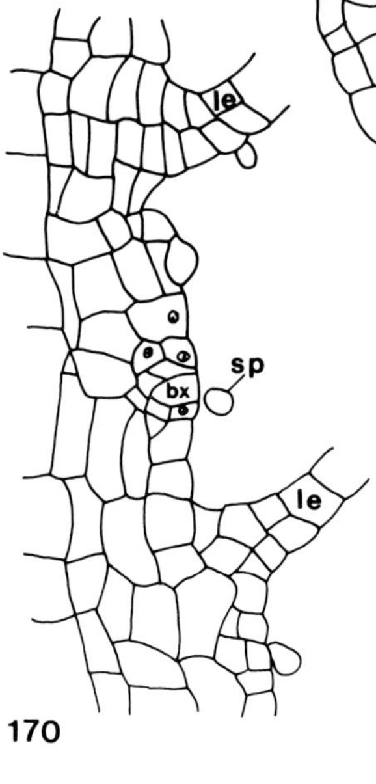

170

50um

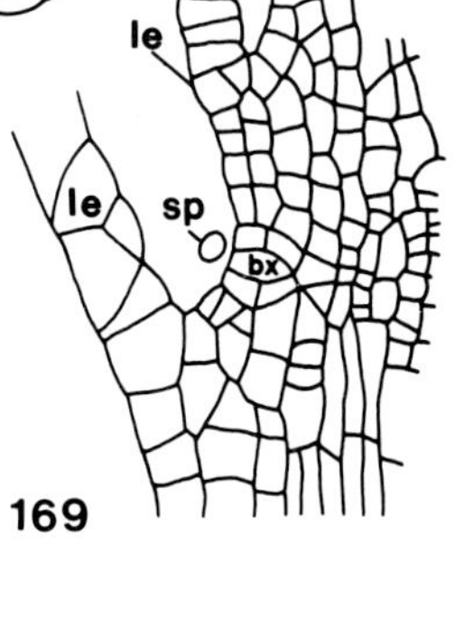

169

FIGURES 171-173. Illustrations of branching in _Haplomitrium_ Nees. -171. Longitudinal section grazing the epidermis of an aerial shoot of _H. mnioides_ (Lindb.) Schust. showing a very young branch in transection. Note the tetrahedral branch apical cell. -172-173. Sections showing the discontinuity between the cells of the main axis and those of the growing branch. 172. Transection through a leafy shoot of _H. hookeri_ (Smith) Nees showing a lateral branch emerging from the main stem. 173. Near median longitudinal section through a portion of the rhizome of _H. mnioides_ bearing an erect branch.

Key: b, branch; bx, branch apical cell; ls, leafy shoot; r, rhizome; sp, slime papilla. (Figs. 171, 173: _Doei s.n._ [ABSH]; 172: _Lorbeer H56_ [PR].)

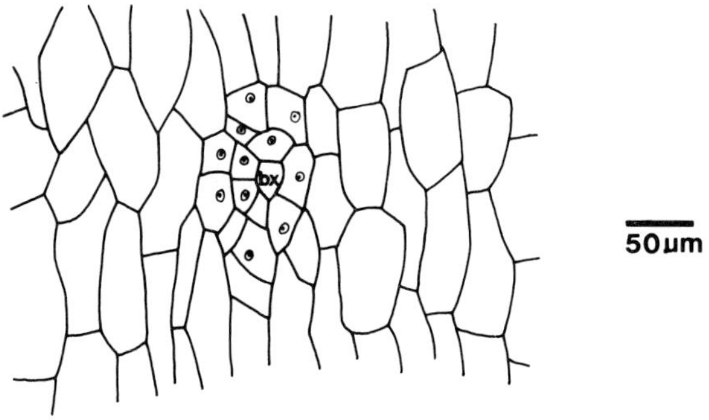

171

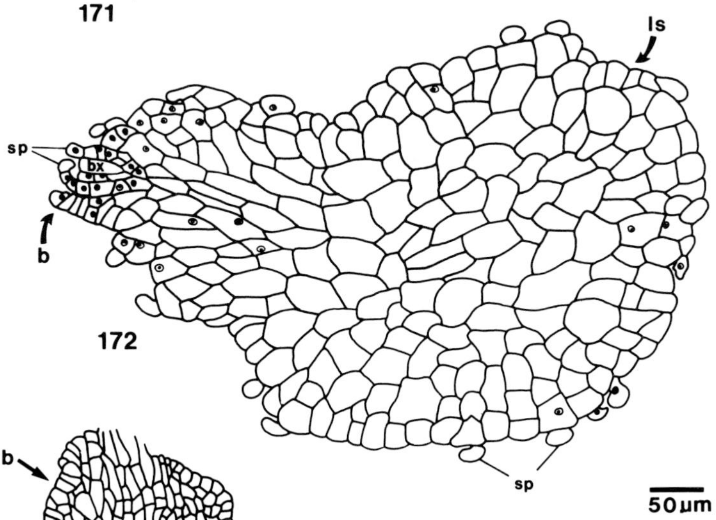

172

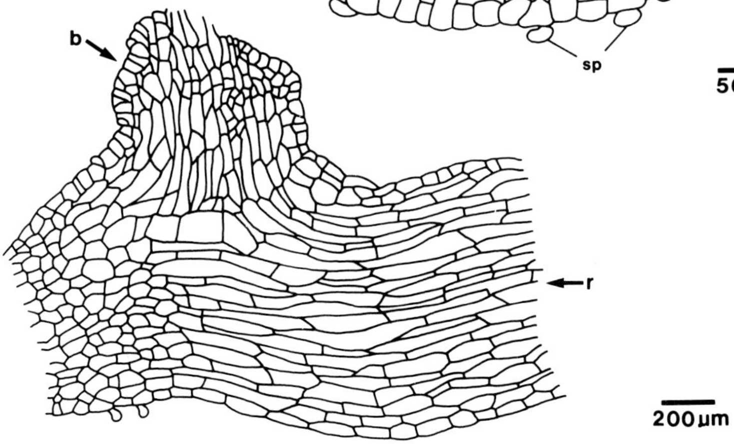

173

FIGURES 174-178. Reproductive gametophytic features of *Haplomitrium hookeri* (Smith) Nees. -174-176. Male plants. 174a-b. Light micrographs. 174a. Shoot with antheridia in a loose apical cluster and in successive series of unmodified bracts posterior to the apex. 174b. Axillary antheridia midway along the shoot. 175-176. Scanning electron micrographs. 175. Apical cluster of antheridia. A perigonial bract has been removed to expose the gametangia. 176a-b. Shoot bearing scattered antheridia. 176a. Distal quarter of the shoot in which the antheridia are in the axils of unmodified bracts (arrow). Note that the surrounding leaves have been removed to expose the antheridia. 176b. Axillary antheridia. -177-178. Female plants. 177. Shoot bearing archegonia in a loose terminal cluster. 178. Apex of a shoot from which a bract has been removed to expose the archegonial cluster.

Key: an, antheridium; ar, archegonium; br, bract. (Figs. 174-176: *Worley & Rawson 12542* [H]; 177: *Hooker s. n.* [neotype; BM]; 178: *Schofield & Spence 83691* [F].)

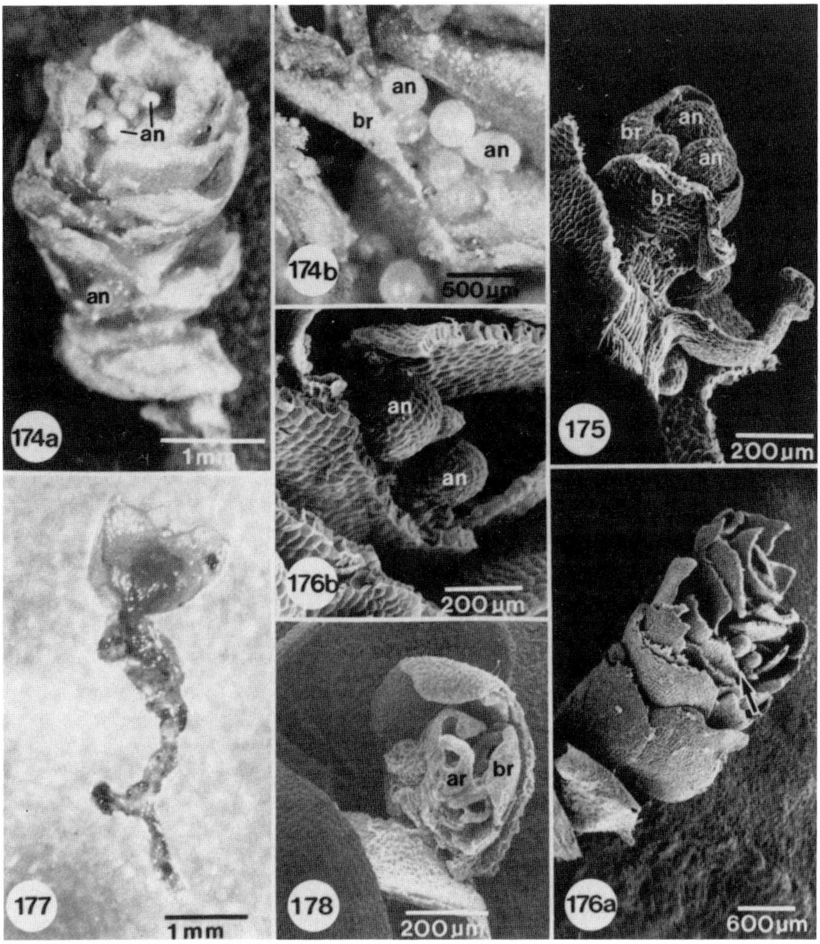
an
an
174a
1 mm
an
br
an
174b
500 µm
br
an
an
br
175
200 µm
an
an
176b
200 µm
177
1 mm
ar
br
178
200 µm
176a
600 µm

FIGURES 179-181. Longitudinal sections made through the fertile regions of *Haplomitrium* Nees shoots showing the gametangia and their associated structures. -179-180. *H. hookeri* (Smith) Nees. 179. Axillary antheridium near the apex of a shoot. Note the slime hair and bractlet. 180. Archegonial cluster located near the shoot apex. Note the bractlet and slime papilla. -181. Axillary antheridia of *H. monoicum* Engel. Note the slime papillae and bractlet.

Key: an, antheridium; ar, archegonium; br, bract; bt, bractlet; sh, slime hair; sp, slime papilla. (Fig. 179: *Loitlesberger 1348* [AD]; 180: *Lorbeer H56* [PR]; 181: *Crosby 14165* [F].)

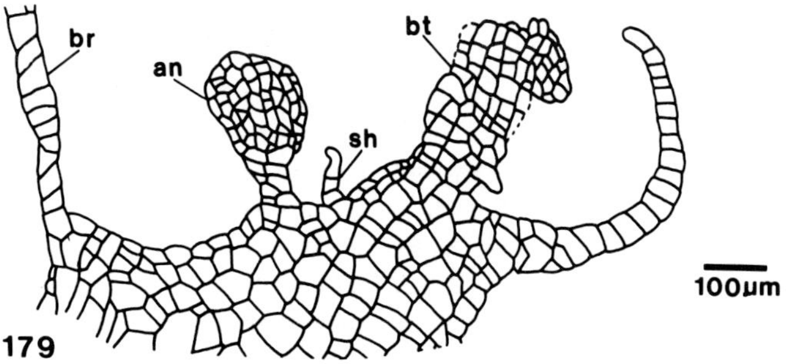

179

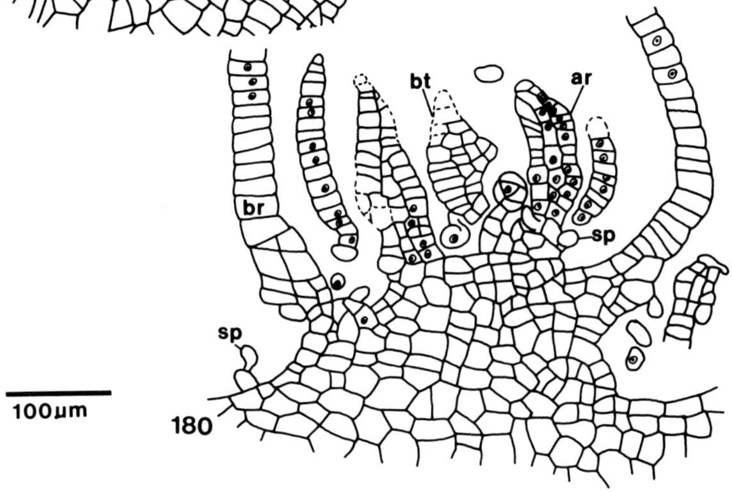

180

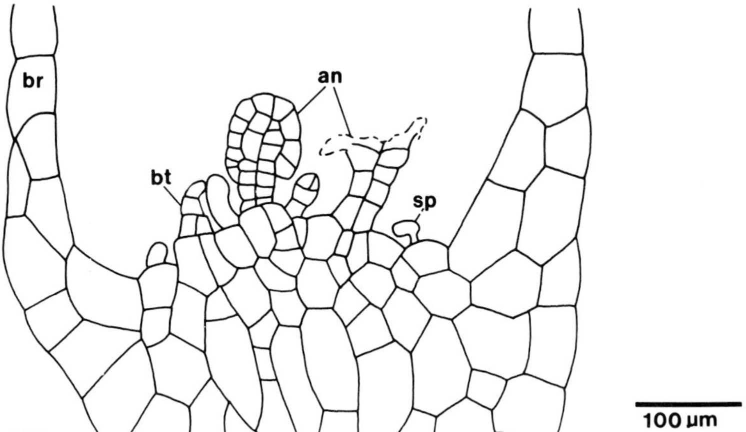

181

FIGURES 182-184. Reproductive gametophytic features of _Haplomitrium monoicum_ Engel. -182-183. Male shoots. 182. Light micrograph showing the shoot habit. Note that the leaves are all similar in size and shape. 183a-b. Scanning electron micrographs. 183a. Portion of a shoot with its antheridia dispersed near the apex. A bract has been removed to expose the antheridia (arrow). 183b. Axillary antheridia. Note that one antheridium is produced per bract. -184. Light micrograph showing the distal portion of a female shoot bearing a loose cluster of archegonia near its apex and one in the fourth series of leaves.

Key: an, antheridium; ar, archegonium; br, unmodified bract. (Figs. 182-184: _Crosby 14165_ [F].)

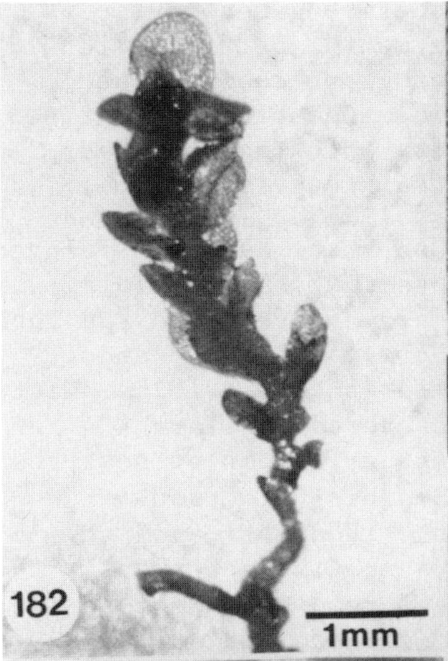
182
1mm

183a
200 µm

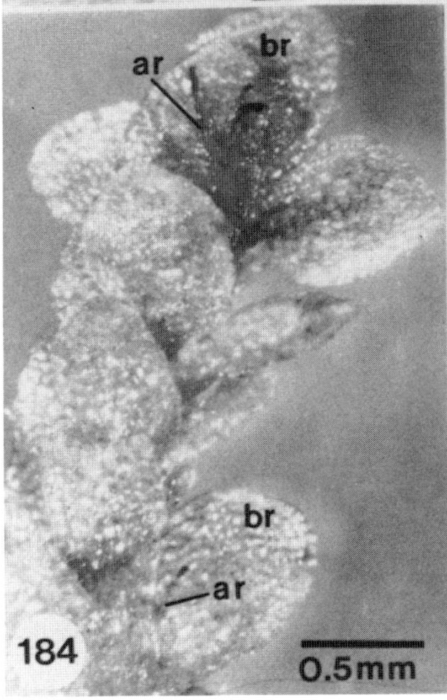
ar
br
br
ar
184
0.5mm

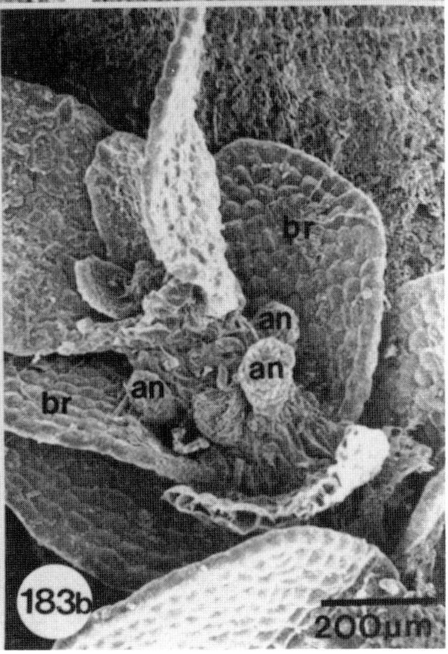
br
an
an
an
br
183b
200µm

FIGURES 185-188. Reproductive gametophytic features of _Haplomitrium intermedium_ Berrie. -185-187. Male shoots. 185. Light micrograph showing the alternating swollen gametangial nodes and vegetative internodal regions of a shoot. Note the enlarged bracts delimiting the androecial rosettes. 186-187. Scanning electron micrographs. 186a. Ill-defined apical androecial rosette. One bract has been removed to expose the antheridia. Note that the remaining peripheral perigonial bracts are enlarged and the shoot apex is slightly dilated. 186b. Antheridia and bractlets of the terminal inflorescence. 187a. Distal portion of a shoot bearing scattered axillary antheridia (arrow). 187b. Antheridia in or near the axils of bracts. -188a-c. Scanning electron micrographs of a female shoot. 188a. Alternating swollen gynoecial nodes and vegetative internodal regions. Some bracts have been removed to expose the archegonial clusters. 188b. Ill-defined apical gynoecial rosette from which a few bracts have been removed to reveal the archegonia. Note the reduced inner perichaetial bract. 188c. Swollen node near the midregion of the shoot.

Key: an, antheridium; ar, archegonium; br, bract; bt, bractlet; ib, reduced inner bract; gn, swollen gametangial node; sp, slime papilla. (Figs. 185, 186, 188: _Berrie 2/62/1_ [SYD]; 187: _Sneddon s.n._ [ABSH].)

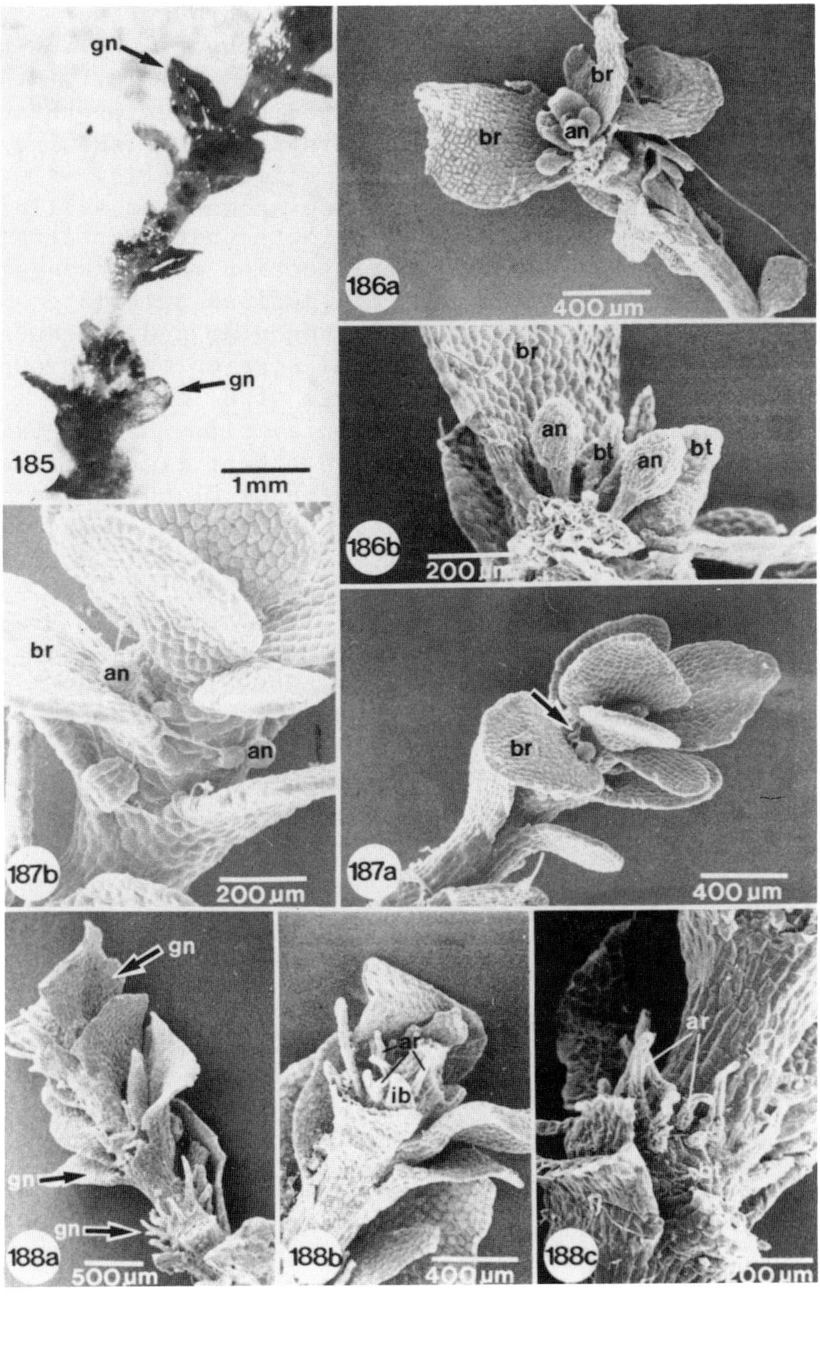
gn
gn
185
1 mm
br
br
an
186a
400 µm
br
an
bt
an
bt
186b
br
an
an
187b
200 µm
br
187a
400 µm
gn
gn
gn
188a
500 µm
ar
ib
188b
400 µm
ar
bt
188c

FIGURES 189-194. Sections made through the fertile regions of *Haplomitrium* Nees shoots showing the gametangia and associated structures. -189-191. *H. intermedium* Berrie. 189. Longitudinal section through a portion of an ill-defined terminal rosette showing a young antheridium, bractlet, and slime papilla. 190. Transection through a dilated node showing an axillary antheridium and slime papillae. Note that bractlets are absent. 191. Longitudinal section made through the peripheral portion of an ill-defined apical gynoecial rosette showing an archegonium and a reduced inner bract. -192. Longitudinal section made through a portion of the gynoecial rosette of *H. gibbsiae* (Steph.) Schust. showing archegonia, slime hairs, slime papillae and bractlets. -193-194. Near transections through mature gametangial rosettes of *H. chilensis* Schust. Note the apical cell in the center of each receptacle. 193. Male rosette. 194. Female rosette.

Key: an, antheridium; ar, archegonium; br, bract; bt, bractlet; ib, reduced inner bract; sp, slime papilla; x, apical cell. (Figs. 189-190: *McKay 3951* [MUCV]; 191: *Selkirk s.n.* [ABSH]; 192: *Engel 16590* [F]; 193: *Engel 11513B* [F]; 194: *Engel 10847* [F].)

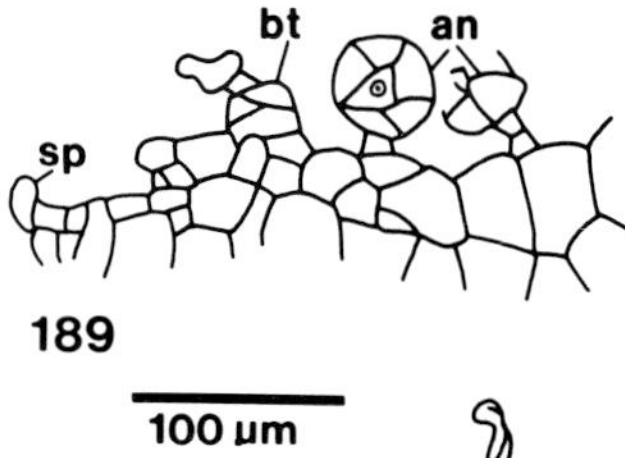

189

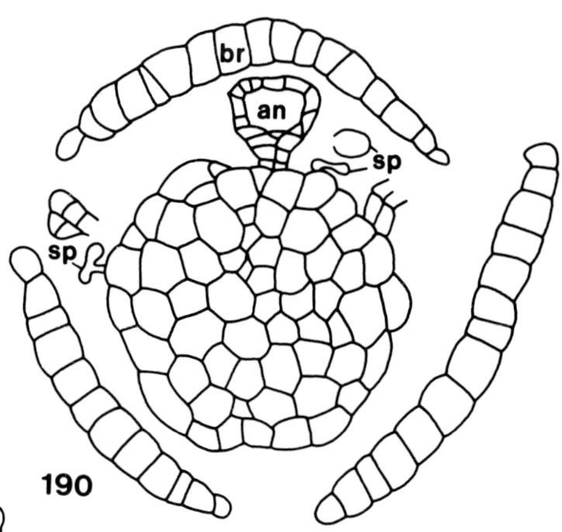

190

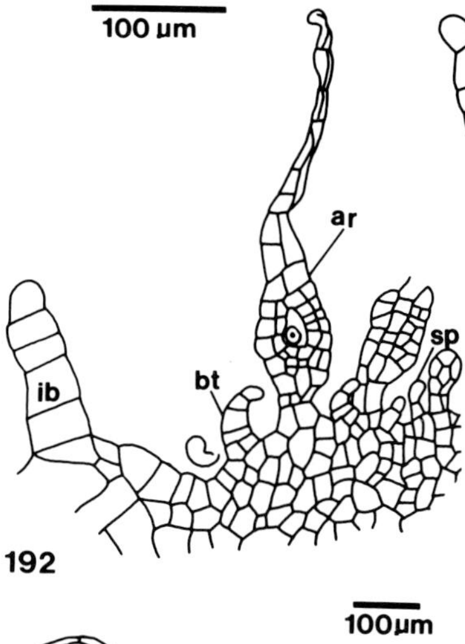

192

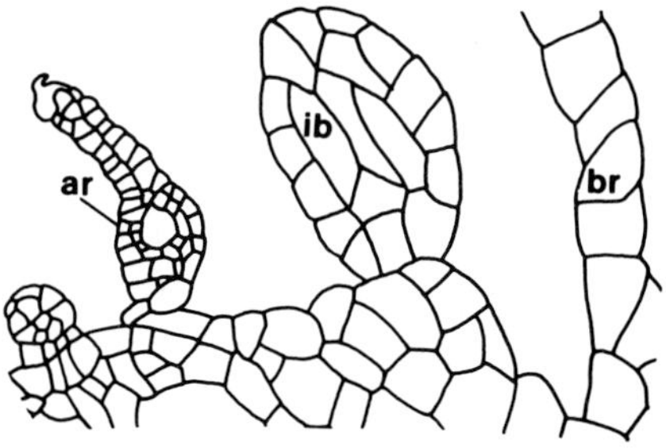

191

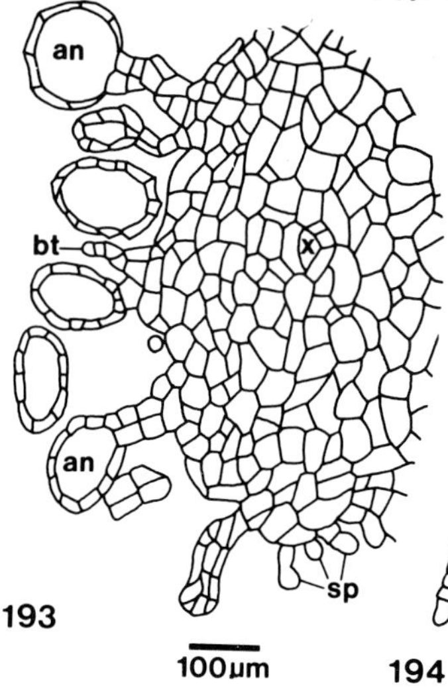

193

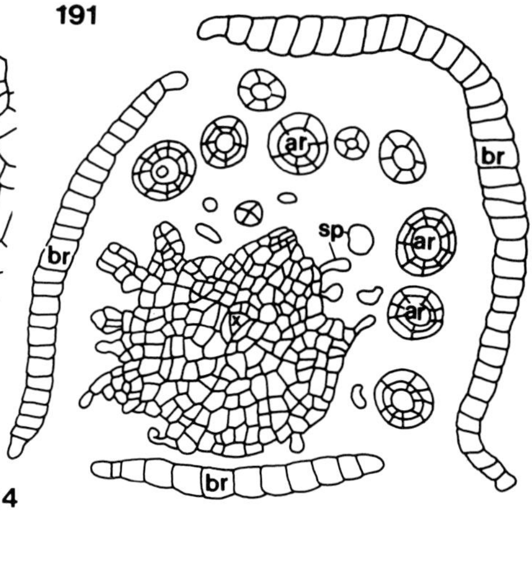

194

FIGURES 195-203. Reproductive gametophytic features of *Haplomitrium gibbsiae* (Steph.) Schust. and *H. chilensis*. Schust. -195-199. Male shoots with terminal androecial rosettes. 195. Light micrograph of a *H. gibbsiae* shoot which is leafy throughout. 196. Light micrograph of a *H. chilensis* shoot with its leaves clustered distally. 197-199. Scanning electron micrographs. 197. Terminal androecial rosette of *H. gibbsiae*. One of the outer bracts has been removed to reveal the antheridia. Note that the leaves of the shoot are clustered distally. 198. Terminal androecial rosette of *H. chilensis*. Two outer bracts have been removed to expose a reduced inner bract and the antheridial cluster. 199. Apical proliferation in *H. gibbsiae*. Note the newly formed shoot with the old antheridia and the outer perigonial bracts basal to it. -200-203. Light micrographs of female shoots with terminal gynoecial rosettes. 200. *H. gibbsiae* shoot which is leafy throughout. 201. *H. gibbsiae* shoot with its leaves clustered distally. 202. *H. chilensis* shoots which are leafy throughout. 203. *H. chilensis* shoot with its leaves clustered distally.

Key: an, antheridium; br, bract; ib, reduced inner bract; ng; new growth. (Figs. 195, 197, 200, 201: *Engel 16590* [F]; 196, 198, 203: *Engel 11537* [F]; 199: *Fife 8453* [ABSH]; 202: *Engel 12133* [F].)

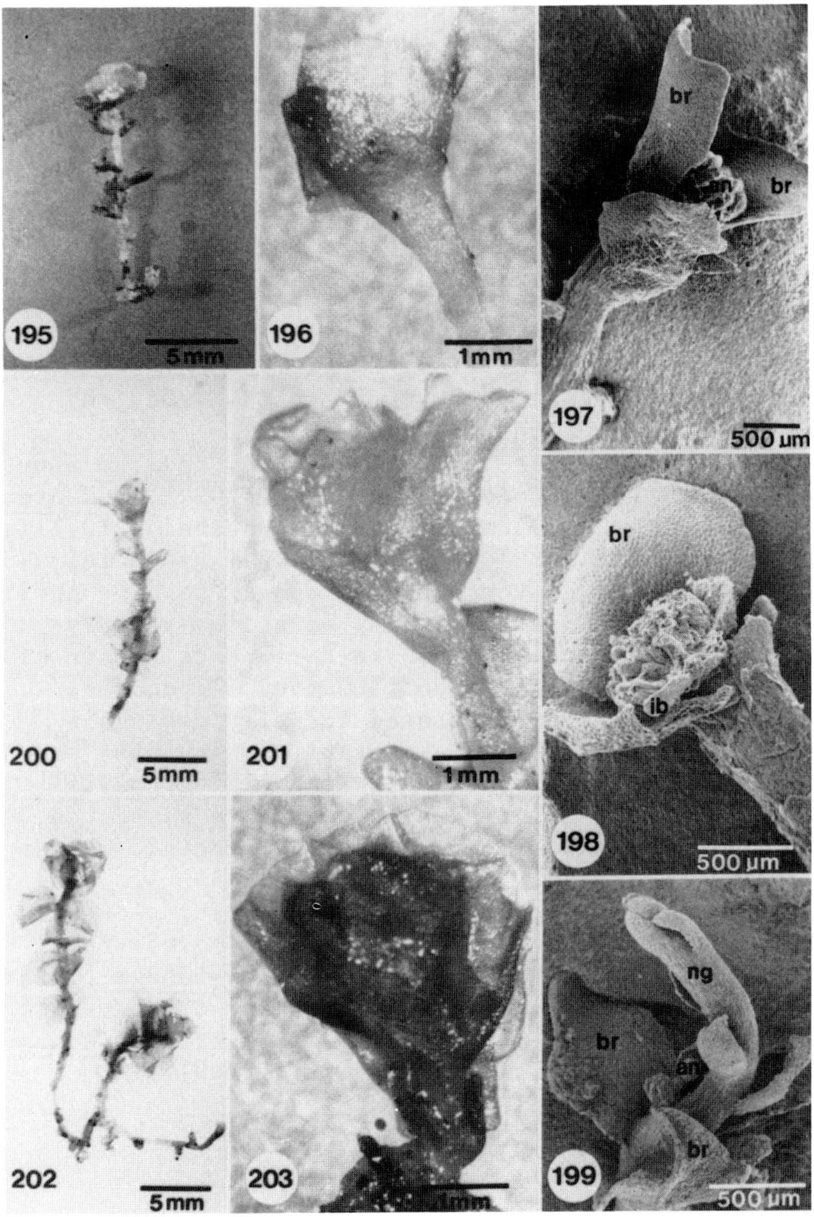
195
5 mm
196
1 mm
br
an
br
197
500 µm
200
5 mm
201
1 mm
br
ib
198
500 µm
202
5 mm
203
ng
br
an
br
199
500 µm

FIGURES 204-212. Reproductive gametophytic features of female plants of *Haplomitrium gibbsiae* (Steph.) Schust., *H. chilensis* Schust., *H. indicum* (Udar & Chandra) Schust., *H. dentatum* (Kumar & Udar) Engel, *H. grollei* Kumar & Udar, and *H. kashyapii* Udar & Kumar. -204-207. Scanning electron micrographs of terminal gynoecial rosettes. Some outer bracts have been removed to reveal the archegonia and inner reduced bracts. 204. *H. gibbsiae*. Note the lingulate-triangular inner bract and the outer enlarged bract. 205. *H. chilensis*. Note the tiny inner bract in front of the archegonia. 206-207. Triangular-shaped reduced inner bracts. 206. *H. gibbsiae*. 207. *H. chilensis*. -208-212. Light micrographs of female plants. 208. *H. indicum* plant bearing terminal gynoecial rosettes. 209. *H. dentatum* plant bearing a terminal gynoecial rosette. 210-211. *H. grollei*. 210. Shoots with terminal gynoecial rosettes. Although these shoots appear to have their leaves clustered distally, they were likely leafy throughout when collected as leaf remnants remain along the entire stem. 211. Shoot bearing a gynoecial rosette which has proliferated apically and again become fertile. 212. *H. kashyapii*. Note that the shoots are leafy throughout and terminal gynoecial rosettes are absent. The archegonia are scattered along the shoots in the axils of the bracts.

Key: ar, archegonium; br, bract; ib, reduced inner bract. (Fig. 204: *Engel 16590* [F]; 205: *Engel 10847* [F]; 206: *Morice, Wylie, & Taylor 135342* [CHR]; 207: *Engel 12133* [F]; 208: *Udar & Kumar C70/5* [LWU]; 209: *Udar & Kumar C-69/4* [LWU]; 210, 211: *Udar & Kumar C72/1* [LWU]; 212: *Udar & Kumar C73/1* [LWU].)

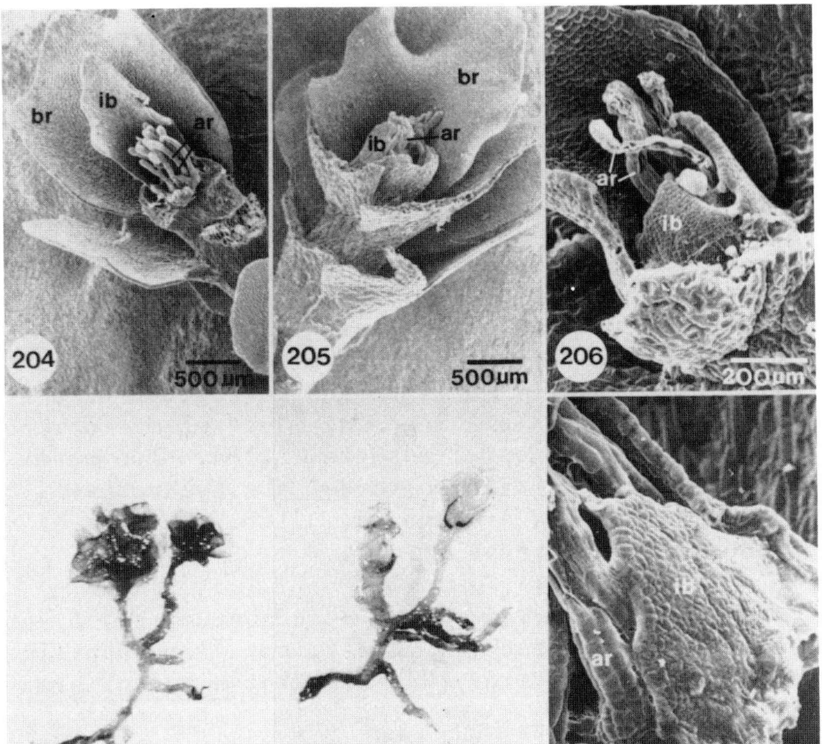

br
ib
ar
204
500 µm
br
ib
ar
205
500µm
ar
ib
206
200µm
ib
ar
207
208
5mm
209
5mm
210
5mm
211
5mm
212
5 mm

FIGURES 213-218. Reproductive gametophytic features of *Haplomitrium mnioides* (Lindb.) Schust. -213. Light micrograph of a male plant with a terminal androecial rosette. Note the distantly spaced vegetative leaves. -214-216. Scanning electron micrographs of male shoots. 214. Young androecial rosette with the bracts overarching the immature antheridia. 215. Antheridia of a mature terminal disc. Note the secondarily produced younger antheridia near the periphery of the cluster. 216. Apical proliferation from an old antheridial receptacle. Note the newly formed vegetative leaves in the center of the antheridial cluster. -217. Light micrograph of a female plant. -218a-b. Scanning electron micrographs of a terminal archegonial rosette. 218a. Two outer bracts have been removed to reveal the two reduced inner bracts and the archegonia. 218b. Smaller of the two reduced inner bracts.

Key: an, antheridium; ar, archegonium; br, bract; ib, reduced inner bract; le, leaf; sp, slime papilla. (Fig. 213: *Lindberg s.n.* [H]; 214-218: *Mizutani s.n.* [ABSH].)

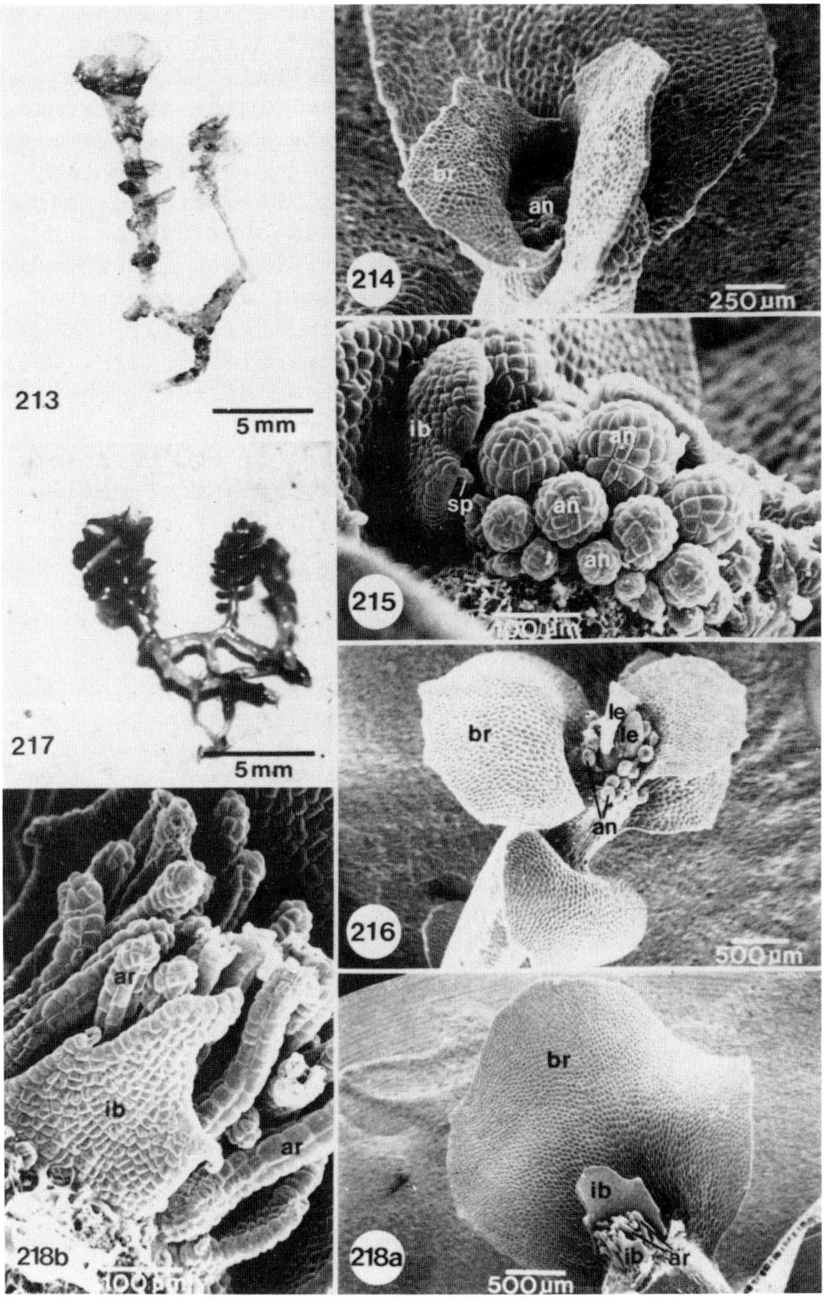
213
5 mm
214
br
an
250 µm
215
ib
sp
an
an
an
an
216
br
le
le
an
500 µm
217
5 mm
218b
ar
ib
ar
218a
br
ib
ib
ar
500 µm

FIGURES 219-222. Sections made through the terminal gynoecial rosettes of _Haplomitrium_ Nees. -219-220. _H. mnioides_ (Lindb.) Schust. 219. Longitudinal section showing a few of the archegonia, slime papillae and bractlets of the rosette. 220a-b. Serial transections. 220a. Archegonia within the outer perichaetial bracts. 220b. Stem tissue below the archegonia (60 µm basal to [a]) showing the centrally located apical cell. -221-222. Longitudinal sections made through the archegonia and associated structures of _H. blumii_ (Nees) Schust. 221. Archegonia, slime papillae, and bractlets. 222. Slime hairs and slime papillae at the base of the archegonia.

Key: ar, archegonium; br, bract; bt, bractlet; sh, slime hair; sp, slime papilla; x, apical cell. (Figs. 219, 220: _Doei s.n._ [ABSH]; 221: _Furuki 384_ [ABSH]; 222: _Verdoorn 61_ [G].)

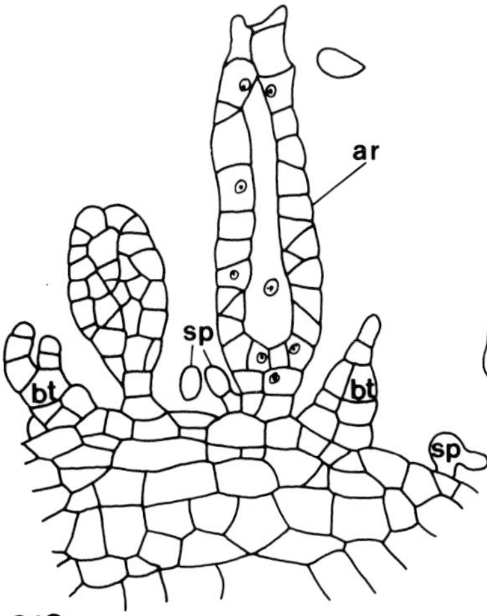

219

100 µm

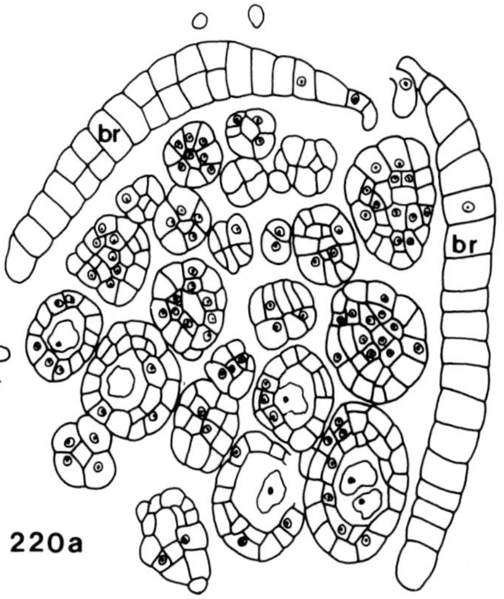

220a

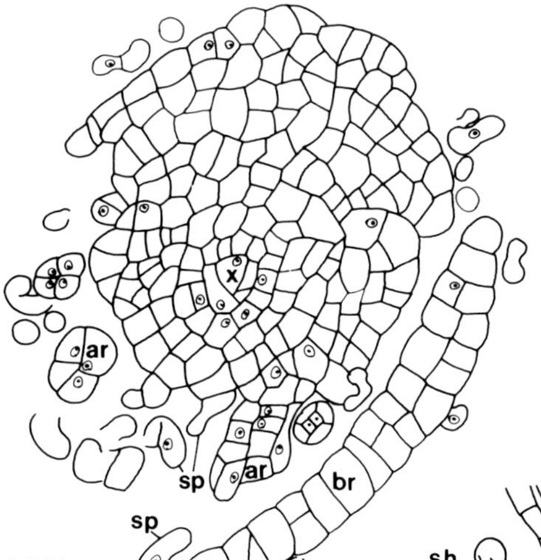

220b

100 µm

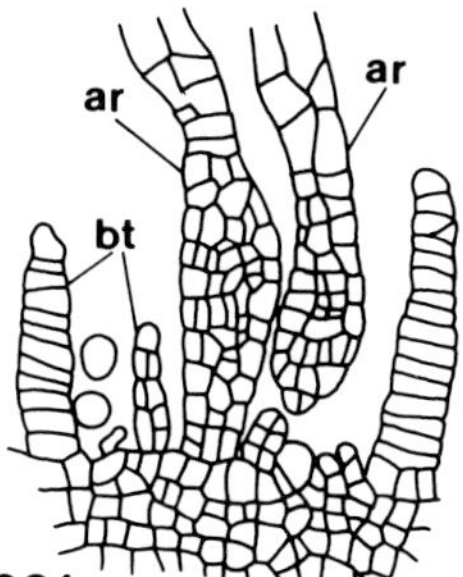

221

100 µm

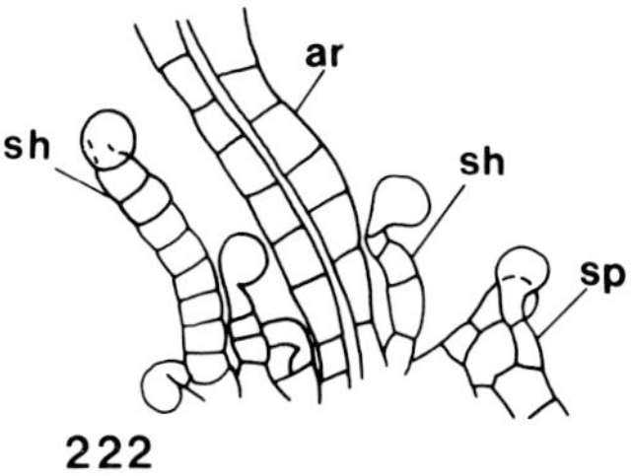

222

FIGURES 223-227. Reproductive gametophytic features of *Haplomitrium blumii* (Nees) Schust. -223-225. Male features. 223. Light micrograph of a male plant bearing an apical androecial rosette. 224. Light micrograph showing the base of an antheridial stalk and the adjacent slime hairs and slime papillae. 225. Scanning electron micrograph showing an antheridial rosette in apical view. Note the three large outer perigonial bracts subtending the antheridial cluster. -226-227. Female features. 226. Light micrograph of a shoot bearing a terminal gynoecial rosette. 227. Archegonial rosette from which an outer bract has been removed to expose a reduced inner bract and the archegonial cluster.

Key: an, antheridium; ar, archegonium; br, bract; ib, reduced inner bract; sh, slime hair; sp, slime papilla. (Fig. 223: *Norris 60313* [H]; 224: *Verdoorn 61* [H]; 225: *Koponen 35568* [G]; 227: *Campbell s.n.* [NY].)

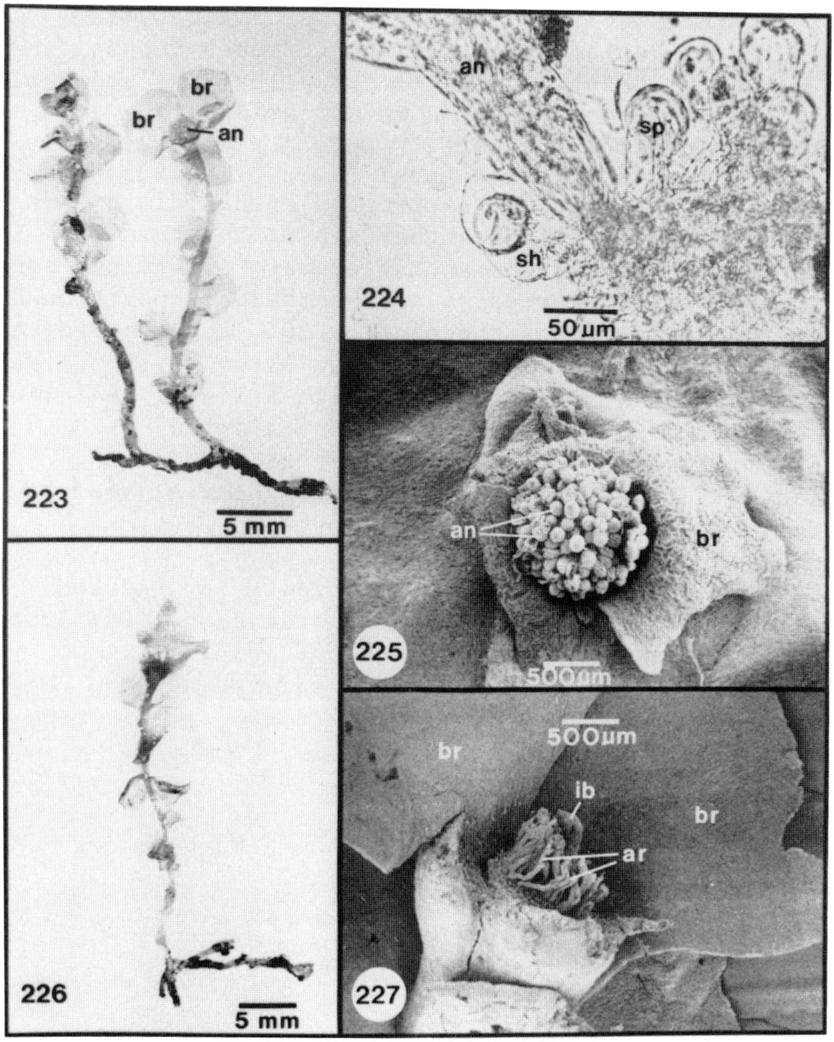

br
br
an
223
5 mm
an
sp
sh
224
50 µm
an
br
225
500µm
226
5 mm
500µm
br
ib
br
ar
227

FIGURES 228-231. Reproductive gametophytic features of *Haplomitrium andinum* (Spruce) Schust. and *H. giganteum* (Steph.) Grolle. -228-230. *H. andinum*. 228. Light micrograph of a male plant bearing a terminal androecial rosette. Dorsal view. 229a-b. Scanning electron micrographs. 229a. Terminal rosette with a bract removed to expose the antheridia. 229b. Reduced inner bract and antheridia of the terminal rosette. Note that the antheridia are of various ages with secondarily formed younger antheridia positioned near the periphery of the disc. 230. Light micrograph of a female plant. Ventral view. Arrow s_1 indicates an unfertilized female shoot, while arrow s_2 indicates a fertilized shoot bearing a shoot-calyptra. -231. Light micrograph showing the habit of a female *H. giganteum* shoot bearing a terminal gynoecial rosette. Dorsal view.

Key: an, antheridium; br, bract; ib, reduced inner bract. (Figs. 228, 229: *Allorge s. n.* [F]; 230: *Spruce s. n.* [NY]; 231: *Elmer 11562* [G]).

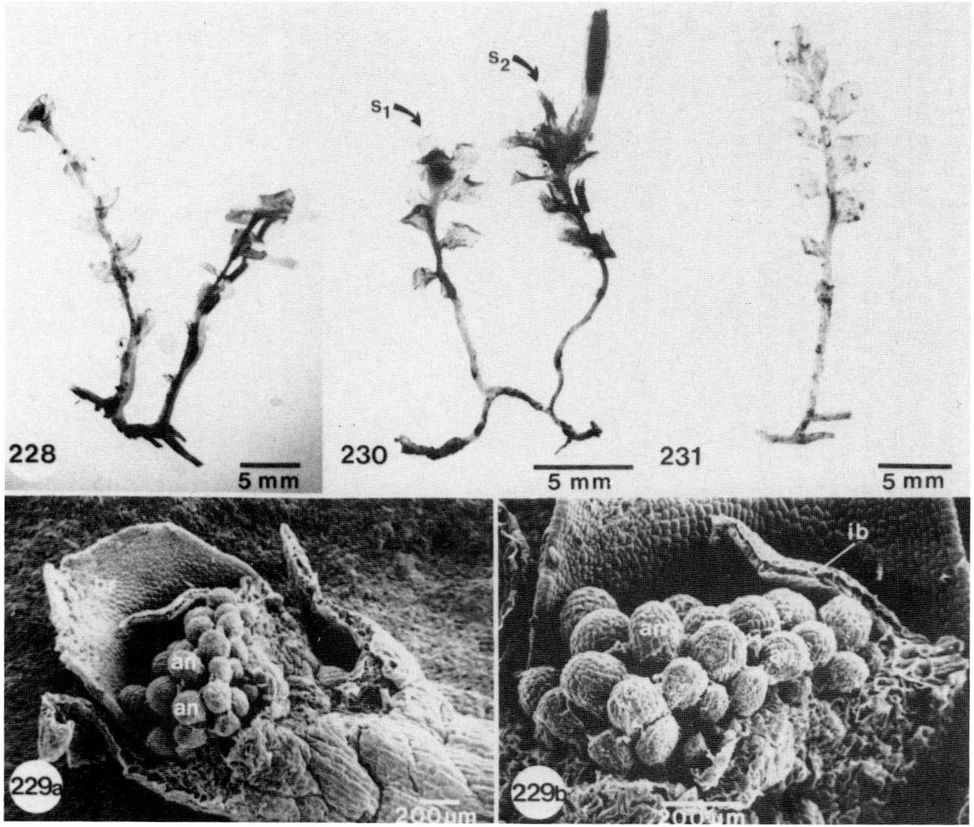

228
5 mm
s1
s2
230
5 mm
231
5 mm
an
an
229a
200 µm
ib
an
229b
200 µm

FIGURES 232-234. Reproductive gametophytic features of the male shoots of _Steereomitrium minutum_ E. O. Campb. -232a-b. Shoots with scattered antheridia. 232a. Habit of shoots. 232b. Axillary and supra-axillary antheridia. -233. Distal portion of a shoot bearing a loose apical cluster of antheridia. Note the axillary antheridia basal to the terminal group (arrow). -234. Vegetative leaf of a fertile shoot.

Key: an, antheridium; br, bract. (Figs. 232-234: _Campbell s.n._ [NY].)

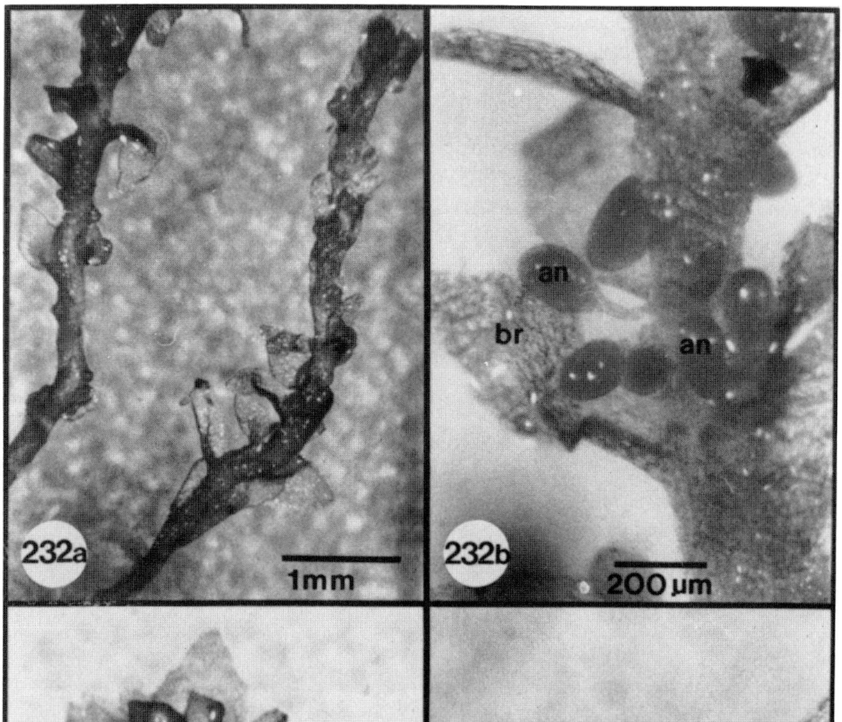
232a
1mm
232b
an
br
an
200 µm
233
an
1mm
234
500

FIGURES 235-236. Longitudinal serial sections made through the apices of male *Haplomitrium* Nees shoots showing antheridial production. Note that the antheridia originate close to the apical cell from the same portion of the merophyte which yields a leaf. -235a-c. *H. mnioides* (Lindb.) Schust. 235b is 20 µm below 235a and 235c is 30 µm below 235a. -236a-b. *H. gibbsiae* (Steph.) Schust. 236b is 10 µm below 236a.

Key: an, antheridium; le, leaf; x, apical cell. (Figs. 235a-c: *Mizutani s.n.* [ABSH]; 236a-b: *Sneddon s.n.* [ABSH].)

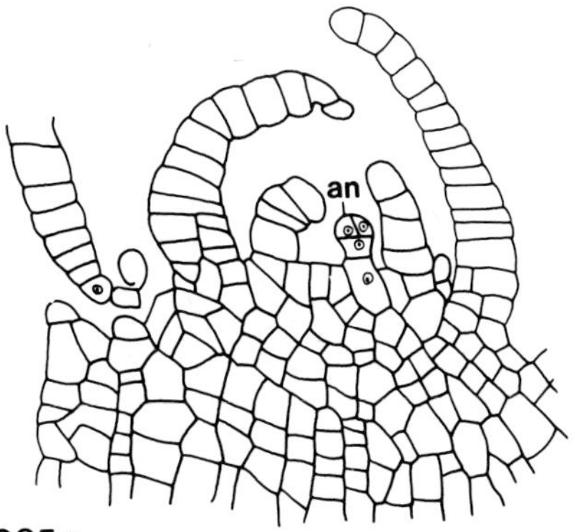

235a

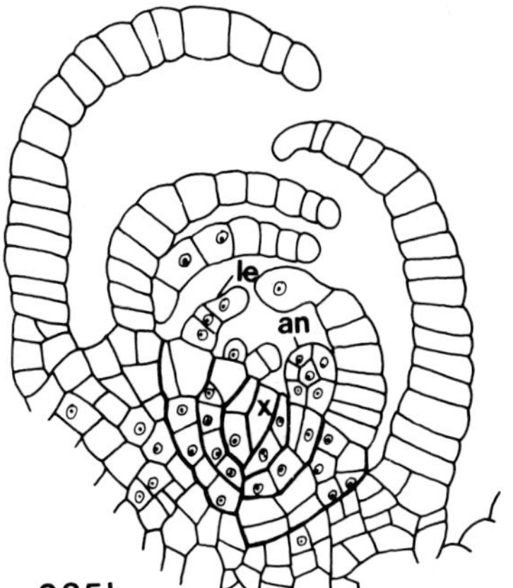

235b

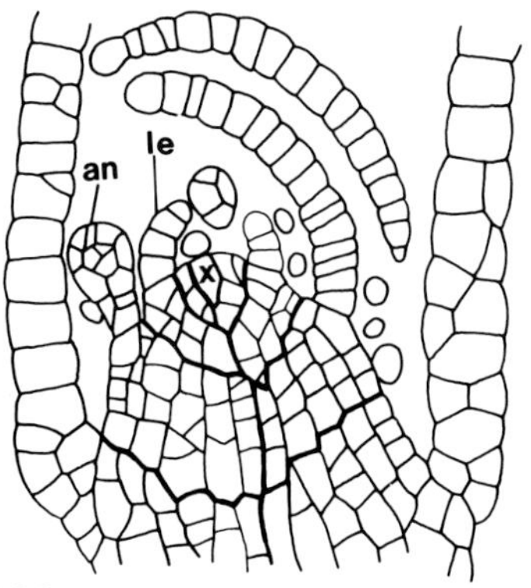

236a

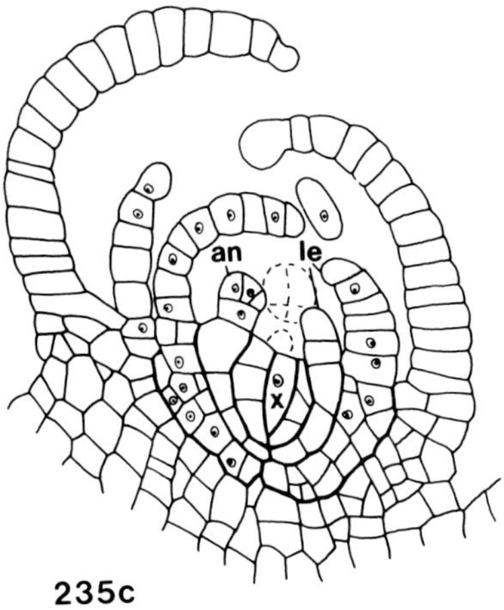

235c

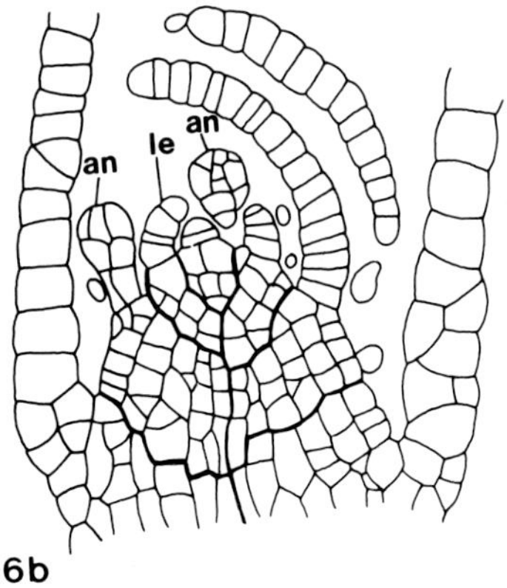

236b

100 µm

FIGURES 237-239. Transections made through the apices of male shoots of *Haplomitrium gibbsiae* (Steph.) Schust. showing antheridial production. Note that the antheridia originate close to the apical cell and may form in place of a leaf or a portion of a leaf. The merophytes are numbered consecutively from oldest to youngest. -237a-b. Serial sections made at 10 µm intervals. 237a. An antheridium has replaced a portion of the leaf in merophyte II so that a leaf, antheridium, and a slime papilla have been produced. 237b. Merophyte VI has given rise to an antheridium and two reduced leaves. -238a-c. Serial sections made at 10 µm intervals. An antheridium has replaced an entire leaf in merophyte I, while merophyte VI has given rise to both an antheridium and a reduced leaf. -239. Section made through an androecial rosette showing the production of three antheridia from a single merophyte.

Key: an, antheridium; sp, slime papilla. (Figs. 237-239: *Sneddon* *s.n.* [ABSH].)

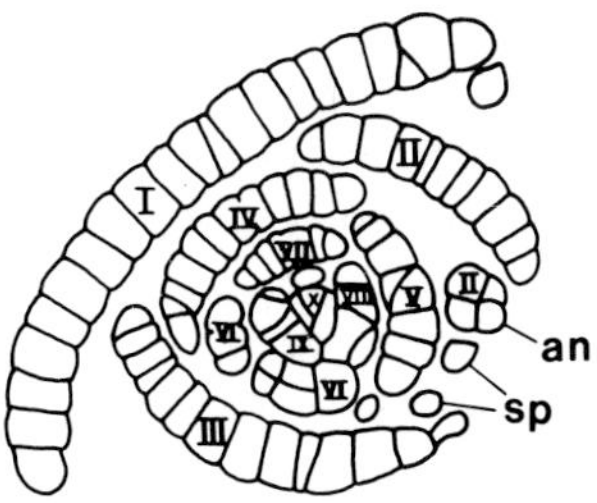

237a

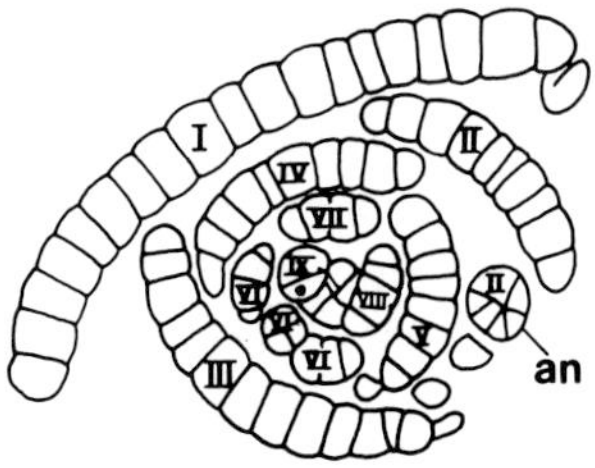

237b

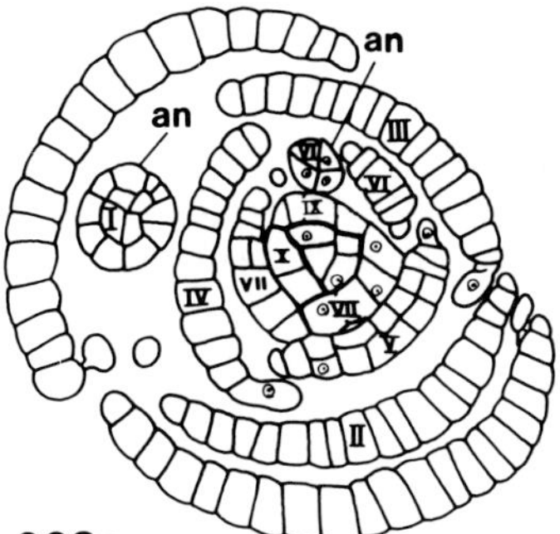

238a

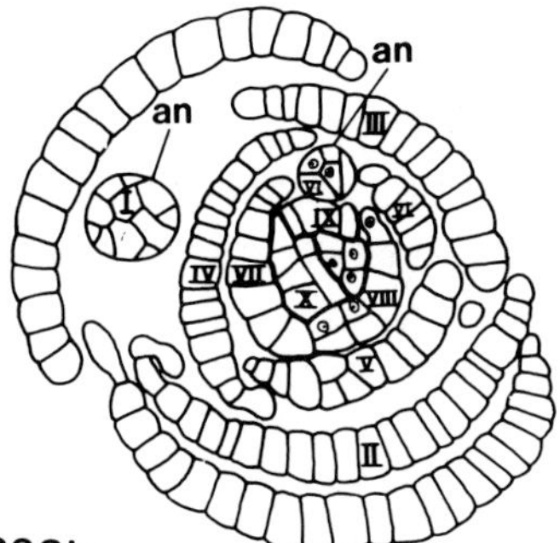

238b

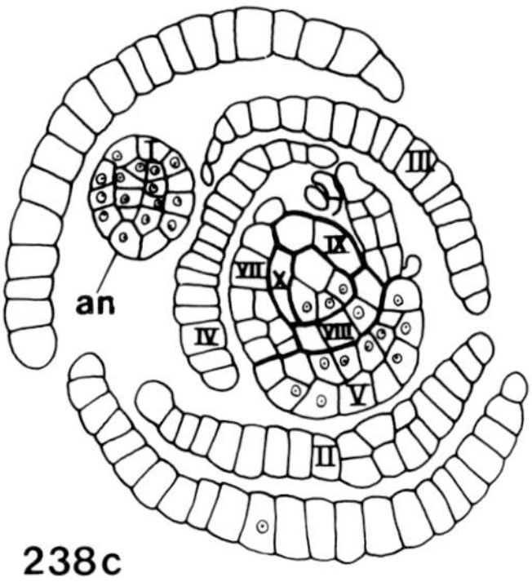

238c

100 µm

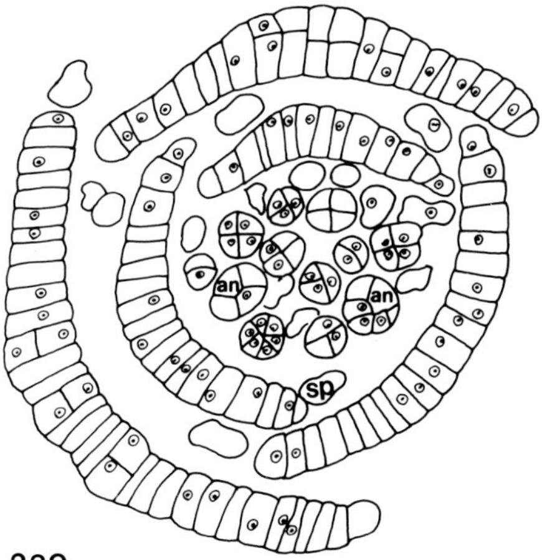

239

FIGURES 240-259. Antheridial ontogeny in *Haplomitrium* Nees. -240a-b. Longitudinal sections. 240a. Swollen superficial cell. 240b. Formation of an outer primary antheridial initial and an inner stalk initial. -241-243. First division of the antheridial initial. 241. Longitudinal view. The stalk initial is undivided. 242. Transection. -243. Longitudinal section. Note the first division of the stalk. -244-248. Division sequences in the antheridial initial which form a central androgonial initial and three jacket initials. -244-247. Two walls which are parallel to the first wall, but oppositely curved. 244, 246, 247. Longitudinal sections. 245. Transection. Note the additional division in one jacket cell. -248. Two curved, vertical divisions which intersect the first wall distally. Longitudinal section. -249. Transection showing the expanded androgonial initial. -250-257. Formation of spermatogenous tissue from the androgonial initial. 250-255. Upper (uc) and lower cell (lc) delimited by a transverse wall, with four quadrants being formed in each tier. 250, 251. Longitudinal sections. 252-255. Transections. 256-257. Repeated synchronous divisions to produce spermatogenous cells. 256. Longitudinal section. Note the tiered antheridial stalk. 257. Transverse section. 258-259. Stalks composed of four cells in transection.

Key: ani, antheridial initial; adi, androgonial initial; bo, body; co, cover cell; j, jacket cell; s, stalk; ss, spermatogenous cell; si, stalk initial. (Figs. 240-244, 246, 248-251, 254, 256, 258, *H. mnioides* (Lindb.) Schust.: *Mizutani s.n.* [ABSH]; 245, 252, 253, 255, 257, 259, *H. gibbsiae* (Steph.) Schust.: *Sneddon s.n.* [ABSH]; 247, *H. monoicum* Engel: *Kitagawa 21694* [Hb. Kitagawa].)

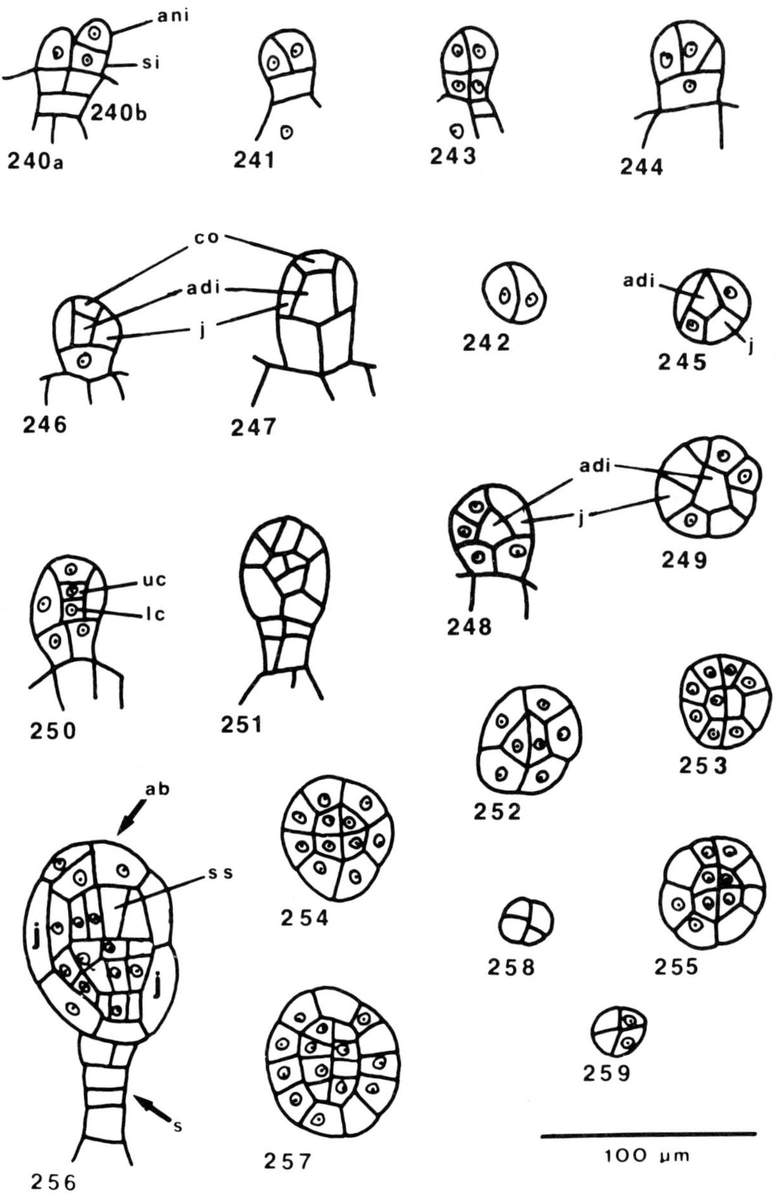
ani
si
240b
240a
241
243
244
co
adi
j
246
247
242
adi
j
245
adi
j
249
248
uc
lc
250
251
252
253
ab
ss
j
j
s
256
254
258
255
259
257
100 µm

FIGURES 260-266. Mature antheridial morphology of *Haplomitrium* Nees. -260-261. *H. monoicum* Engel. Note the very abbreviated stalk and the ovoid to ellipsoid body. 260. Light micrograph. 261. Scanning electron micrograph. -262-266. Scanning electron micrographs. 262. *H. hookeri* (Smith) Nees. Note the ovoid bodies. In this view the stalks are obscured by bract remnants. 263. *H. intermedium* Berrie. 264. *H. mnioides* (Lindb.) Schust. 265. *H. gibbsiae* (Steph.) Schust. Note the tiered, rectangular jacket cells of the ovoid bodies, and the very long stalks. 266. *H. chilensis* Schust. Note the tiered, rectangular jackets cells of the ellipsoid bodies.

Key: bo, antheridial body; s, antheridial stalk. (Figs. 260, 261: *Crosby 14165* [F]; 262: *Worley & Rawson 12542* [H]; 263: *Berrie 2/62/1* [SYD]; 264: *Mizutani s.n.* [ABSH]; 265: *Fife 8453* [ABSH]; 266: *Engel 11537* [F].)

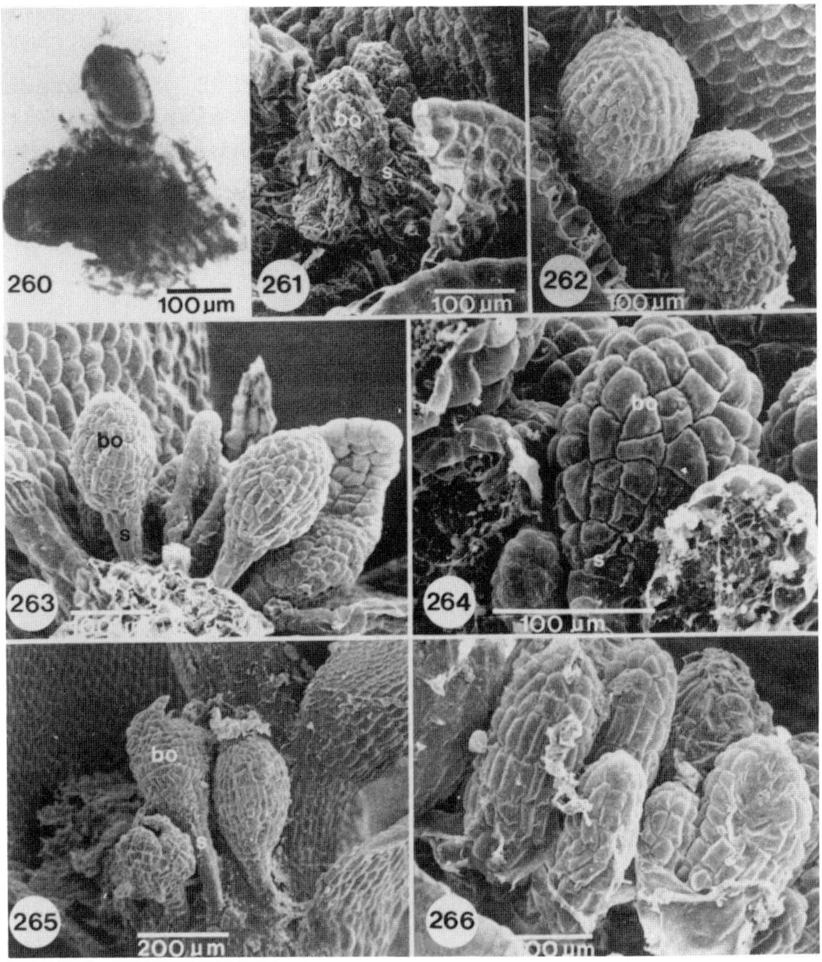
260
100 µm
261
bo
s
100 µm
262
100 µm
263
bo
s
264
bo
s
100 µm
265
bo
s
200 µm
266

FIGURES 267-272. Mature antheridia of *Haplomitrium* Nees and *Steereomitrium* E. O. Campb. -267-268. *H. blumii* (Nees) Schust. 267. Antheridium with an ovoid body. 268. Antheridium with an ellipsoid body. -269. *H. andinum* (Spruce) Schust. Note the ellipsoid body. -270. *S. minutum* E. O. Campb. -271. Mature spermatozoids of *H. gibbsiae* (Steph.) Schust. -272. Dehisced *H. blumii* antheridium.

Key: bo, antheridial body; s, antheridial stalk. (Figs. 267, 272: *Verdoorn 61* [G]; 268: *Touw 15228* [L]; 269: *Spruce s. n.* [NY]; 271: *Fife 8453* [ABSH]).

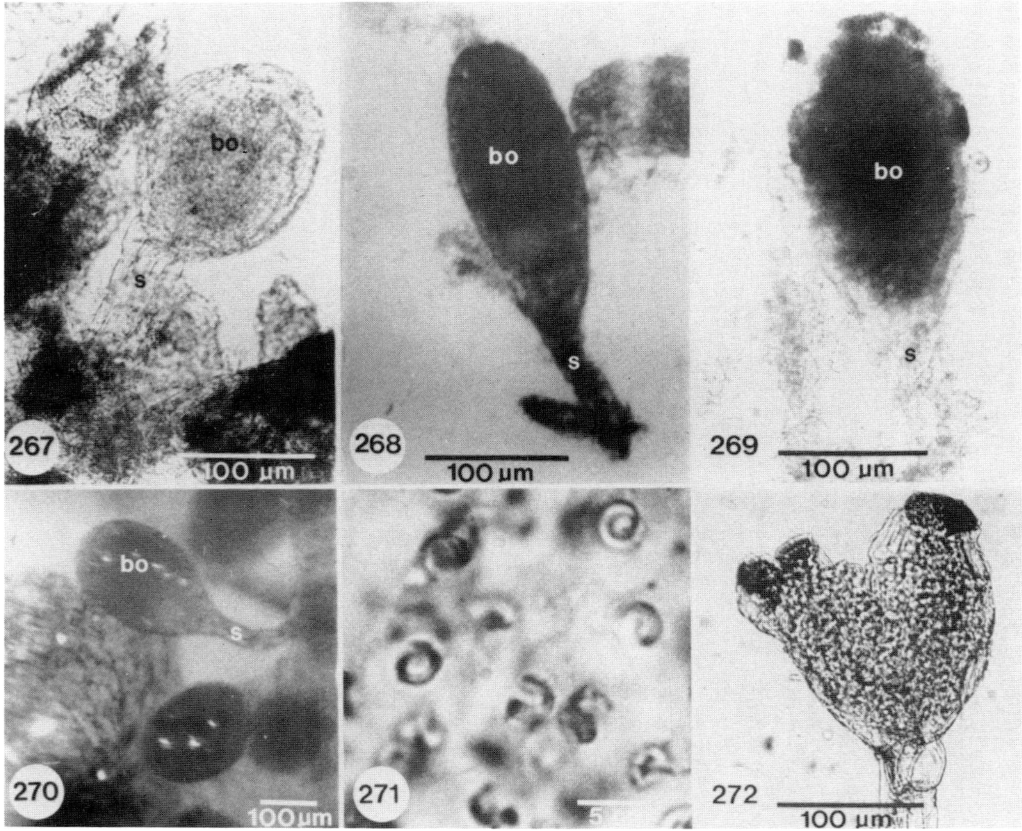
267
bo
s
100 µm
268
bo
s
100 µm
269
bo
s
100 µm
270
bo
s
100µm
271
272
100 µm

FIGURES 273-275. Fine structure of the antheridia of *Haplomitrium* Nees. -273a-b. Two-tiered androgonial region and unistratose jacket cells of a young *H. gibbsiae* (Steph.) Schust. antheridium. 273a. Androgonial region. 273b. Portion of the jacket. -274. Jacket cells of a more mature *H. mnioides* (Lindb.) Schust. antheridium. Note the starch-containing plastids. -275. Spermatogenous cells of *H. mnioides*.

Key: k, karyosome; l, lipid; m, mitochondrion; n, nucleus; nu, nucleolus; ob, oil body; p, plastid; sg, starch grain; v, vacuole. (Figs. 273: *Sneddon s.n.* [ABSH]; 274, 275: *Doei s.n.* [ABSH].)

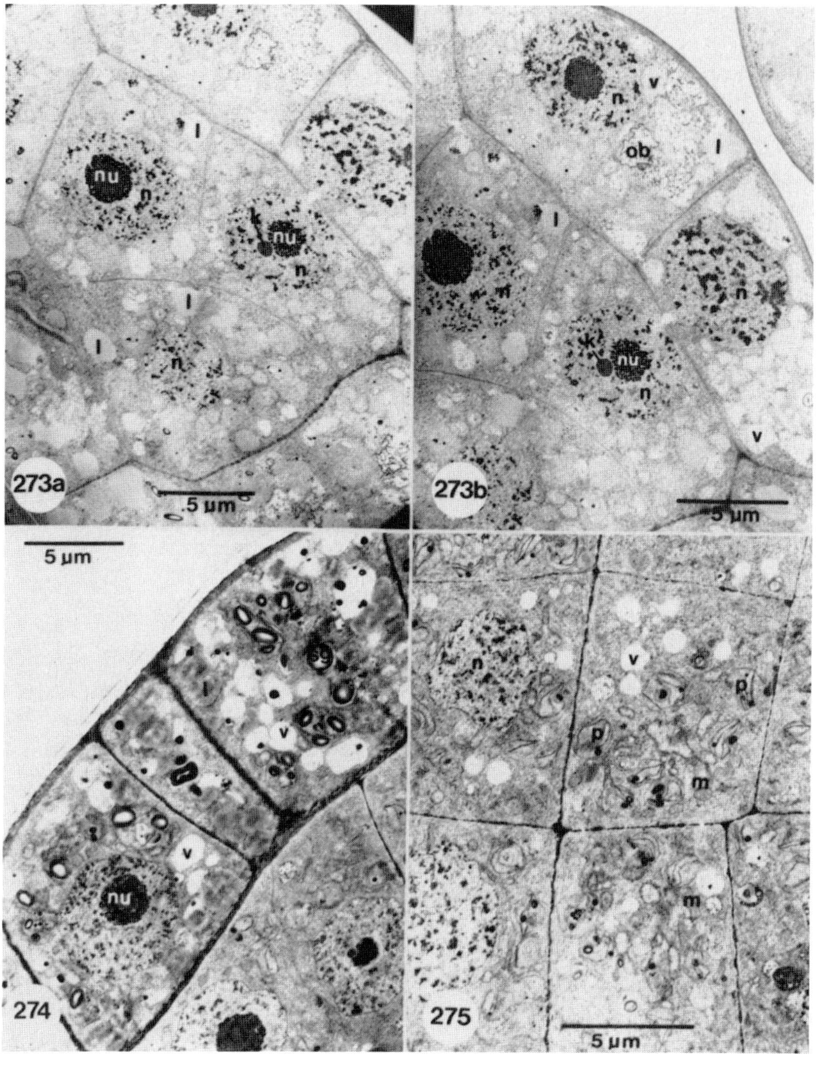

273a
nu
n
l
5 µm
273b
ob
v
k
5 µm
274
5 µm
sg
v
nu
275
n
v
p
m
5 µm

FIGURES 276-277. Serial sections made through the apices of female *Haplomitrium* Nees shoots showing archegonial production. Note that the archegonia originate close to the apical cell from the same portion of the merophyte which yields a leaf. Thus, the archegonia may form in place of a leaf or a portion of a leaf. -276a-c. *H. gibbsiae* (Steph.) Schust. Serial longitudinal sections made at 10 µm intervals. -277a-c. *H. hookeri* (Smith) Nees. Serial transections made at 10 µm intervals. The merophytes are numbered consecutively from oldest (I) to youngest (XI). Note that merophyte I has given rise to an archegonium and two reduced leaves, while only an archegonium and a leaf have formed from merophyte II.

Key: ar, archegonium; ib, inner bract; sp, slime papilla; x, apical cell. (Fig. 276: *Sneddon s.n.* [ABSH]; 277: *Lorbeer H56* [PR].)

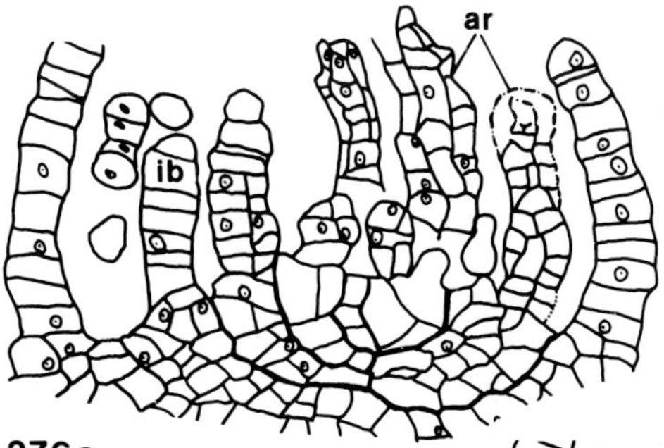

276a

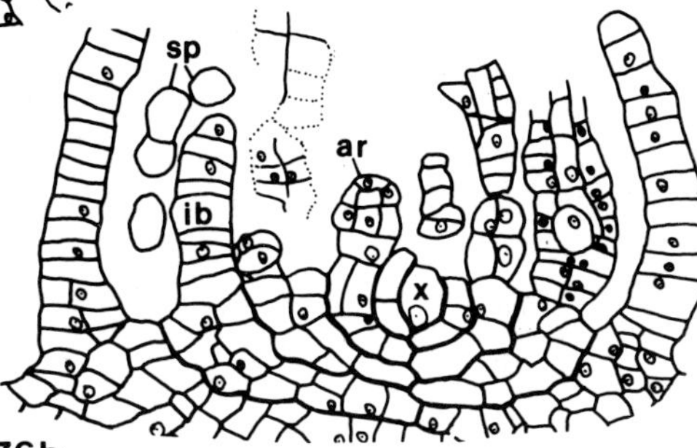

276b

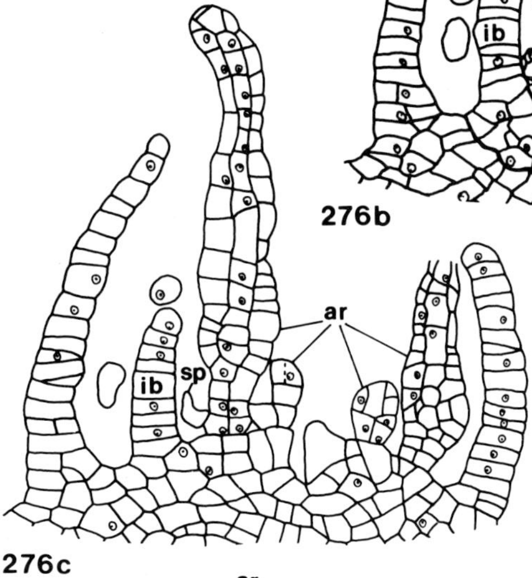

276c

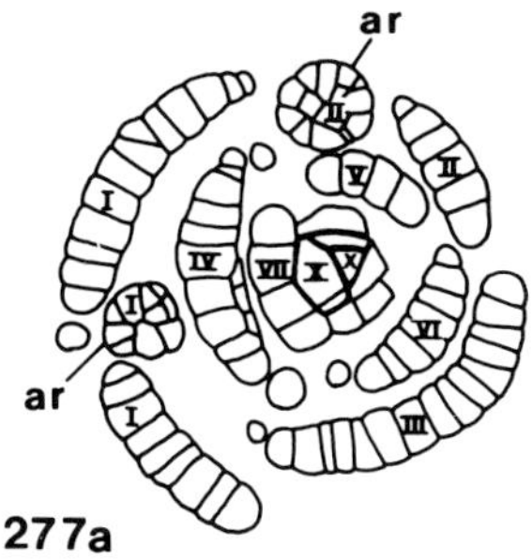

277a

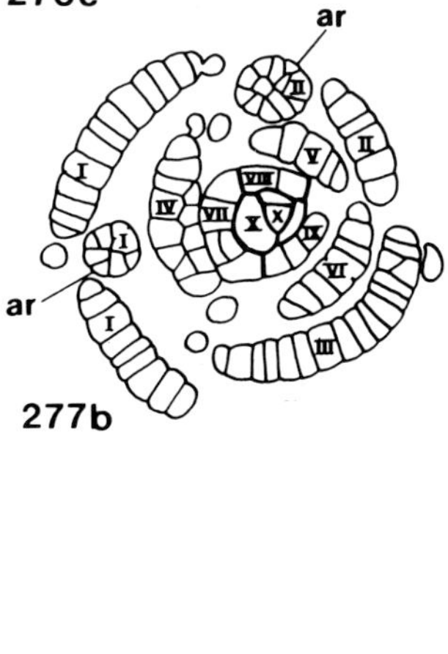

277b

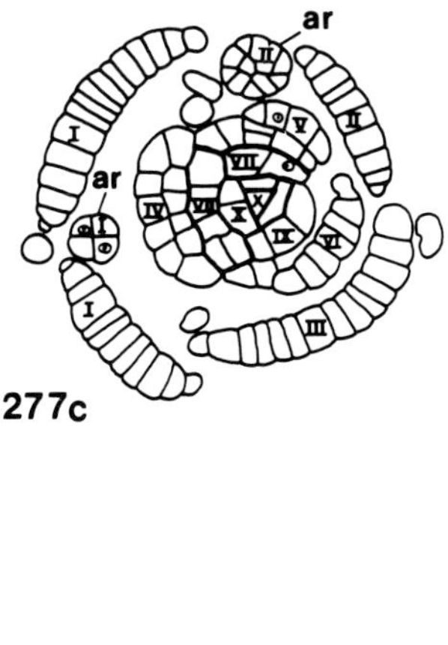

277c

100 µm

FIGURES 278-279. Sections made through the gynoecia of Haplomitrium Nees. -278a-c. Serial transections made through the terminal archegonial receptacle of H. chilensis Schust. showing the centrally located apical cell. 278a-c are each cut at 10 µm intervals with 278a the most terminal section and 278c the most basal. -279. Longitudinal section showing a portion of the terminal archegonial rosette of H. blumii (Nees) Schust. Note the synchronous development of the archegonia.

Key: ar, archegonium; bt, bractlet; sp, slime cell; x, apical cell. (Fig. 278: Engel 10847 [F]; 279: Furuki 384 [ABSH].)

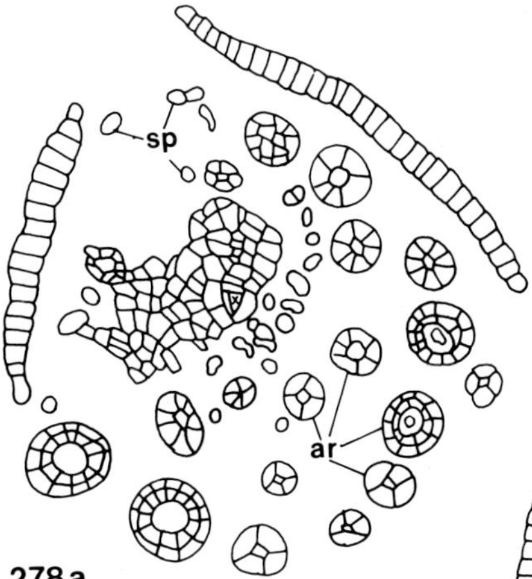

278a

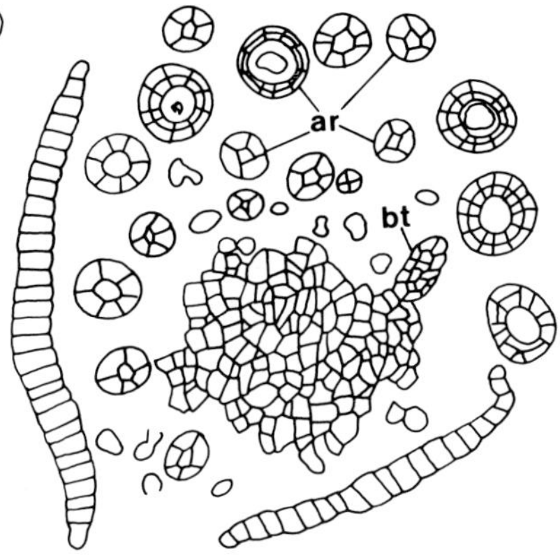

278b

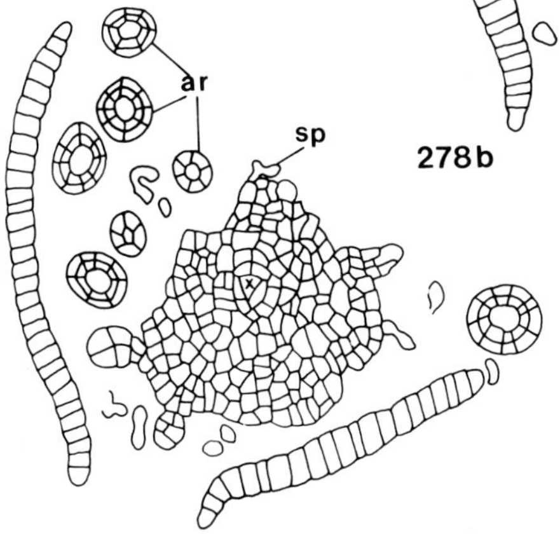

278c

100µm

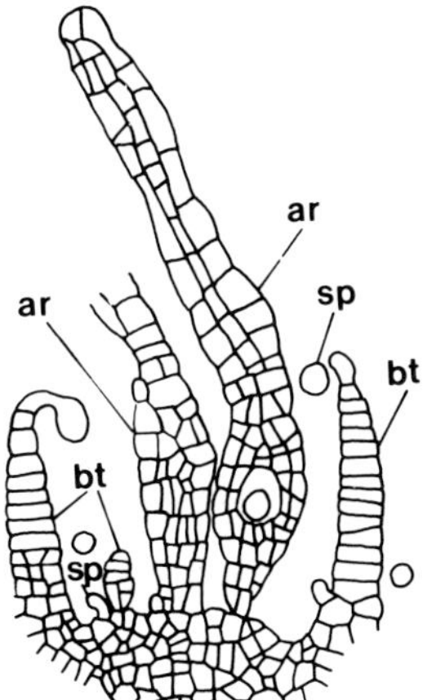

279

FIGURES 280-294. Archegonial ontogeny in Haplomitrium Nees. -280. Swollen superficial cell delimited from the axis by a transverse wall. Longitudinal section. 281. Formation of an outer primary archegonial initial and an inner stalk initial. -282-284. First division of the archegonial initial. 282. Longitudinal view. The stalk initial is undivided. 283. Transection. -284. Longitudinal section. Note the first division of the stalk initial. -285-289. Division sequences in the archegonial initial which form a pyramidal central cell and three outer cells. 285-287. Two divisions which are parallel to the first wall, but oppositely curved. 285-286, 288. Longitudinal sections. 287. Transection. 289. Two curved, vertical divisions which intersect the first wall distally. Note the division in the outer cell to yield four cells. Longitudinal section. -290-291. Division of the expanding central cell to form a primary neck canal cell and a primary ventral canal cell. Repeated divisions in the outer cells compensate for the enlarging central region. -292a-b. Archegonial necks in transection. 292a. Midregion composed of four cells. 292b. Apical portion composed of five cells. -293. Transection made through the mature bistratose venter. -294. Stalk composed of four cells in transection.

Key: ari, archegonial initial; co, cover cell; pc, pyramidal central cell; ni, primary neck canal cell; oc, outer cell; vi, primary ventral canal cell. (Figs. 280, 282, 285, 287-289, H. mnioides (Lindb.) Schust.: Doei s.n. [ABSH]; 284, H. blumii (Nees) Schust.: Furuki 384 [ABSH]; 283, 286, 290, 292-294, H. gibbsiae (Steph.) Schust.: Sneddon s.n. [ABSH]; 291, H. hookeri (Smith) Nees: Lorbeer H56 [PR].)

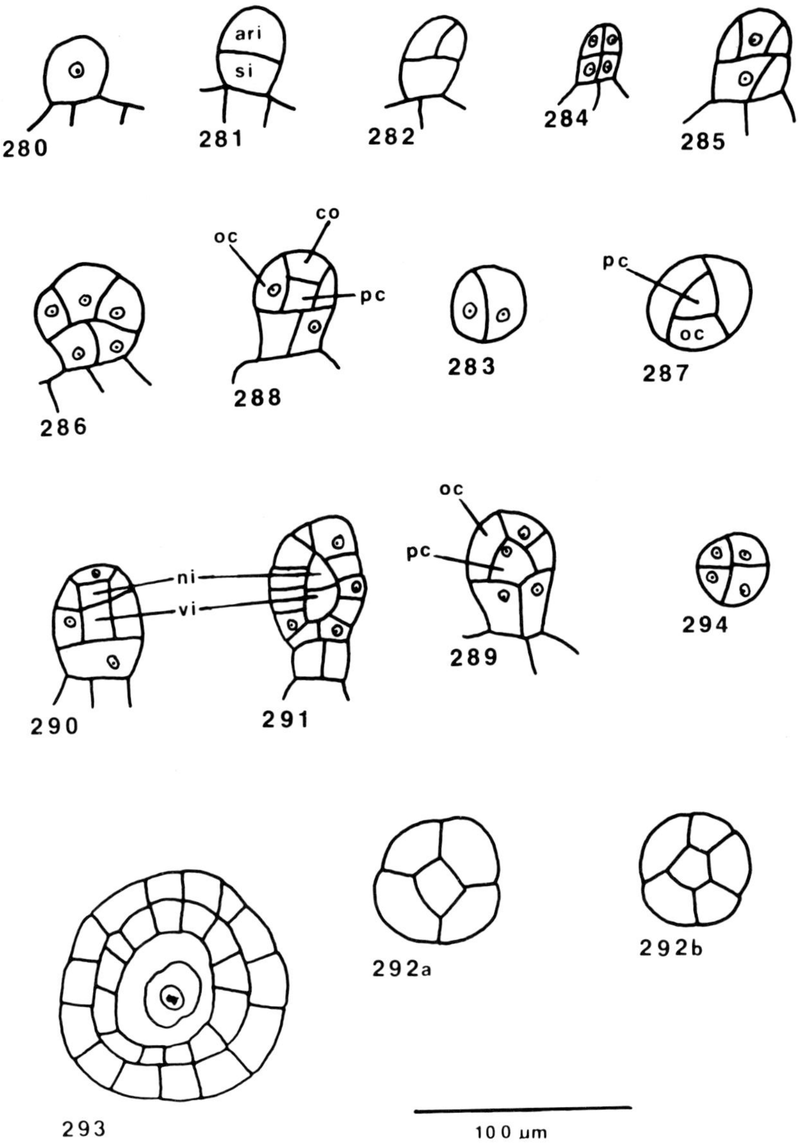

280
ari
si
281
282
284
285
oc
co
pc
286
288
283
pc
oc
287
ni
vi
oc
pc
290
291
289
294
293
292a
292b
100 µm

FIGURES 295-297. Fine structure of the immature archegonium of *Haplomitrium hookeri* (Smith) Nees. -295a-c. Longitudinal sections. 295a. Entire archegonium. 295b. Apex. Note the cover cell, unistratose neck cells, and the inner row of neck canal cells. 295c. Ventral canal cell within the venter. -296. Neck canal cells of an immature archegonium. Note the vesicles which line the inner wall of the neck cells (arrows) and the prominent autophagic vacuoles and plastids of the neck canal cells. -297. Autophagic vacuoles and plastids in the neck canal cells. Note the plasmodesmata which link the neck canal cells.

Key: av, autophagic vacuole; co, cover cell; n, nucleus; nc, neck canal cell; p, plastid; pd, plasmodesma; v, vacuole; vlc, ventral canal cell. (Figs. 295-297: *Lorbeer* *H56* [PR].)

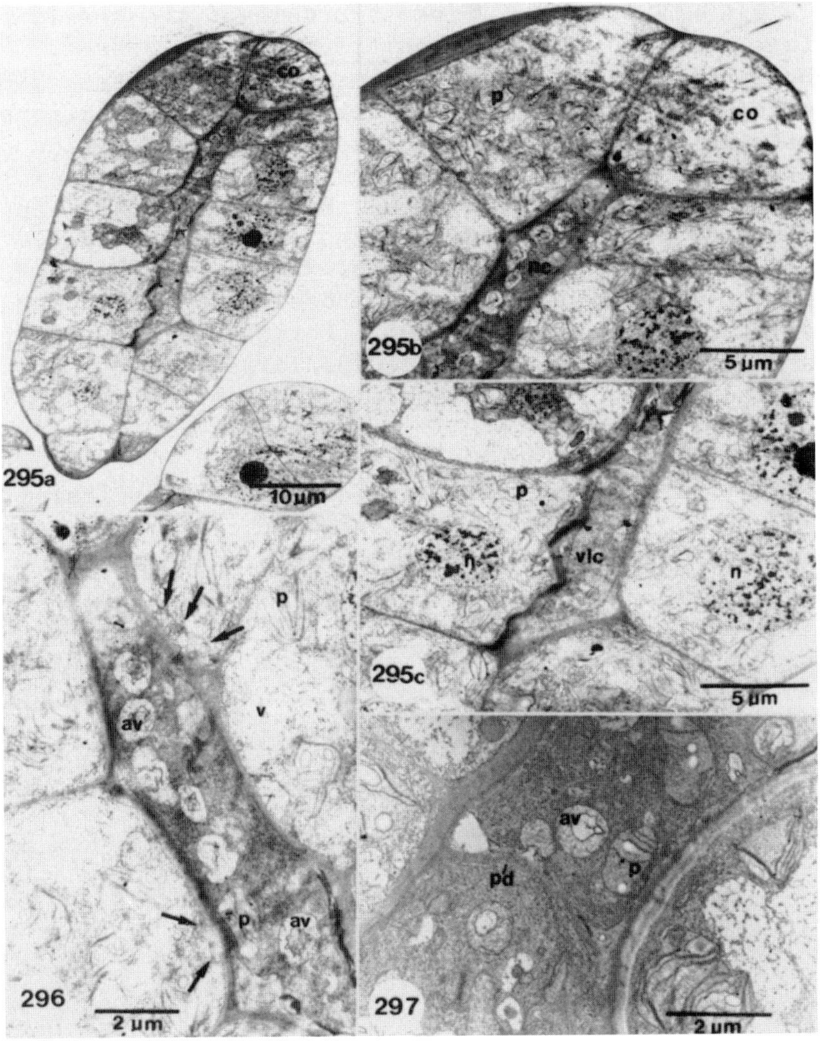

295a
co
10 µm
295b
p
co
5 µm
295c
p
n
vlc
n
5 µm
296
p
v
av
p
av
2 µm
297
av
pd
p
2 µm

FIGURES 298-299. Fine structure of the nearly mature archegonium of Haplomitrium Nees. -298a-c. Longitudinal sections. H. hookeri (Smith) Nees. 298a. Apex of the neck showing the cover cell, and portions of the neck and neck canal cells. 298b. Midregion of the neck showing a neck canal cell and a portion of the neck cells. 298c. Portion of the adjacent neck canal cells. Note the plasmodesmata which link the cells. -299a-b. Venter in transection. H. blumii (Nees) Schust. 299a. Note the bistratose wall surrounding the ventral canal cell. 299b. Peripheral portion of the ventral canal cell. Note the large zone of material separating the ventral canal cell and the venter cells.

Key: co, cover cell; er, endoplasmic reticulum; g, dictyosome; m, mitochondrion; n, nucleus; nc, neck canal cell; ne, neck cell; p, plastid; pd, plasmodesma; sg, starch grain; v, vacuole; vlc, ventral canal cell. (Fig. 298: Lorbeer H56 [PR]; 299: Furuki 384 [ABSH].)

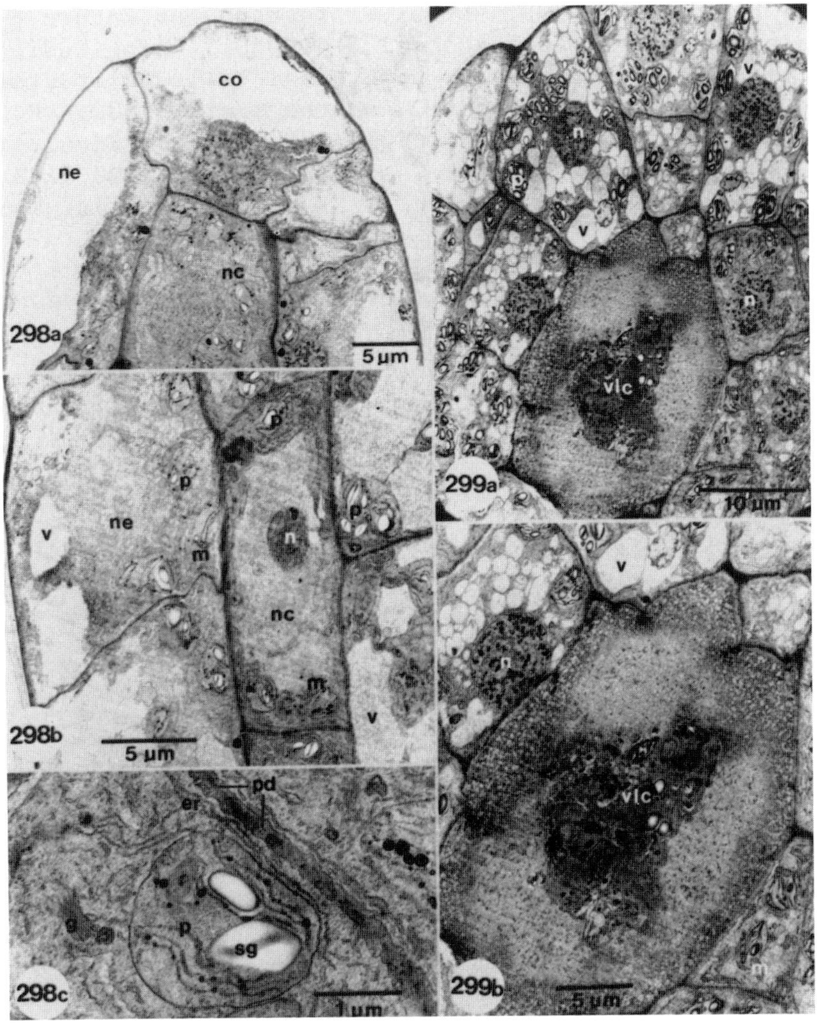
co
ne
nc
298a
5 µm
p
v
ne
m
n
nc
m
v
298b
5 µm
pd
er
p
sg
298c
1 µm
v
n
v
n
vlc
299a
10 µm
v
n
vlc
m
299b
5 µm

FIGURES 300-305. Light micrographs of mature *Haplomitrium* Nees archegonia. -300-301. Longitudinal sections. Note the neck, neck canal cells, bistratose venter, ventral canal cell, and the stalk. 300. *H. gibbsiae* (Steph.) Schust. 301. *H. monoicum* Engel. -302. *H. indicum* (Udar & Chandra) Schust. -303. *H. dentatum* (Kumar & Udar) Engel. -304. *H. grollei* Kumar & Udar. -305. *H. blumii* (Nees) Schust.

Key: ne, neck; nc, neck canal cell: s, archegonial stalk; sp, slime papilla; ve, venter; vlc, ventral canal cell. (Fig. 300: *Sneddon s. n.* [ABSH]; 301: *Crosby 14165* [F]; 302: *Udar & Kumar* C70/5 [LWU]; 303: *Udar & Kumar C72/1* [LWU]; 305: *Furuki 384* [ABSH].)

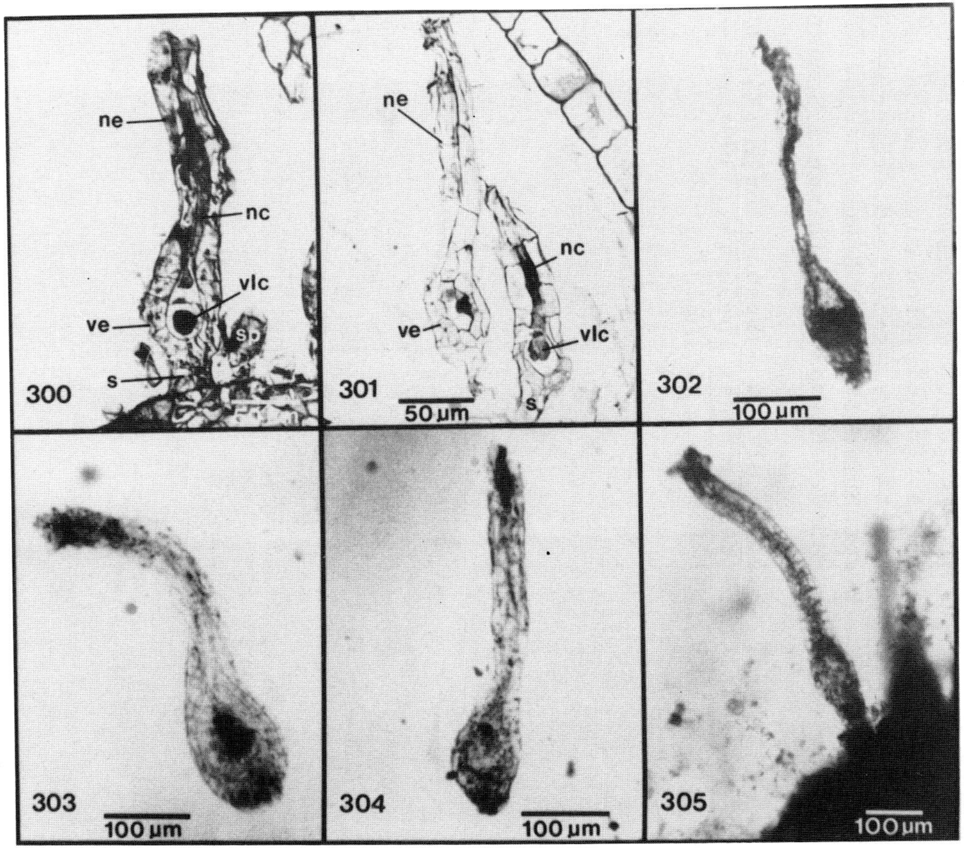
300
ne
nc
vlc
ve
sp
s
301
ne
nc
vlc
ve
s
50 µm
302
100 µm
303
100 µm
304
100 µm
305
100 µm

FIGURES 306-310. Archegonial morphology of *Haplomitrium* Nees. Scanning electron micrographs. -306. *H. hookeri* (Smith) Nees. -307. *H. intermedium* Berrie. -308. *H. gibbsiae* (Steph.) Schust. -309. *H. chilensis* Schust. -310. *H. mnioides* (Lindb.) Schust.

Key: (Fig. 306: *Schofield & Spence 83691* [F]; 307: *Berrie 2/62/1* [SYD]; 308: *Engel 16590* [F]; 309 from *Engel 10847* [F]; 310: *Doei s.n.* [ABSH].)

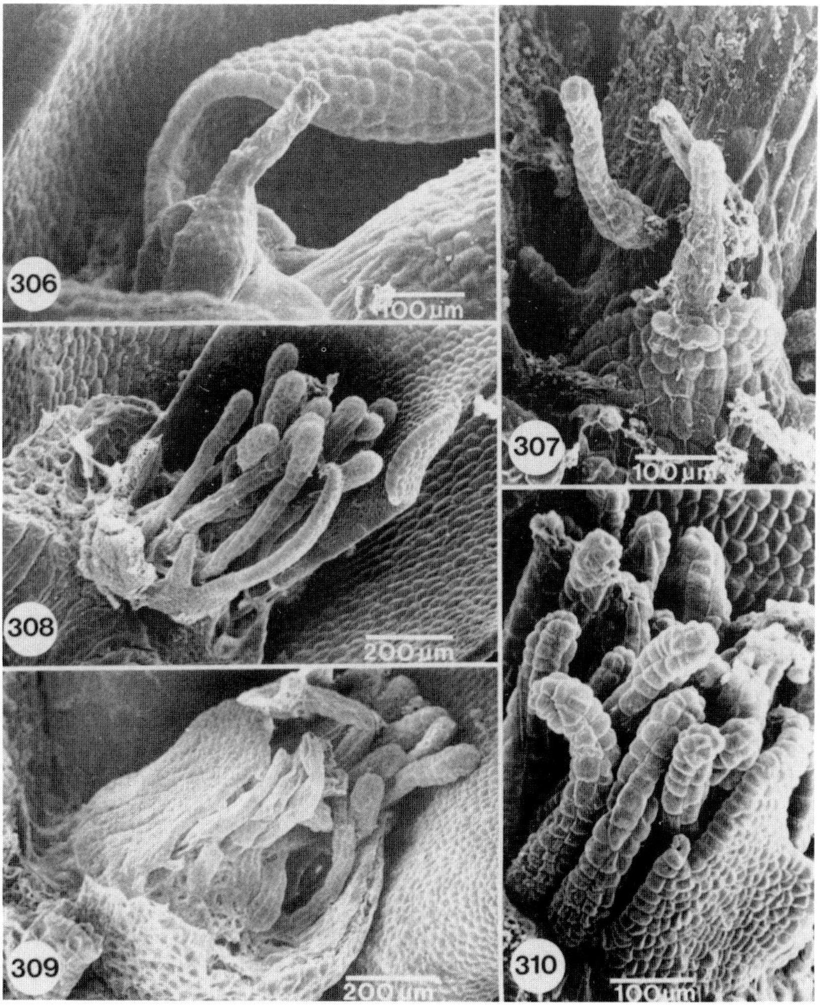
306
100 µm
307
100 µm
308
200 µm
309
200 µm
310
100 µm

FIGURES 311a-e. Diagrams of the ontogenetic patterns responsible for the sporophyte protective structures of *Haplomitrium* Nees and *Steereomitrium* E. O. Campb. -311a. True calyptra of *H. hookeri* (Smith) Nees, *S. minutum* E. O. Campb., *H. ovalifolium* Schust., and *H. monoicum* Engel. Only archegonial meristematic activity occurs. The calyptra is smooth. -311b. Shoot-calyptra of *H. intermedium* Berrie and some individuals of *H. andinum* (Spruce) Schust. Stem activity is only extensive enough to raise a few structures slightly. The elevated inner bracts become lingulate. -311c. Shoot-calyptrae of *H. andinum*, *H. blumii* (Nees) Schust., and *H. mnioides* (Lindb.) Schust. Two possibilities exist. (1) Stem activity carries the available intra-gynoecial structures aloft, but not within the active archegonial meristematic zone. The raised structures are restricted to the basal regions of the shoot-calyptra. (2) Stem activity is extensive enough and appropriately timed to place the raised structures within the archegonial meristematic zone which carries them to the apex. -311d. Perigynium of *H. chilensis* Schust., the Indian endemics, and rarely in *H. gibbsiae* (Steph.) Schust. Stem activity occurs in a small zone beneath the fertilized archegonium. The perigynium bears no foliar outgrowths except a rare bractlet. -311e. Perigynium typical of *H. gibbsiae*. An extensive axial zone lifts numerous intra-gynoecial structures including foliar ones which are then elaborated.

Key: Stippling indicates zone of meristematic activity.

Λ bractlet

P slime papilla or slime hair

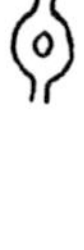 unfertilized archegonium

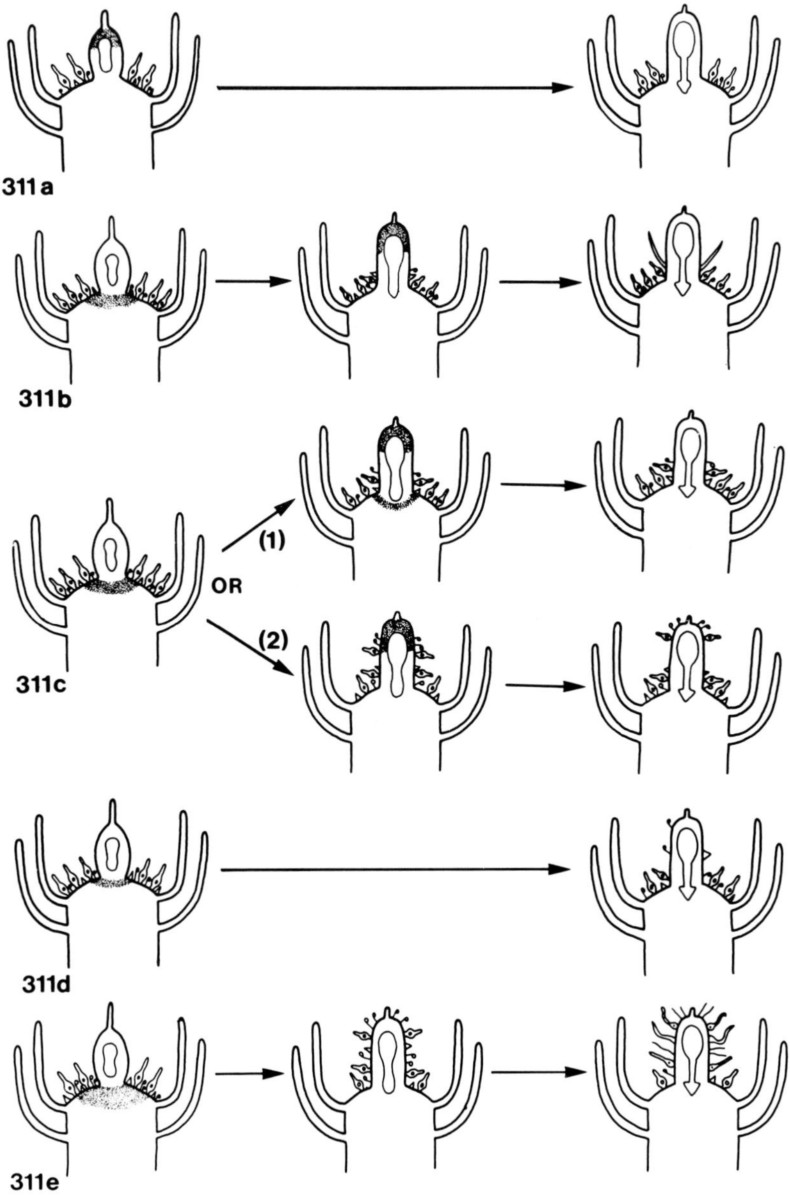

311a

311b

311c

311d

311e

FIGURES 312-318. Calyptrae of *Haplomitrium hookeri* (Smith) Nees, *H. monoicum* Engel, and *Steereomitrium minutum* E. O. Campb. -312-315. *H. hookeri*. 312. Light micrograph of calyptra. 313-315. Scanning electron micrographs. 313. Immature calyptra subtended by an unfertilized archegonium. Note the small, meristematic cells at the apex of the venter (arrow a_1) and the more elongate cells near the base (arrow a_2). 314. Mature calyptra. Note that it is devoid of outgrowths. 315. Transection through a paraffin embedded calyptra and enclosed sporophyte. Note the calyptra wall is 2-4-stratose. -316. Light micrograph of a *S. minutum* shoot bearing an immature calyptra. The bracts have been folded back to reveal the smooth surface of the calyptra. -317-318. *H. monoicum*. 317. Light micrograph of a plant bearing a calyptra. 318. Scanning electron micrograph showing the smooth calyptra. Note the cellular collapse of the apical region (arrow) due to it being composed of small, delicate meristematic cells.

Key: ar, archegonium; cl, calyptra; cpw, capsule wall. (Figs. 312, 314, 315: *Long 9930* [BM]; 313: *Schofield & Spence 83691* [F]; 316: *Campbell s.n.* [NY]; 317, 318: *Crosby 14165* [F].)

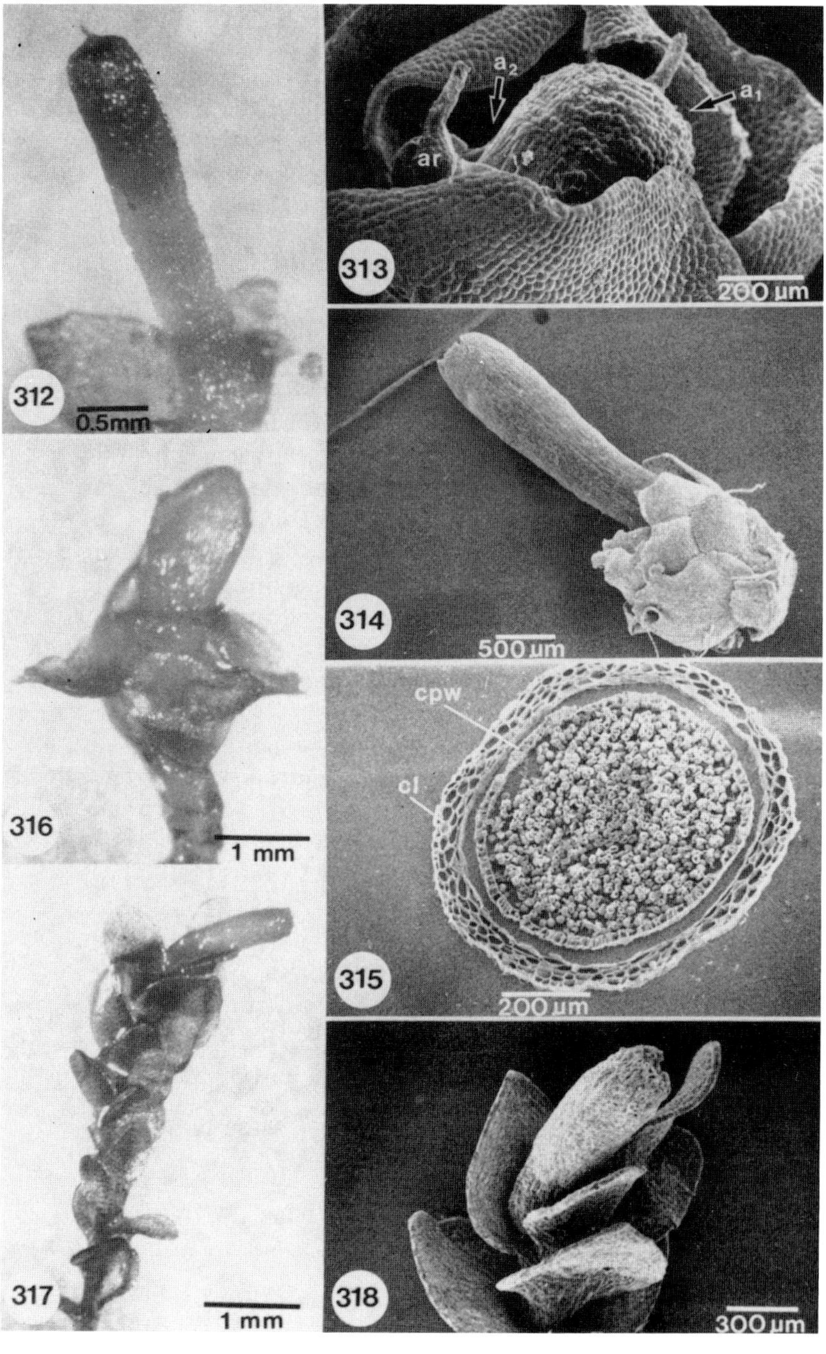
312
0.5mm
313
a2
a1
ar
200 µm
314
500 µm
315
cpw
cl
200 µm
316
1 mm
317
1 mm
318
300 µm

FIGURES 319-323. Longitudinal sections made through the sporophyte protective structures of *Haplomitrium hookeri* (Smith) Nees, *H. andinum* (Spruce) Schust. and *H. blumii* (Nees) Schust. -319. Young calyptra of *H. hookeri*. Note that meristematic activity occurs exclusively in the archegonium and consequently, the accessory structures associated with the fertilized archegonium are not elevated. -320. Distal portion of a *H. andinum* shoot-calyptra showing elevated slime papillae and the start of archegonial meristematic activity (arrows). Note the old archegonial neck. -321-323. *H. blumii*. 321. Immature shoot-calyptra. Note the meristematic activity in the stem below the fertile archegonium. 322. Older stage of development showing the continuation of axial activity, the extent of stem derived tissue (arrows) and the elevation of some intra-gynoecial structures. 323. Apical portion of an older shoot-calyptra in which archegonial meristematic activity has begun. Note the small, quadrate cells of the meristematic zone at the apex of the archegonium and the deep insertion of the foliar outgrowths.

Key: bt, bractlet; cpw, capsule wall; fg, foliar outgrowth; ne, archegonial neck; sp, slime papilla. (Fig. 319: *Long 9930* [BM]; 320: *Spruce s.n.* [NY]; 321, 323: *Norris 61631* [H]; 322: *Furuki 384* [ABSH].)

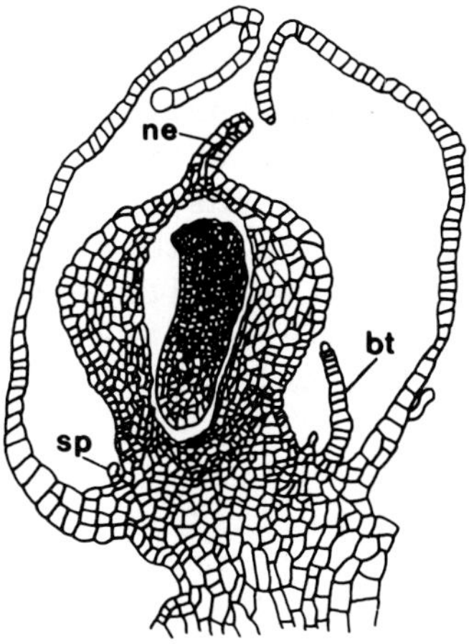

319

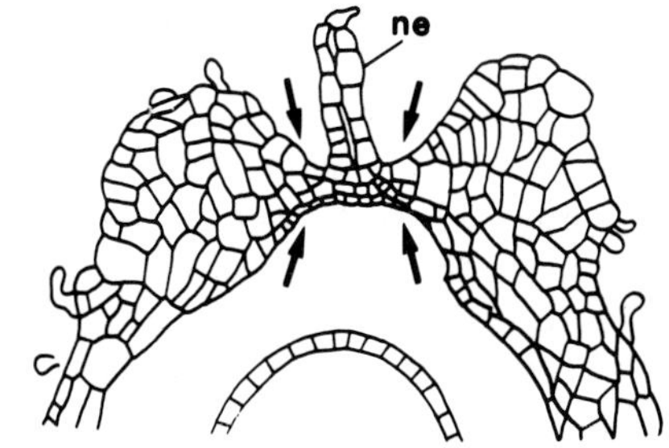

320

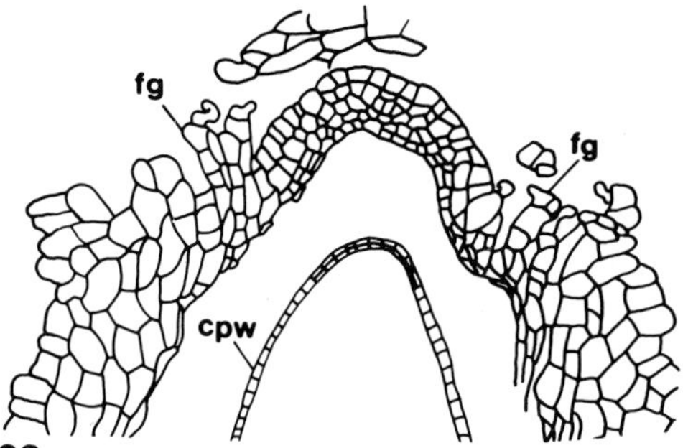

323

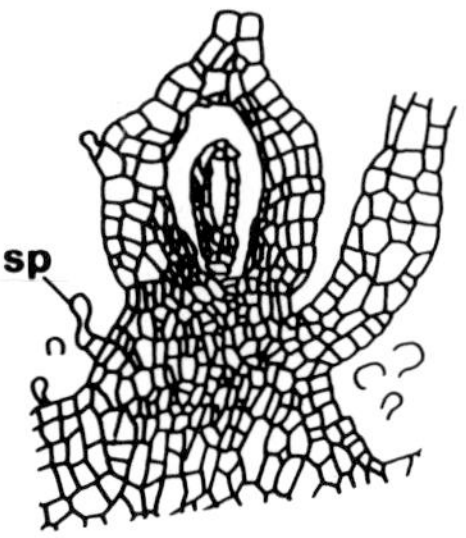

321

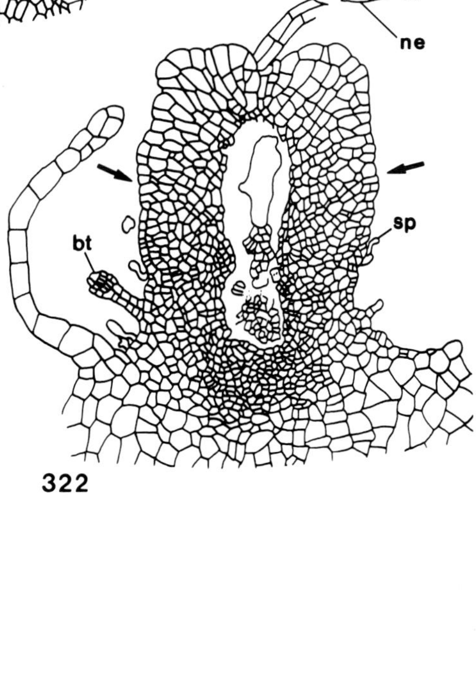

200 um

322

FIGURES 324-325. Shoot-calyptrae of *Haplomitrium intermedium* Berrie. -324a-c. Light micrographs. 324a-b. Opposite views of a female plant bearing a shoot-calyptra. Note the raised, lingulate inner bracts, the branch which originates from the gynoecium, and the unmodified perichaetial bracts. 324c. Secondarily elongated inner bracts which are inserted on the shoot-calyptra. -325. Scanning electron micrograph of a mature shoot-calyptra with its elevated, lingulate bracts.

Key: b, branch; br, outer perichaetial bract; ib, elevated, secondarily elongated inner bract; sc, shoot-calyptra. (Figs. 324-325: *Berrie 2/62/1* [SYD].)

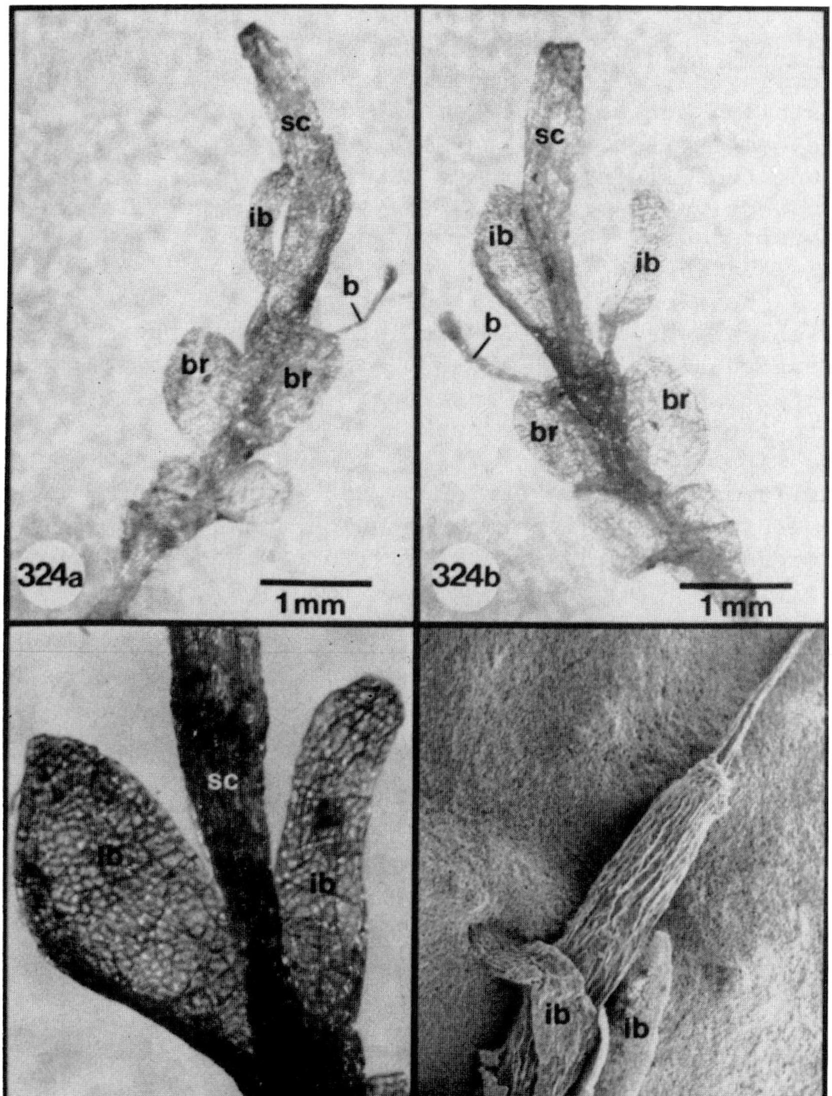
sc
ib
b
br
br
324a
1 mm
sc
ib
ib
b
br
br
324b
1 mm
sc
ib
ib
b
324c
0.5 mm
ib
ib
325
600 µm

FIGURES 326-330. Shoot-calyptrae of Haplomitrium andinum (Spruce) Schust. -326-328. Light micrographs. 326a. Shoot-calyptra which is smooth distally, but bears raised lingulate inner bracts in its basal fifth. 326b. Elevated, secondarily elongated, lingulate inner bracts. 327a. Shoot-calyptra with bractlets and slime papillae scattered over the entire surface. 327b. Apex with an elevated bractlet and slime papillae. 328. Base of a shoot-calyptra with the bracts folded back to reveal an unelaborated bractlet which has been elevated slightly. -329-330. Scanning electron micrographs. 329a. Distal portion of a female shoot with the bracts folded back to reveal the entire shoot-calyptra. 329b. Apex showing the collapsed meristematic region (arrow). 330. Distal portion of a mature shoot-calyptra from which the sporophyte has emerged. Note its irregularly torn mouth and smooth walls.

Key: br, bract; bt, bractlet; ib, raised, secondarily elongated inner bract; sc, shoot-calyptra. (Figs. 326, 328-330: Spruce s.n. [NY]; 327: Spruce s.n. [BM].)

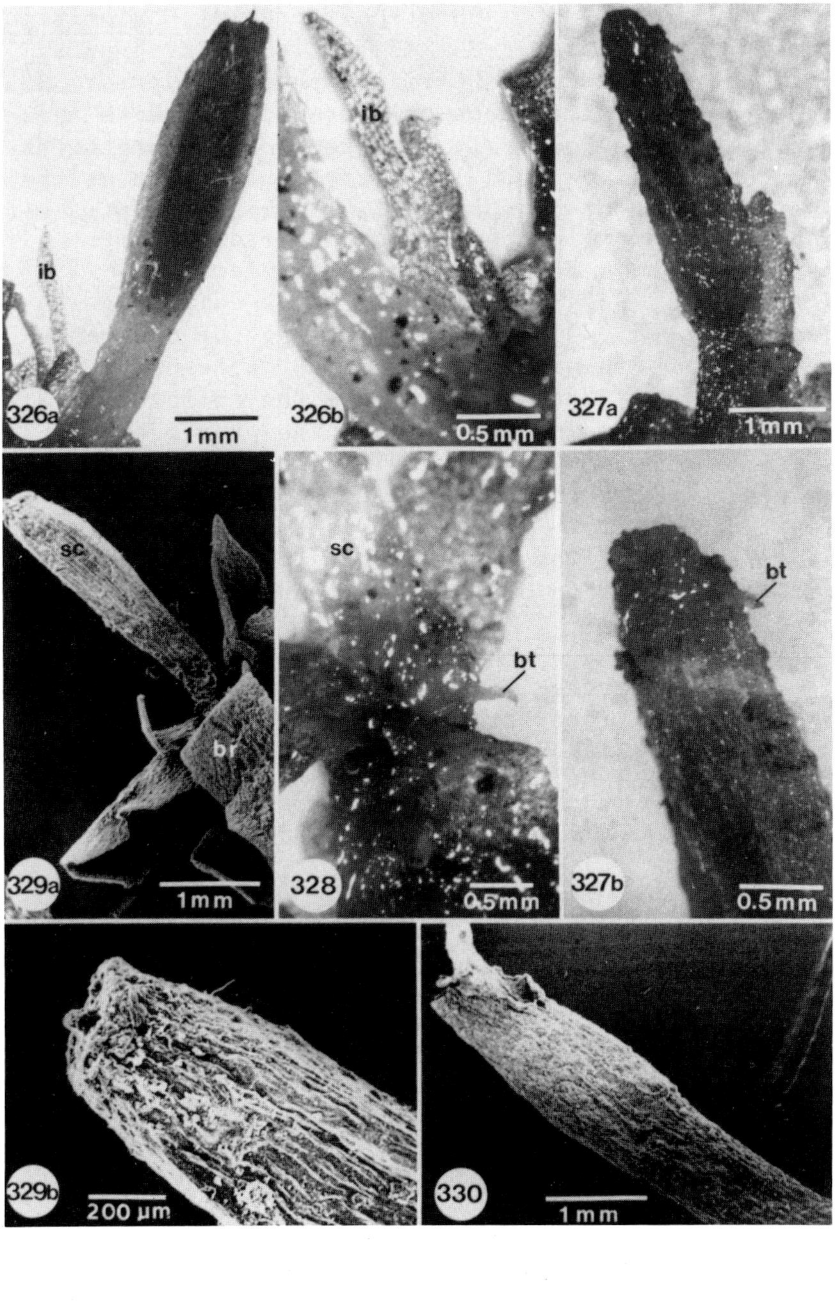
326a
ib
1 mm
326b
ib
0.5 mm
327a
1 mm
329a
sc
br
1 mm
328
sc
bt
0.5 mm
327b
bt
0.5 mm
329b
200 µm
330
1 mm

FIGURES 331-336. Shoot-calyptrae of *Haplomitrium blumii* (Nees) Schust. -331-334. Light micrographs. 331. Habit of a plant bearing a shoot-calyptra. 332. Apical portion of a shoot-calyptra covered with raised bractlets and secondarily produced foliar outgrowths. 333. Slightly elevated lingulate inner bract at the base of the shoot-calyptra. 334. Shoot-calyptra with archegonia and slime papillae scattered over its entire length. -335-336. Scanning electron micrographs. 335a-c. Immature shoot-calyptra with bracts removed. 335a. Habit showing the ornamented apex, the smooth midregion, and the archegonia in the basal region. 335b. Apex. Note the foliar outgrowths beginning to form and the presence of raised bracts. 335c. Base with archegonia and bractlets. Note that some of the archegonia are elevated slightly. 336. Old shoot-calyptra in which the apical foliar outgrowths and elevated apical bractlets have been destroyed by the emergent sporophyte.

Key: ar, archegonium; bt, bractlet; fg, foliar outgrowth; ib, inner bract. (Figs. 331, 334: *Furuki 384* [ABSH]; 332: *Schiffner 398* [G]; 333: *Verdoorn 60* [G]; 335: *Norris 61631* [H]; 336: *Campbell s.n.* [NY].)

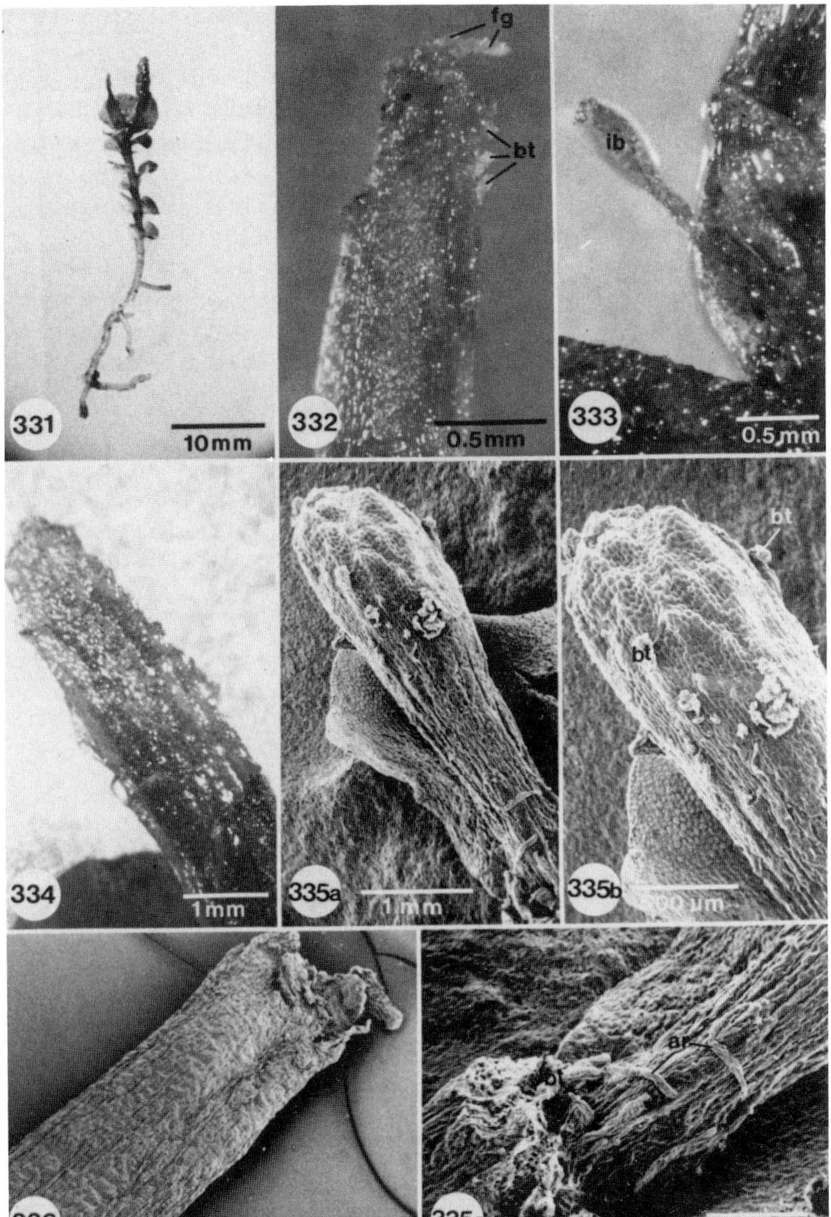
331
10 mm
332
fg
bt
0.5 mm
333
ib
0.5 mm
334
1 mm
335a
1 mm
335b
bt
bt
336
1 mm
335c
ar

FIGURES 337-340. Shoot-calyptrae of *Haplomitrium mnioides* (Lindb.) Schust. -337-338. Light micrographs. 337. Shoot bearing an immature shoot-calyptra. 338a-b. Shoot-calyptra with archegonia and bractlets raised only in basal one-fifth. 338a. Smooth distal portion. 338b. Bracts folded back to expose the basal archegonia, some of which are raised slightly. -339-340. Scanning electron micrographs. 339a-b. Immature shoot-calyptra removed from the plant. 339a. Note the slightly elevated basal bractlets and archegonia, the smooth midregion, and the ornamented apex. 339b. Basal region showing elevated bractlets and archegonia. 340a-b. Mature shoot-calyptra from which the sporophyte has emerged. The outer perichaetial bracts have been removed. 340a. Entire protective structure showing accessory structures at the apex and base, and a smooth midregion. 340b. Distal portion with elevated bractlets and archegonia.

Key: ar, archegonium; br, bract; bt, bractlet; sc, shoot-calyptra; sp, slime papilla. (Figs. 337, 339-340: *Horikawa 62* [H]; 338: *Furuki 7408* [ABSH].)

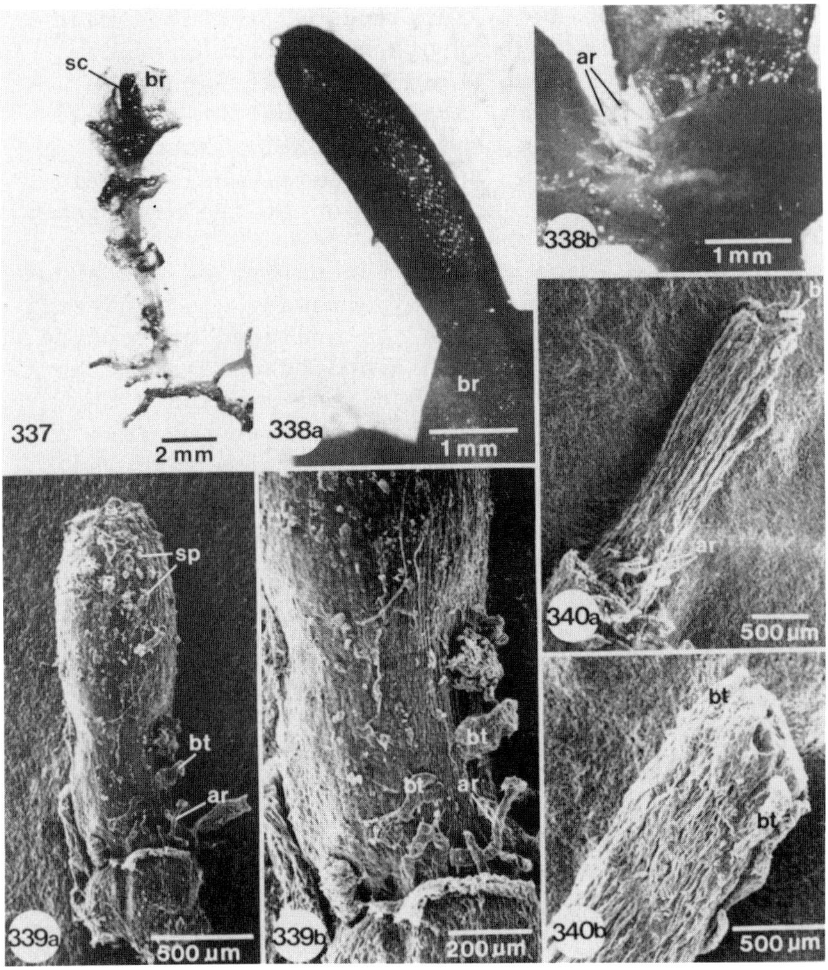
sc
br
337
2 mm
338a
br
1 mm
ar
c
338b
1 mm
340a
ar
500 µm
339a
sp
bt
ar
500 µm
339b
bt
bt
ar
200 µm
340b
bt
bt
500 µm

FIGURES 341-344. Longitudinal sections made through the sporophyte protective structures of *Haplomitrium mnioides* (Lindb.) Schust. and *Haplomitrium gibbsiae* (Steph.) Schust. -341-343. Shoot-calyptrae of *H. mnioides*. 341. Young developmental stage. Note the axial meristematic activity and the slight elevation of the archegonia, slime papillae and bractlets. 342. Older developmental stage showing the extent of the shoot derived tissue (indicated by arrows). 343. Nearly mature shoot calyptra with its enclosed sporophyte. Note the elevated archegonia and bractlets in the basal region and the smooth distal portion. The small, quadrate cells at the apex mark the zone of later archegonial meristematic activity (arrows). -344. Perigynium of *H. gibbsiae*. Note the thick perigynial walls which still appear meristematic and the elevated, elaborated slime hairs, slime papillae and bractlets.

Key: ar, archegonium; br, bract; bt, bractlet; cpw, capsule wall; ne, neck; sh, slime hair; sp, slime papilla. (Fig. 341: *Hattori 101* [H]; 342, 343: *Doei s.n.* [ABSH]; 344: *Engel 16590* [F].)

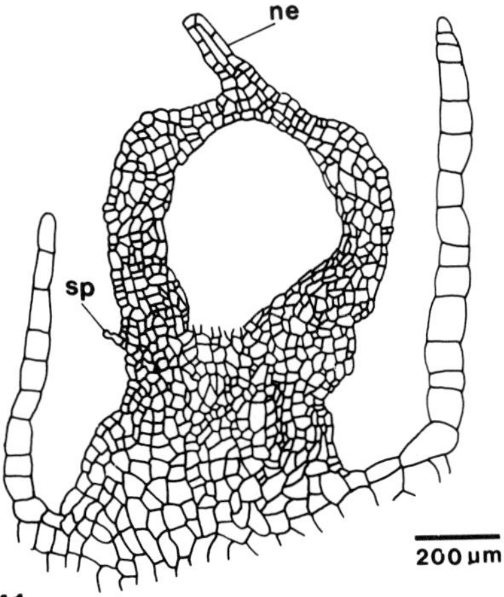

341

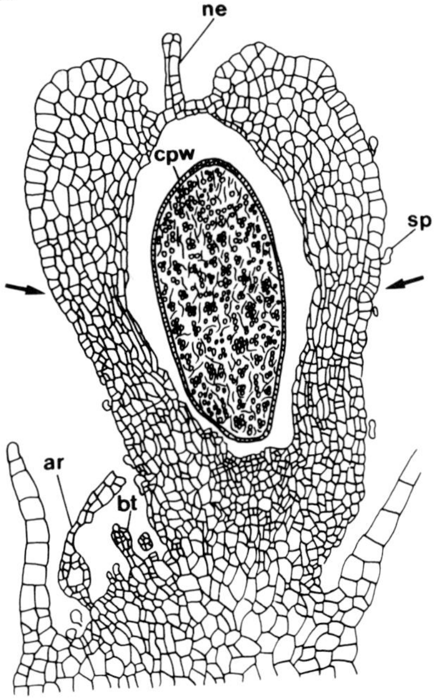

342

200 µm

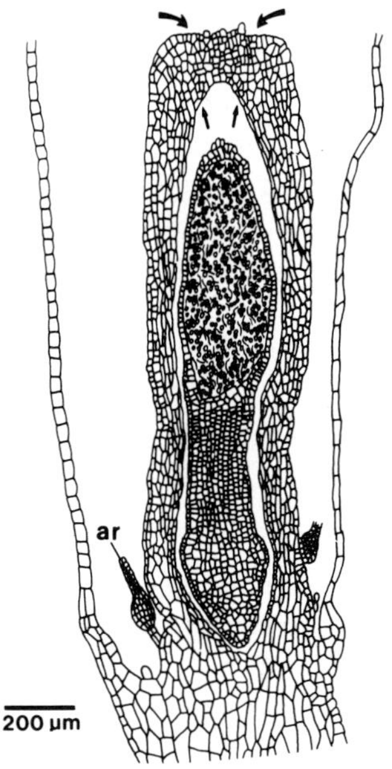

343

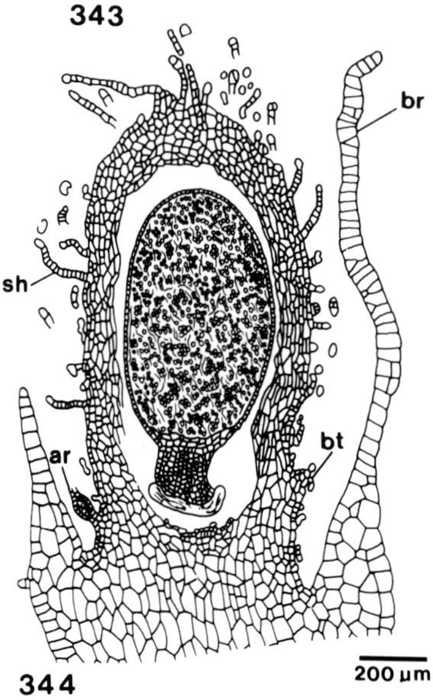

344

FIGURES 345-351. Perigynia of *Haplomitrium gibbsiae* (Steph.) Schust. and *H. chilensis* Schust. -345-348. *H. gibbsiae*. 345-347. Light micrographs. 345. Two plants each bearing a perigynium. 346a-b. Perigynium devoid of foliar structures. 346a. Entire perigynium. 346b. Apex. 347. Perigynium with a single, raised bractlet. -348. Scanning electron micrograph of the apex of a perigynium which is covered with elevated archegonia and elaborated bractlets. -349-351. Light micrographs of *H. chilensis* perigynia. 349. Plant bearing a perigynium. 350. Perigynium with a single, raised bractlet. 351a-b. Perigynium lacking foliar outgrowths. 351a. Archegonia on swollen receptacle and some which are elevated slightly onto the base of the perigynium. 351b. Apex.

Key: ar, archegonium; bt, bractlet; sp, slime papilla. (Figs. 345, 348: *Sneddon s.n.* [ABSH]; 346: *Burrell, Scott, & Taylor 136036* [AD]; 347: *Allison H2264* [AD]; 349-351: *Engel 11513B* [F].)

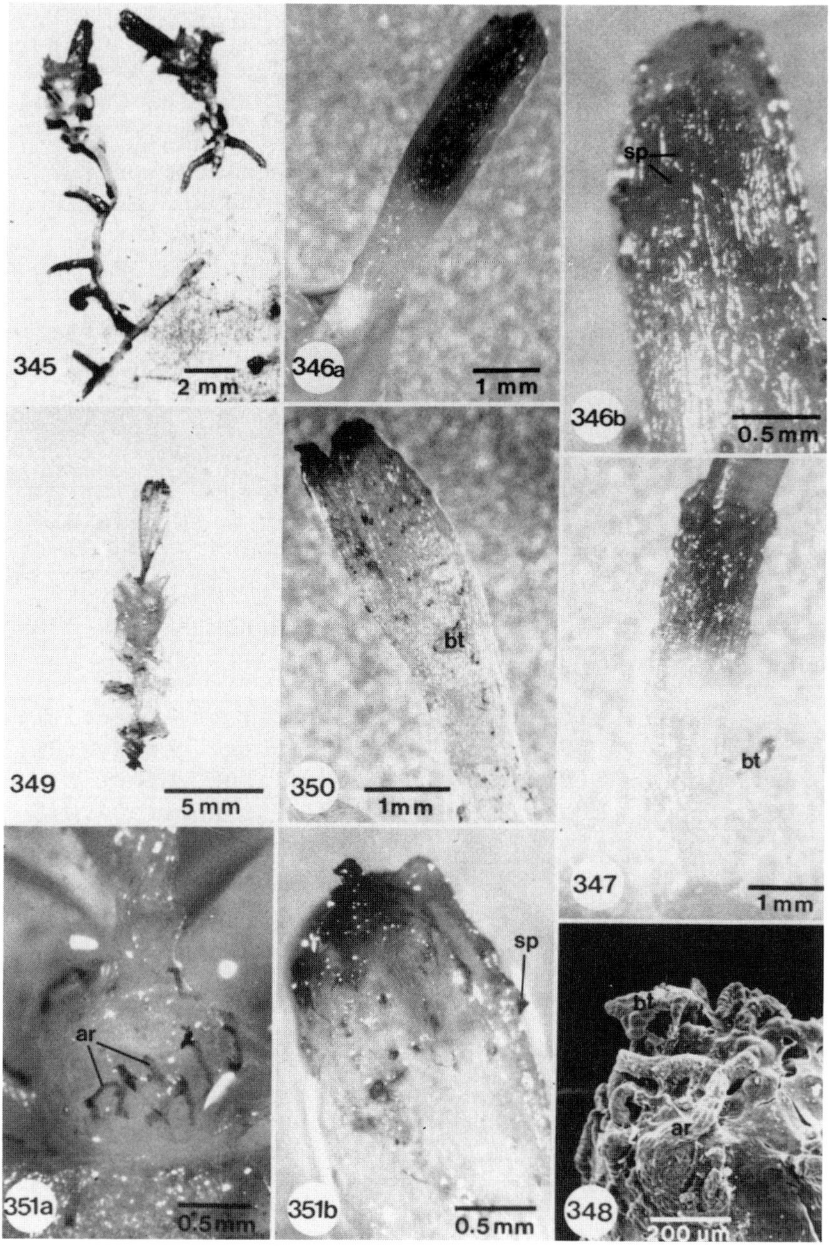
345
2 mm
346a
1 mm
sp
346b
0.5 mm
349
5 mm
bt
350
1mm
bt
347
1 mm
ar
351a
0.5 mm
sp
351b
0.5 mm
bt
ar
348
200 µm

FIGURES 352-356. Perigynia of *Haplomitrium gibbsiae* (Steph.) Schust., *H. indicum* (Udar & Chandra) Schust., and *H. grollei* Kumar & Udar. -352-353. Scanning electron micrographs of a mature *H. gibbsiae* perigynium bearing foliar outgrowths. 352*a*-b. Immature perigynium. A bract has been removed to expose the elaborate ornamentation. 352a. Entire perigynium. 352b. Elaborated slime hairs and bractlets. 353a-b. Apex of an old perigynium. 353a. Note the large foliar structure (= elaborated bractlet). 353b. Wedge-shaped initial responsible for the secondary growth of the foliar structure. -354-356. Light micrographs. 354-355. *H. indicum*. 354. Plant with one unfertilized and one fertilized female shoot bearing an immature perigynium. 355a. Immature perigynium. Note the absence of foliar structures. 355b. Distal portion of a perigynium covered only by slime papillae. 356a-b. *H. grollei*. 356a. Base of a perigynium with the bracts folded back. 356b. Basal fifth of the perigynium bearing elevated bractlets.

Key: ar, archegonium; br, bract; bt, bractlet; fx, apical cell of foliar structure; pg, perigynium; sh, slime hair. (Figs. 352, 353: *Allison, Scott & Taylor 135503* [MUCV]; 354, 355: *Udar & Kumar C-70/5* [LWU]; 356: *Udar & Kumar C 72/1* [LWU].)

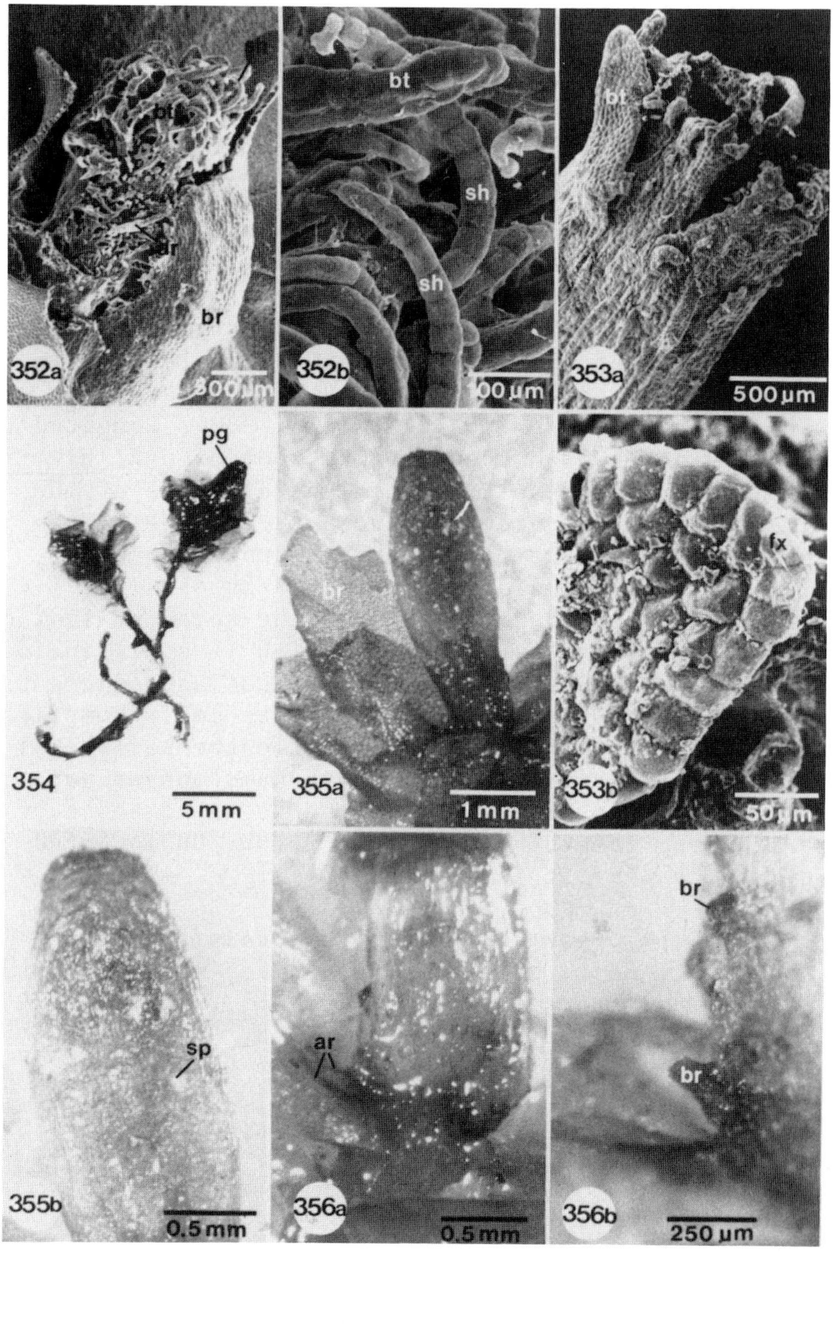
bt
ar
br
352a
300 µm
bt
sh
sh
352b
100 µm
bt
353a
500 µm
pg
354
5 mm
br
355a
1 mm
fx
353b
50 µm
sp
355b
0.5 mm
ar
356a
0.5 mm
br
br
356b
250 µm

FIGURES 357-363. Sporophytic features of *Haplomitrium* Nees. -357-358. Light micrographs. 357. Young sporophyte just emerging from the calyptra of *H. hookeri* (Smith) Nees. 358. Elongate sporophyte of *H. mnioides* (Lindb.) Schust. bearing a cylindrical capsule and hyaline seta. -359. Scanning electron micrograph of a longitudinal section made through a paraffin embedded immature sporophyte of *H. blumii* (Nees) Schust. which is still enclosed within the shoot-calyptra. The foot is inserted in the gametophyte to the level of the first series of bracts. -360-362. Light micrographs. 360. Mature capsule of *H. hookeri*. Note the blunt apex. 361. Mature *H. andinum* capsule with a beaked apex. 362. Longitudinal section made through the embedded foot of a *H. blumii* sporophyte. Note the small peripheral cells of the foot and of the adjacent gametophytic tissue. This marks the sporophyte-gametophyte transfer zone. -363a-b. Scanning electron micrographs of a longitudinal section made through the seta and foot of a young sporophyte of *H. blumii* which is still enclosed within the shoot-calyptra. 363a. Foot and adjacent gametophytic tissue. Larger, quadrate cells comprise the central region of the foot, while the peripheral cells are smaller. The sporophyte-gametophyte juncture is marked by the tiny cells of the foot and those of the adjacent gametophytic zone. 363b. Unelongated seta. Note that it is nonarticulate.

Key: bk, beak; br, bract; cl, calyptra; cp, capsule; ft, foot; ga, gametophyte; sc, shoot-calyptra; se, seta; tz, transfer zone. (Figs. 357, 360: *Long 9930* [BM]; 358: *Doei s.n.* [ABSH]; 359, 362, 363: *Furuki 384* [ABSH]; 361: *Spruce s.n.* [BM].)

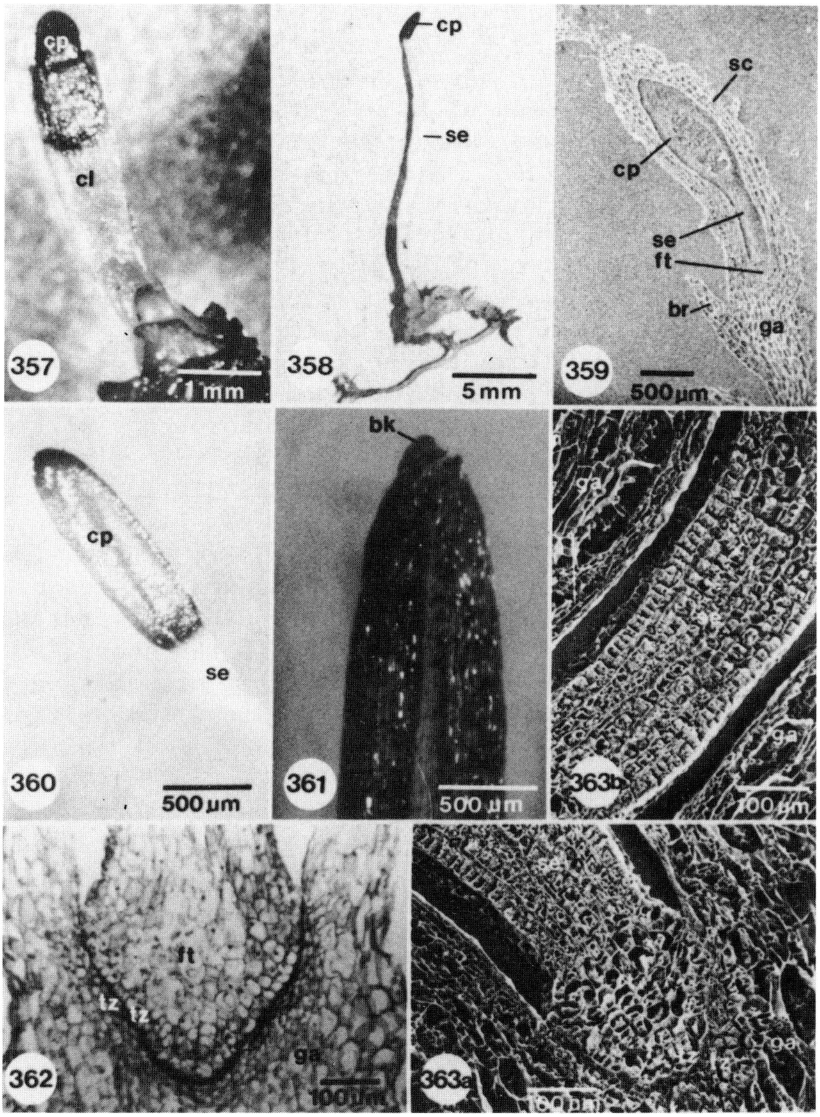

cp
cl
357
1 mm
cp
se
358
5 mm
sc
cp
se
ft
br
ga
359
500 µm
cp
se
360
500 µm
bk
361
500 µm
ga
ga
363b
100 µm
ft
tz
tz
ga
362
100 µm
se
ga
363a
100 µm

FIGURES 364-370. Features of the sporophyte capsules and setae of *Haplomitrium* Nees. -364-368. Sections made through setae. 364-367. Transections. 364. Light micrograph of an elongated seta of *H. blumii* (Nees) Schust. Note that it is of Douin's generalized type. 365. Scanning electron micrograph of an elongated *H. hookeri* (Smith) Nees seta showing a central zone of enlarged cells. 366. Light micrograph of an unelongated seta of *H. mnioides* (Lindb.) Schust. showing the hyaline cells comprising the central region. 367. Scanning electron micrograph of a paraffin embedded elongated seta of *H. mnioides* with a hollow central zone. -368. Scanning electron micrograph of a longitudinal section through a portion of a paraffin embedded elongated seta of *H. mnioides* showing the hollow central zone. -369-370. Longitudinal sections through capsules. Light micrographs. 369a-b. *H. hookeri*. 369a. Apical portion. Note the darkly stained thickening band of each cell of the unistratose wall. 369b. Bistratose composition of the capsule base. Note the short, stout elaters within the capsule. 370. Apex of a beaked *H. gibbsiae* capsule. Note the multistratose composition of the wall.

Key: bk, beak; cl, calyptra; cpw, capsule wall; el, elater; se, seta; sr, spore; tb, thickening band. (Fig. 364: *Furuki 384* [ABSH]; 365: *Long 9930* [BM]; 366-368: *Furuki 7408* [ABSH]; 369: *Jensen 1347* [AD]; 370: *Fife 5737* [CHR].)

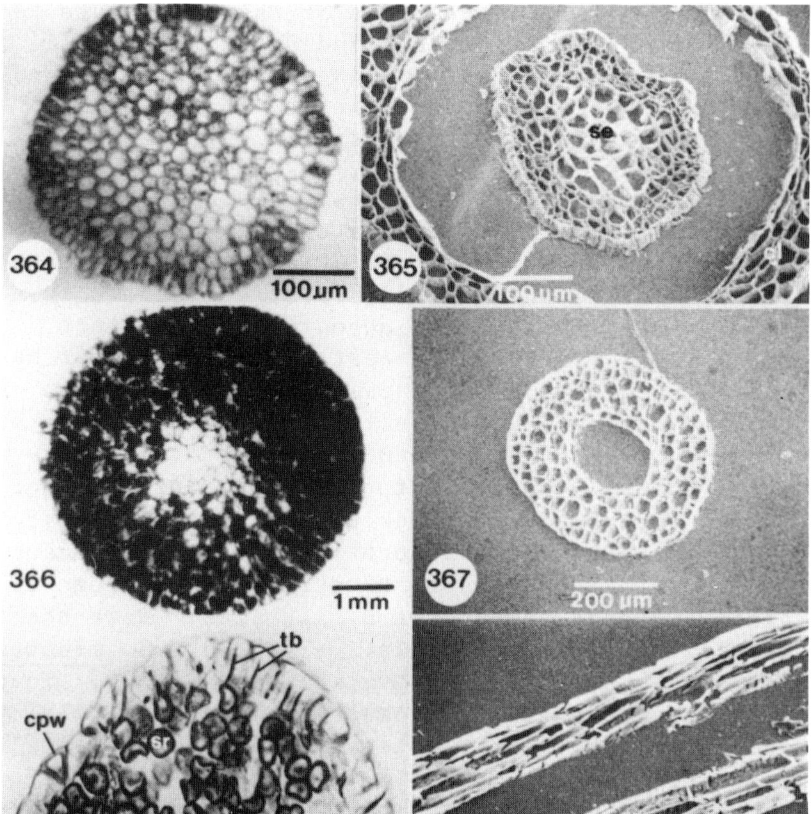
364
100 µm
365
se
100 µm
366
1 mm
367
200 µm
tb
cpw
sr
sr
369a
50 µm
368
200 µm
sr
el
el
369b
se
50 µm
bk
370
500 µm

FIGURES 371-379. Features of the sporophyte capsules of *Haplomitrium* Nees and *Steereomitrium* E. O. Campb. -371-375. Capsule walls of *Haplomitrium*. 371. Light micrograph showing the single, longitudinal annular thickening band traversing each capsule wall cell. *H. blumii* (Nees) Schust. 372-373. Scanning electron micrographs of the inner surface of the capsule wall of *H. hookeri* (Smith) Nees. 372. Viewed from directly above. Note the dark appearance of the longitudinal thickening bands. 373. Viewed at an angle. The thickening band appears as a distinct protrusion only because the rest of the cell wall has collapsed in around it as a result of the fixation procedure. 374. Scanning electron micrograph of the external surface of the capsule wall of *H. gibbsiae* (Steph.) Schust. Note that the cell walls have not collapsed in around the thickening bands and therefore, the bands do not appear as protrusions. 375. Scanning electron micrograph of a transection made through a paraffin embedded capsule of *H. gibbsiae*. Note the thickening bands of the cells. -376-379. *S. minutum* E. O. Campb. 376-378. Light micrographs. 376. Young sporophyte just emerging from the calyptra. 377. Elongated sporophyte. The capsule has opened along two opposite dehiscence lines. 378. Capsule wall cells each bearing a single, longitudinal annular thickening band. 379. Scanning electron micrograph showing the unistratose composition of the capsule wall.

Key: cl, calyptra; cp, capsule; cpw, capsule wall; se, seta; tb, thickening band; w, cell wall. (Fig. 371: *Furuki 384* [ABSH]; 372: *Long 9930* [BM]; 373: *Jensen 1347* [AD]; 374: *Allison, Scott & Taylor 135503* [MUCV]; 375: *Fife 5737* [CHR]; 376-379: *Campbell s.n.* [NY].)

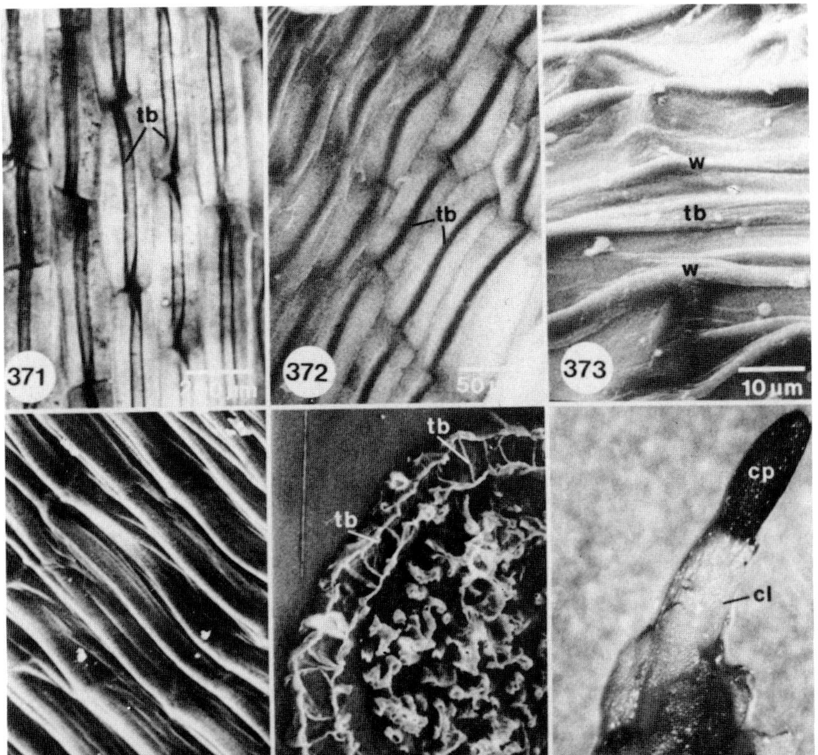
371
tb
372
tb
373
w
tb
w
10 µm
374
375
tb
tb
376
cp
cl
5 mm
379
cpw
378
tb
377
cp
se
5 mm

FIGURES 380-384. Fine structure of the immature capsule wall cells of *Haplomitrium mnioides* (Lindb.) Nees. -380-381. Longitudinal sections showing the generalized deposition of material toward the interior of the cell walls (arrows). Note that organellar degeneration has begun. -382. Tangential walls of adjacent capsule cells. The fibrous middle lamella is distinct from the granular cell wall. Note the more opaque layer of extra material being deposited on the cell walls toward the cell interior by the vesicles which have crossed the plasmalemma. -383. Plasmodesmata which have been plugged by the deposition of the extra wall material. -384. Thickening bands being formed toward the interior of the cell on the radial walls of conterminous cells.

Key: cd, plugged plasmodesma; l, lipid droplet; ml, middle lamella; ob, oil body; p, plastid; pl, plasma membrane; tb, thickening band; tm, extra wall material; vs, vesicle; w, cell wall. (Figs. 380-384: *Furuki 7408* [ABSH].)

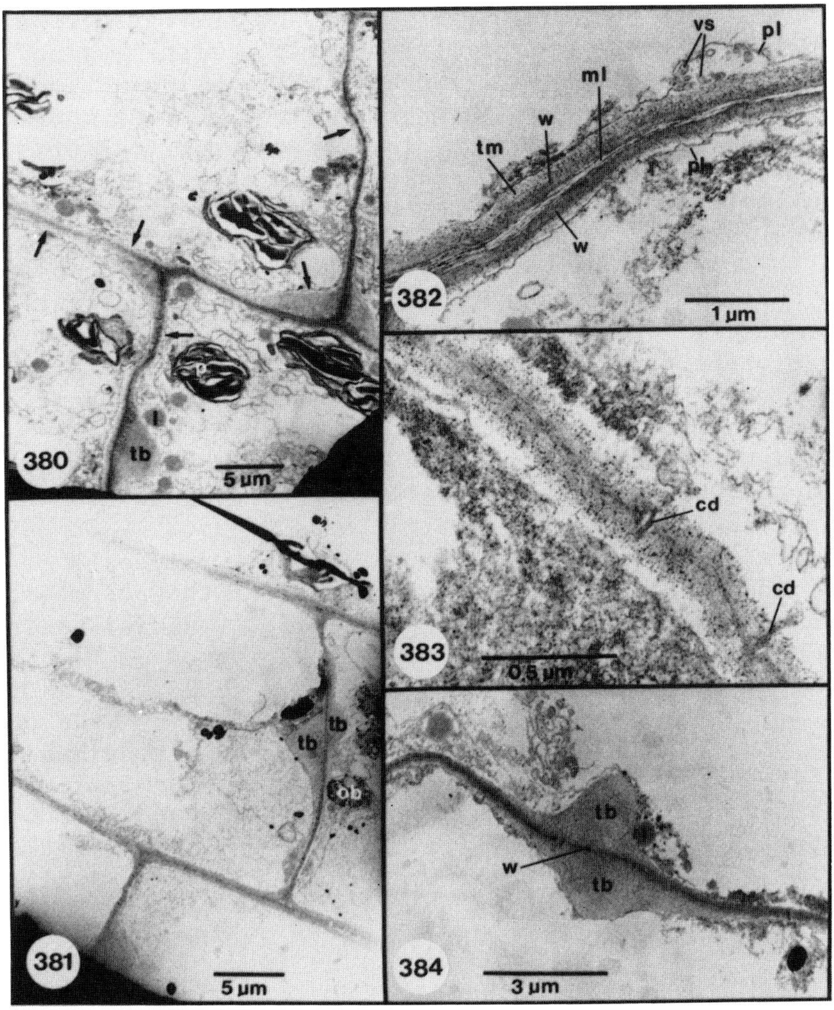

380
tb
5 µm
381
tb
tb
cb
5 µm
382
vs
pl
ml
w
tm
pl
w
1 µm
383
cd
cd
0.5 µm
384
tb
w
tb
3 µm

FIGURES 385-389. Development of the capsule wall thickening band in _Haplomitrium_ Nees. Transmission electron micrographs. -385-386. _H. mnioides_ (Lindb.) Schust. 385. Radial thickening bands forming toward the cell interior from the localized deposition of material by the Golgi vesicles which have crossed the plasmalemma. Note that the younger thickening band (tb_2) is not directly adjacent to the older band (tb_1) on the conterminous cell wall. 386. Thickening band on a radial wall. Note that it is on the inner surface of the cell wall, but external to the plasmalemma. A dictyosome and vesicles are visible. -387. External wall of a capsule cell bearing a mature radial band. Note the more osmiophilic nature of the band as compared to the younger ones shown in Figures 384-386, and the thicker external cell wall as compared to the inner cell walls. _H. blumii_ (Nees) Schust. -388-389. Mature thickening bands of the capsule wall cells of _H. mnioides_. Longitudinal sections. Note the absence of organelles in these mature cells. 388. Radial walls of conterminous cells. 389. Thickening on a tangential wall.

Key: cd, plugged plasmodesma; ew, external wall; g, dictyosome; l, lipid droplet; ob, oil body; p, plastid; pl, plasmalemma; tb, thickening band; tm, layer of thickening material; vs, vesicle; w, cell wall. (Figs. 385, 386, 388, 389: _Furuki 7408_ [ABSH]; 387: _Furuki 384_ [ABSH].)

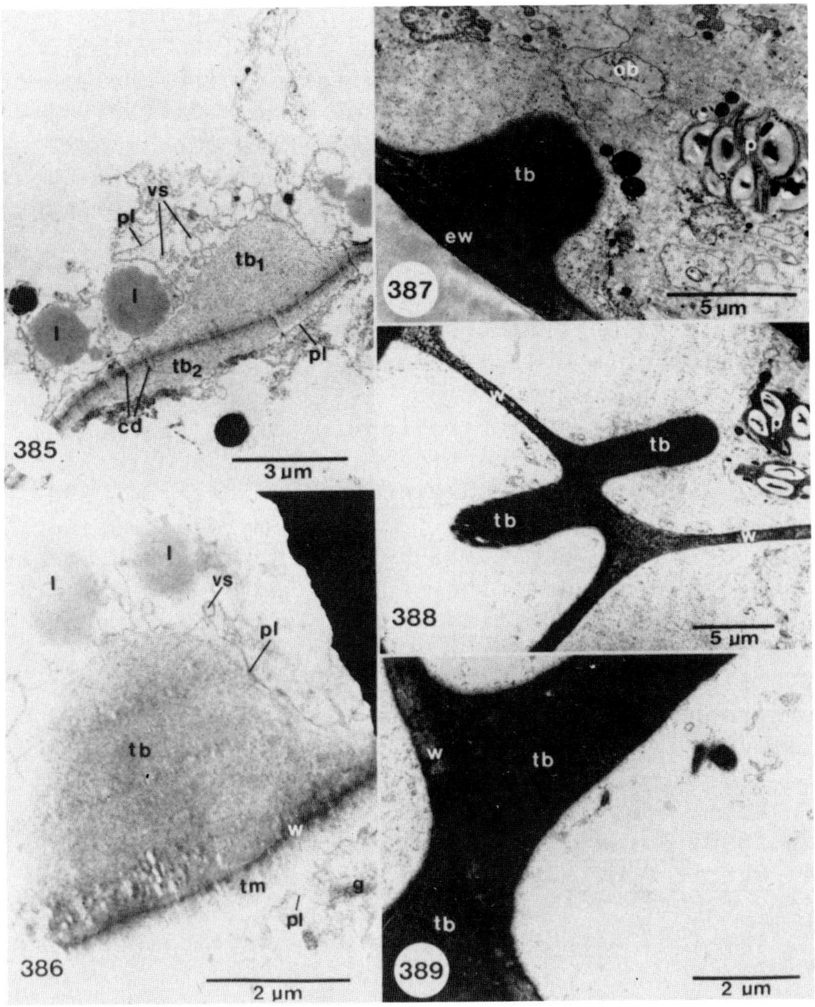
385
vs
pl
tb1
l
l
tb2
pl
cd
3 µm
386
l
l
vs
pl
tb
w
tm
g
pl
2 µm
387
ab
p
tb
ew
5 µm
388
w
tb
tb
w
5 µm
389
w
tb
tb
2 µm

FIGURES 390-401. Elater morphologies of *Haplomitrium* Nees and *Steereomitrium* E. O. Campb. -390. Scanning electron micrograph of the sporogenous tissue of an immature capsule of *H. blumii* (Nees) Schust. which is still enclosed within the shoot-calyptra. Longitudinal section through a paraffin embedded capsule. The elater thickenings have not yet formed and the spore walls are not yet ornamented. -391-393. Mature elaters of *H. hookeri* (Smith) Nees. 391-392. Light micrographs. 391. Portion of a long, attenuate elater which is bispiralled with unispiralled tips. 392. Short, stout elater with a branched tip. 393. Scanning electron micrograph of the unispiralled tip of an attenuate elater. -394-399. Scanning electron micrographs of mature elaters. 394. *H. intermedium* Berrie. 395. *H. gibbsiae* (Steph.) Schust. 396. *H. mnioides* (Lindb.) Nees. 397. *H. blumii*. 398-399. *H. andinum* (Spruce) Schust. 398. Midportion. 399. Tip. -400-401. Mature elaters of *S. minutum* E. O. Campb. Note the unispiralled median portion and the bispiralled tips. 400. Light micrograph. 401. Scanning electron micrograph. (Arrow marks the point where the band spilts.)

Key: cpw, capsule wall; el, elater; sr, spore. (Fig. 390: *Furuki 384* [ABSH]; 391-393: *Long 9930* [BM]; 394: *Berrie 3/62* [SYD]; 395: *Allison, Scott & Taylor 135503* [MUCV]; 396: *Doei s.n.* [ABSH]; 397: *Norris 61631* [H]; 198-199: *Spruce s.n.* [G]; 400, 401: *Campbell s.n.* [NY].)

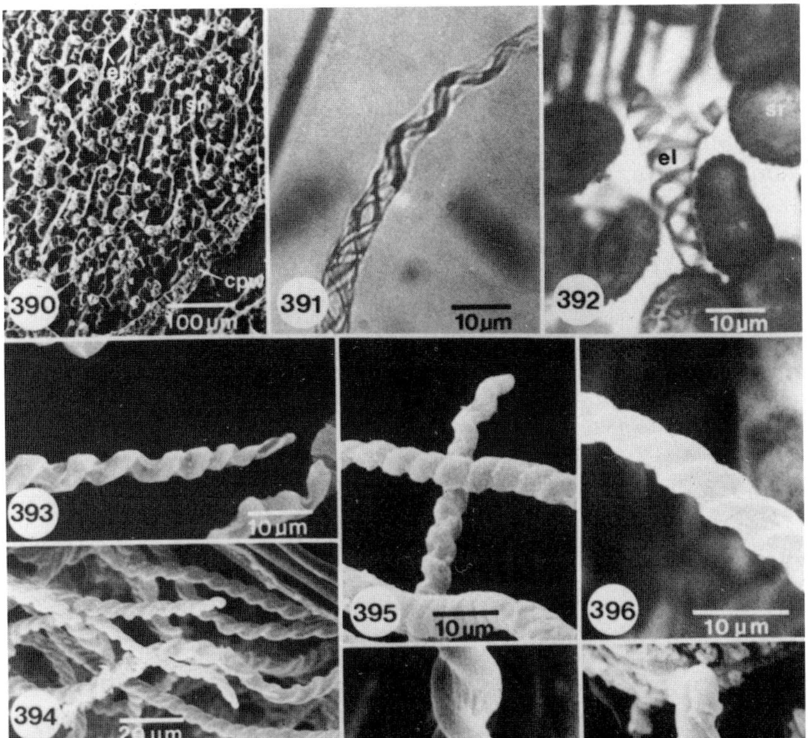
390
el
sr
cpw
100 µm
391
10 µm
392
sr
el
10 µm
393
10 µm
394
20 µm
395
10 µm
396
10 µm
397
10 µm
398
5 µm
399
5 µm
400
cpw
30 µm
401
cpw
30 µm

FIGURES 402-405. Fine structure of developing elaters of *Haplomitrium* Nees. -402a-b. Portion of an immature *H. mnioides* (Lindb.) Schust. elater in transection. 402a. Note the opaque spiral thickening bands, the presence of large lipid droplets and starch grains in the degenerating cytoplasm, and the lipids adhering to the outside of the elater wall. 402b. Spiral thickening band formed toward the interior of the cell. The large lipid adhering to the outside of the elater wall is not membrane bound. -403. Transection through a young elater. Note the generalized deposition of extra wall material toward the interior of the cell, its accumulation into opaque spiral thickening bands, and the large starch grains and lipid droplets in the degenerating cytoplasm. -404-405. Mature *H. blumii* (Nees) Schust. elaters. The thickening bands are densely osmiophilic. 404. Longitudinal section. The cytoplasm is devoid of organelles. 405. Transection.

Key: do, degenerating organelles; l, lipid droplet; ol, external lipid droplet; sg, starch grain; tb, thickening band; w, cell wall. (Figs. 402, 403: *Furuki 7408* [ABSH]; 404, 405: *Furuki 384* [ABSH].)

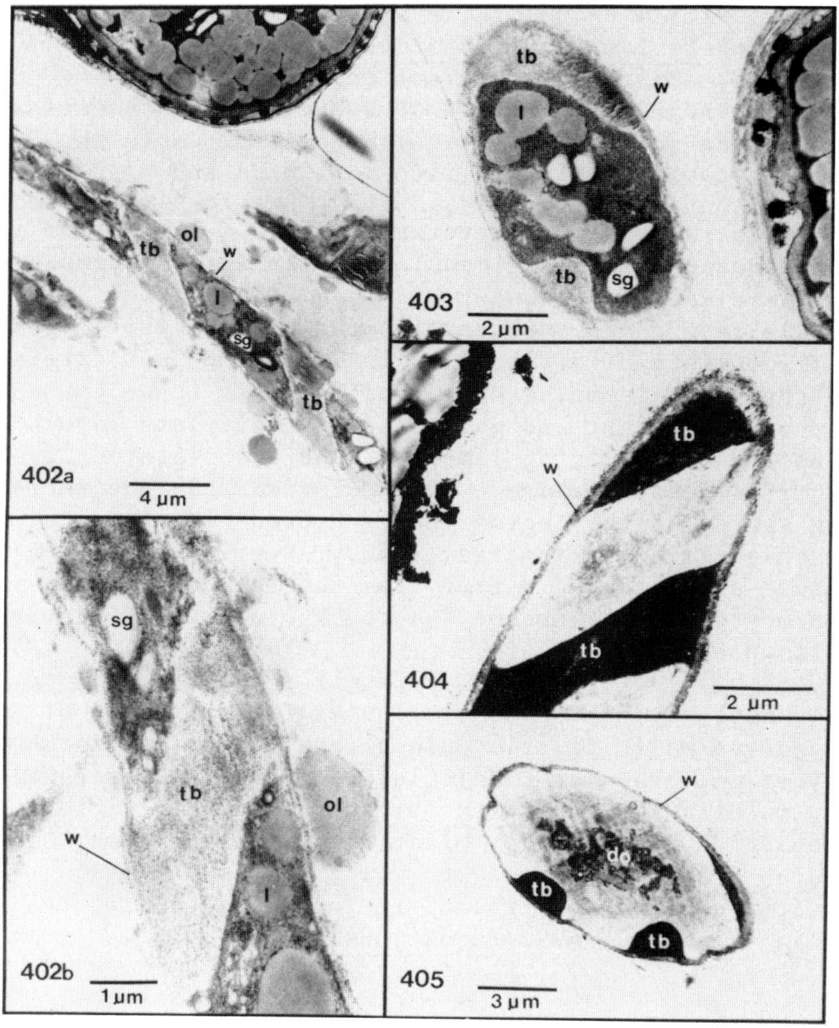

402a
tb
ol
w
l
sg
tb
4 µm
403
tb
w
l
tb
sg
2 µm
404
tb
w
tb
2 µm
402b
sg
tb
ol
w
l
1 µm
405
w
do
tb
tb
3 µm

FIGURES 406-409. Fine structure of developing spores of *Haplomitrium mnioides* (Lindb.) Schust. -406-408. Spores in the late tetrad stage of development. 406. Tetrad of spores which are enclosed within the sporocyte wall. Transection. Note the dense cytoplasm which is rich in lipids and starch-containing plastids. 407. Distinctly layered inner tetrad and outer sporocyte walls. The latter is composed of an inner flocculent layer which encompasses the lamellae of the outer exine, a finely granular middle layer, and an outer fibrous layer. The tetrad wall consists of a fibrillar intine lying just external to the plasmalemma, a thin, multilaminar inner exine of alternating light and dark bands, a foveolate layer composed of low lying lamellae, and a sculptured layer of trilaminar lamellae. 408. Proximal surface where the spores of the tetrad meet. Sporoderm ornamentation in this region is limited. -409. Section through a mature spore which is free from the tetrad. The sporocyte wall is absent and the sporoderm sculpturing is complete.

Key: cy, cytoplasm; f, outer fibrous layer of sporocyte wall; fl, flocculent layer; fo, foveolate layer; gm, granular middle layer of the sporocyte wall; ie, multilaminar inner exine; in, intine; l, lipid droplet; n, nucleus; p, plastid; pl, plasmalemma; sg, starch grain; sw, sporocyte wall; sx, sexine; tl, sculptured layer of trilaminar lamellae; tw, tetrad wall. (Figs. 406-409: *Furuki 384* [ABSH].)

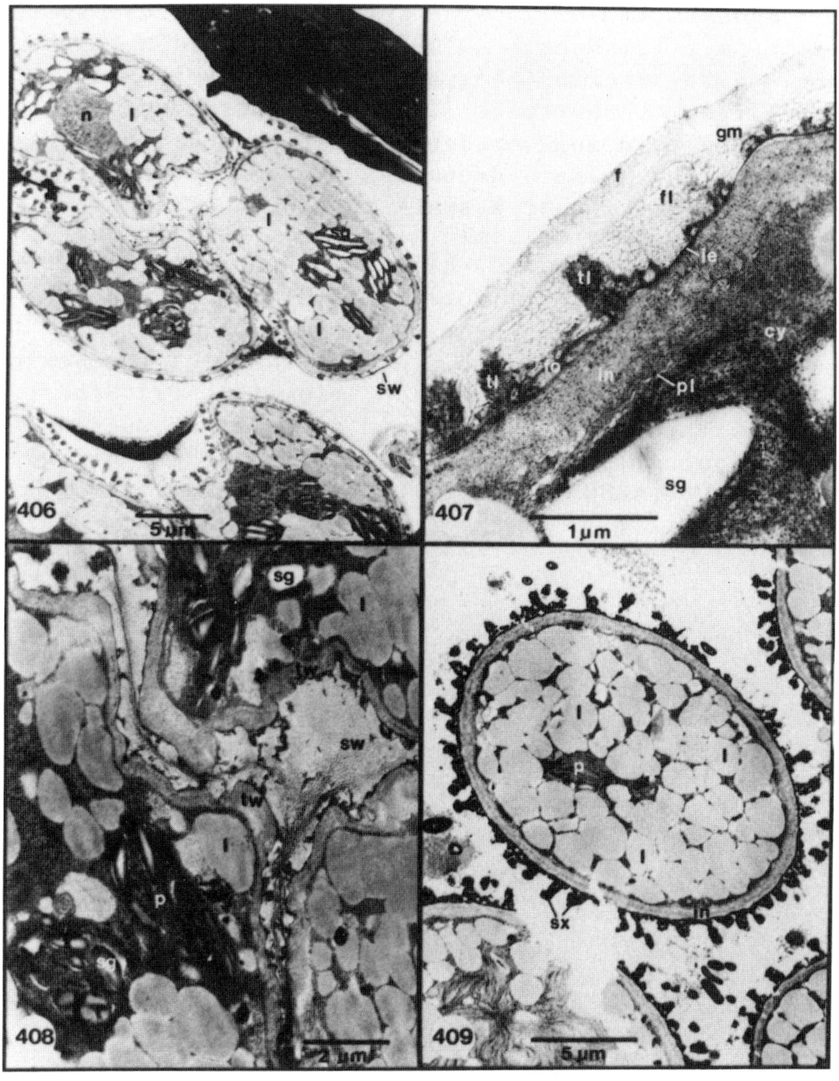
406
n
l
l
l
sw
5 µm
407
gm
f
fl
le
tl
cy
tl
to
in
pl
sg
1 µm
408
sg
l
sw
tw
l
p
sg
2 µm
409
l
p
l
l
sx
in
5 µm

FIGURES 410-413. Fine structure of mature spores of *Haplomitrium* Nees. -410-411. *H. mnioides* (Lindb.) Nees. 410. Section showing the cytoplasmic composition of the spore. A large central nucleus is surrounded by a zone of dense plastids and a peripheral region of large lipid droplets. 411. Section of the peripheral portion of a spore showing the large, tightly packed cytoplasmic lipid droplets and the sculptured spore wall. -412-413. Spore walls in longitudinal section. Note the trilaminar composition of the exine, the electron dense sporopollenin sculpturing, and the areolate appearance of the sexine. 412. *H. mnioides*. 413a-b. *H. blumii* (Nees) Schust.

Key: in, intine; l, lipid droplet; n, nucleus; nu, nucleolus; p, plastid; sx, sexine. (Figs. 410-412: *Furuki 7408* [ABSH]; 413: *Furuki 384* [ABSH].)

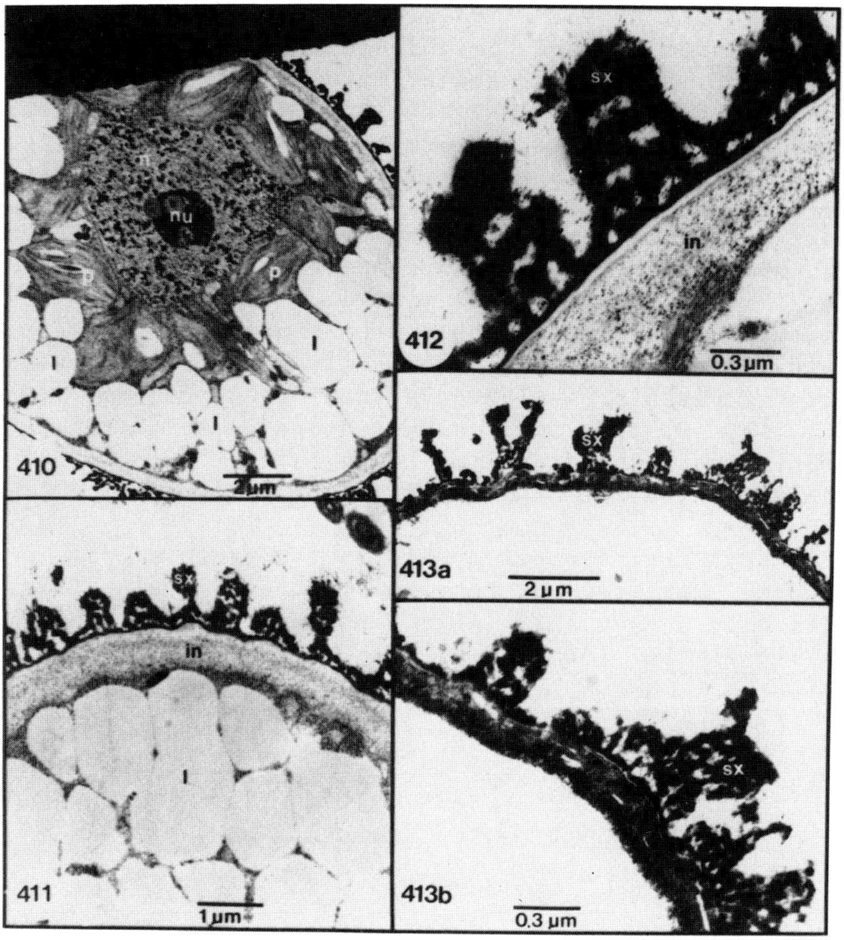

410
nu
p
p
l
l
l
2 µm
411
sx
in
l
1 µm
412
sx
in
0.3 µm
413a
sx
2 µm
413b
sx
0.3 µm

FIGURES 414-417. Mature spore morphologies of _Haplomitrium hookeri_ (Smith) Nees, _Steereomitrium minutum_ E. O. Campb., and _H. mnioides_ (Lindb.) Schust. Scanning electron micrographs. -414-415. _H. hookeri_. Note the verruculate sculpture. 414. Tetrad. 415a. Free spore in which the lightly sculptured proximal surface is visible. 415b. Verruculate ornamentation. Note that some verruculae are partially fused with one another. One verrucula which bears a central aperture is visible. -416a-b. _S. minutum_. This spore is virtually indistinguishable from those of _H. hookeri_. -416a. Spore in which the lightly sculptured proximal face is visible. -416b. Verruculate sporoderm. -417a-b. _H. mnioides_. 417a. Spore in which the lightly sculptured proximal face is visible. 417b. Dense tuberculate ornamentation.

Key: ap, central aperture of verrucula; ps, proximal surface; ta, tubercula; va, verrucula. (Figs. 414, 415: _Jensen 1347_ [AD]; 416: _Campbell s.n._ [NY]; 417: _Doei s.n._ [ABSH].)

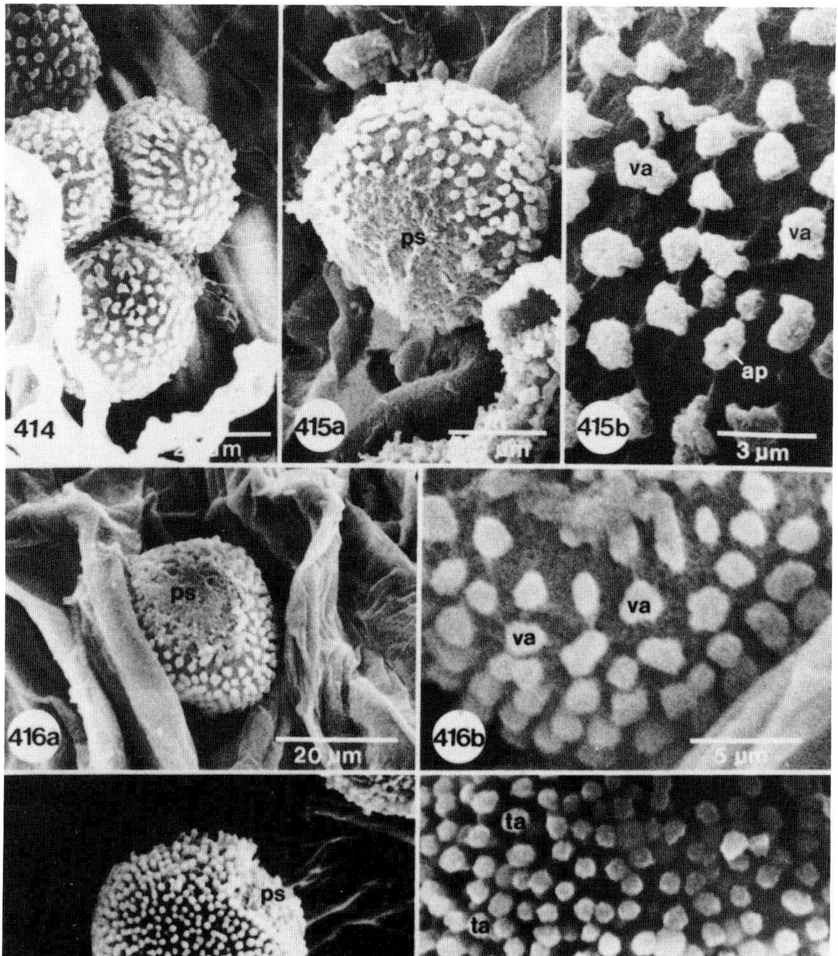
414
415a
ps
415b
va
va
ap
3 µm
416a
ps
20 µm
416b
va
va
5 µm
417a
ps
10 µm
417b
ta
ta

FIGURES 418-421. Mature spore morphologies of *Haplomitrium gibbsiae* (Steph.) Schust. and *H. intermedium* Berrie. Scanning electron micrographs. -418-420. *H. gibbsiae*. 418. Tetrad. Irregularly roughened verrucae ornament the spores. 419. Free spore. Smooth verrucae ornament the spore. 420. Lateral view of smooth verrucae. Note the coarse foveolate layer evident on the proximal surface. -421a-b. *H. intermedium*. 421a. Spore ornamented with dense vermiform ridges. 421. Ridged sporoderm.

Key: ps, proximal surface; vm, vermiform ridge; vr, verruca. (Figs. 418-420: *Allison, Scott & Taylor 135503* [MUCV]; 421: *Berrie 3/62* [SYD].)

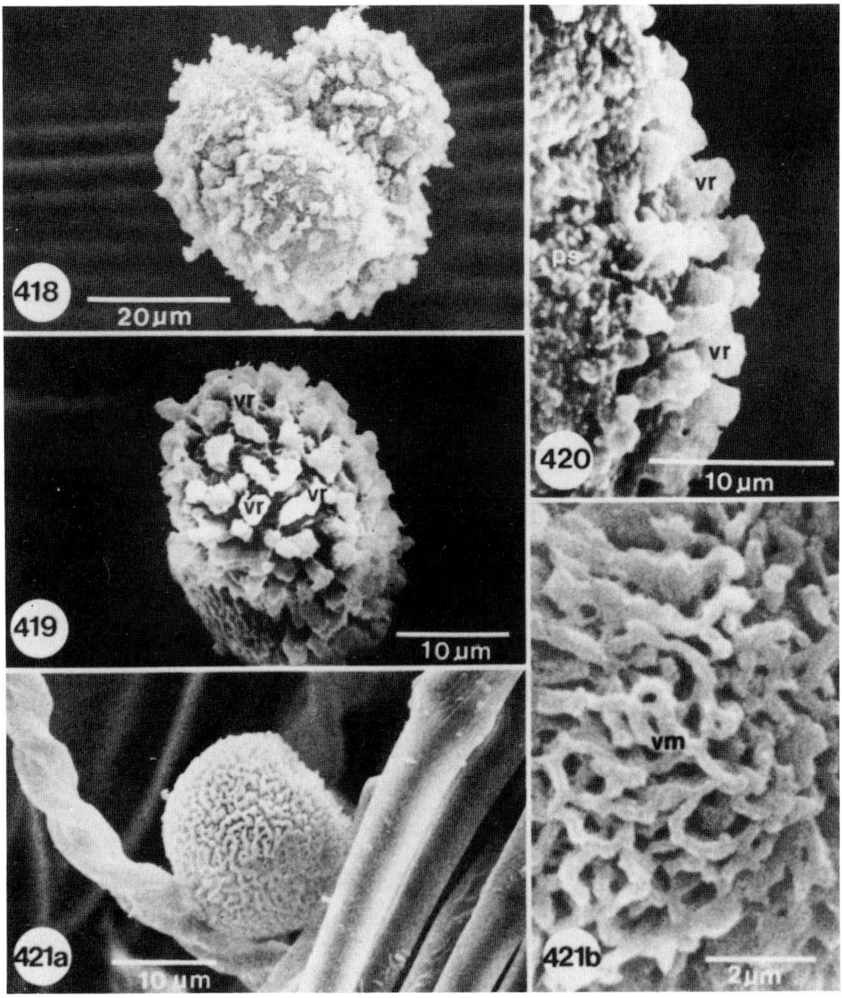
418
20 µm
vr
ps
vr
420
10 µm
vr
vr
vr
419
10 µm
vm
421a
10 µm
421b
2 µm

FIGURES 422-424. Mature spore morphologies of *Haplomitrium blumii* (Nees) Schust. and *H. andinum* (Spruce) Schust. Scanning electron micrographs. -422-423. *H. blumii*. 422a. Spore. 422b. Sporoderm composed of relatively short ridges which often anastomose. Note the reticulate nexine. 423. Spore in which the proximal surface is visible. -424a-b. *H. andinum*. Note its similar morphology to the *H. blumii* spores in Figures 422-423. 424a. Spore in which the proximal surface is visible. 424b. Sporoderm composed of short ridges which may anastomose. Note the reticulate nexine.

Key: nx, nexine; ps, proximal surface; rd, ridge. (Figs. 422, 423: *Norris 61631* [H]; 424: *Spruce s.n.* [NY].)

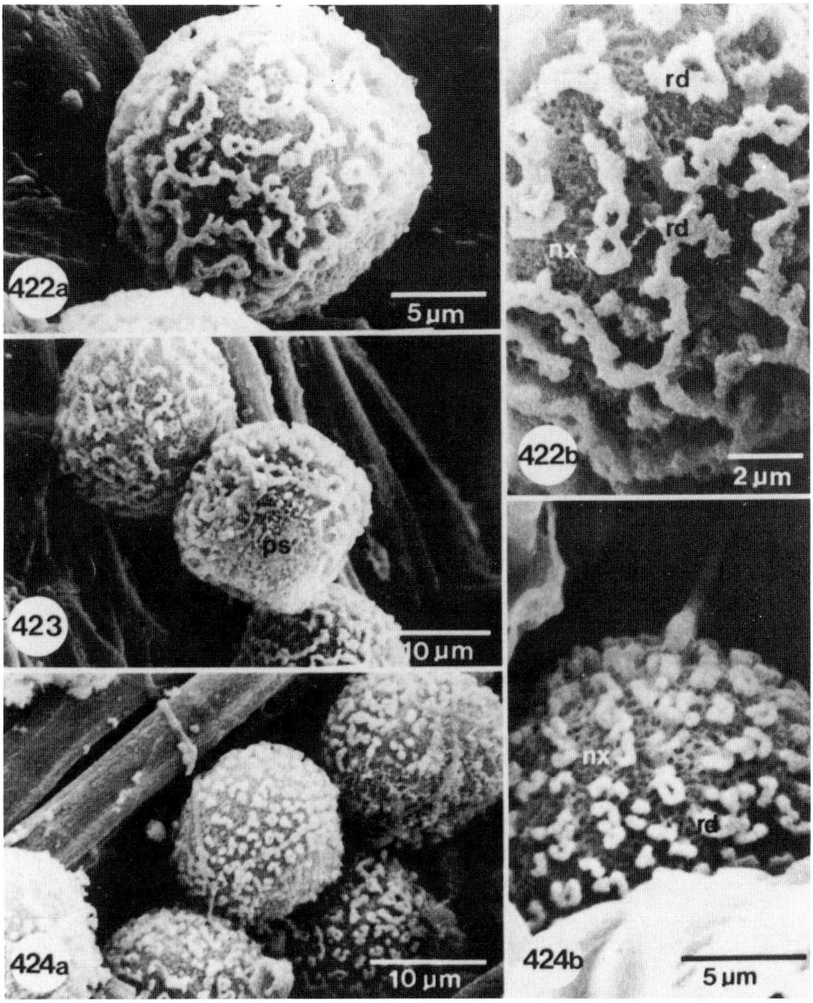

422a
5 µm
rd
rd
nx
422b
2 µm
ps
423
10 µm
nx
rd
424a
10 µm
424b
5 µm

FIGURES 425-428. Dehiscence in *Haplomitrium* Nees. Scanning electron micrographs. -425. *H. hookeri* (Smith) Nees capsule which has opened along two opposite suture lines into two valves with blunt apices. One valve retains a row of specialized dehiscence cells (arrow). Note the elaters which adhere to the apices of the open valves. -426a-c. Dehisced *H. gibbsiae* (Steph.) Schust. capsule. 426a. Capsule with one complete suture (s_1) and a second partial suture (s_2). An untorn row of dehiscence cells is positioned along the valve opposite to the partial suture. 426b. Multistratose apex of capsule. Note the row of special dehiscence cells bearing transverse rods which is positioned along the valve opposite to the suture. 426c. Row of specialized dehiscence cells bearing transverse rods. -427. Dehisced *H. mnioides* (Lindb.) Schust. capsule. -428. Portion of a row of specialized dehiscence cells on the edge of an open valve of an *H. hookeri* capsule.

Key: el, elater; tr, transverse rod; sd, specialized dehiscence cell. (Fig. 425: *Long 9930* [BM]; 426: *Allison, Scott & Taylor 135503* [MUCV]; 427: *Doei s.n.* [ABSH]; 428: *Jensen 1347* [AD].)

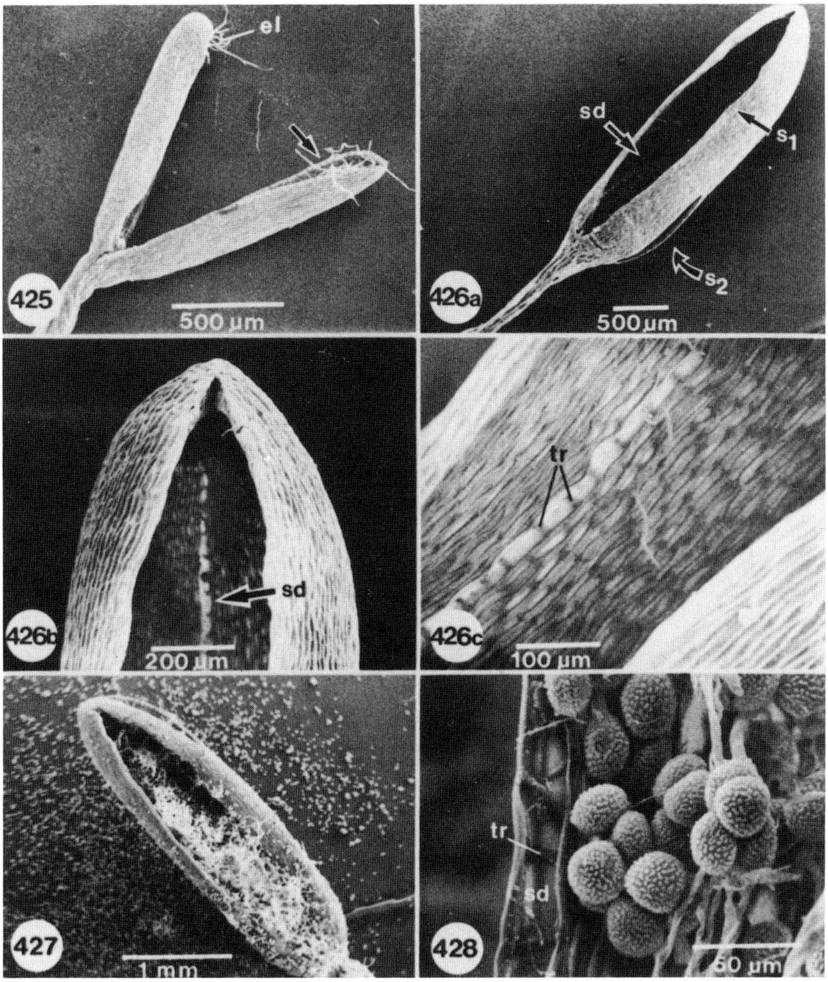

el
425
500 µm
sd
s1
s2
426a
500µm
sd
426b
200 µm
tr
426c
100 µm
427
1 mm
tr
sd
428
50 µm

FIGURES 429-439. Spores and protonemata of _Haplomitrium_ _mnioides_ (Lindb.) Schust. -429. Swollen spore protruding through the ruptured exospore. -430. Initial transverse division delimiting a terminal cell and a basal cell. -431a-b. Two views of the same protonema showing the first longitudinal division of the terminal cell. 431a. Lateral view. 431b. Near apical view. -432a-b. Two views of the same protonema showing the second wall lying perpendicular to the first division in the terminal cell. 432a. Lateral view. 432b. Apical view. -433a-c. Three views of the same protonema showing the quadrant stage. 433a. Lateral view. 433b. Basal view. 433c. Apical view. -434a-c. Three views of the same protonema showing the initiation of tier production. 434a-b. Lateral views with (a) oriented at a 90° angle to (b). 434c. Apical view. -435. Lateral view showing two tiers of cells, each consisting of four cells. -436a-b. Two different views of the same protonema showing two tiers of cells. Note the spore wall on the basal cell. 437a. Basal view. 437b. Lateral view. -438a-b. Different views of the same protonema showing generalized divisions. 438a. Lateral view. 438b. Near apical view. -439a-b. Two different lateral views of the same massive, generalized protonema.

Key: bc, basal cell; e, exospore; lt, lower tier; tc, terminal cell; ut, upper tier.

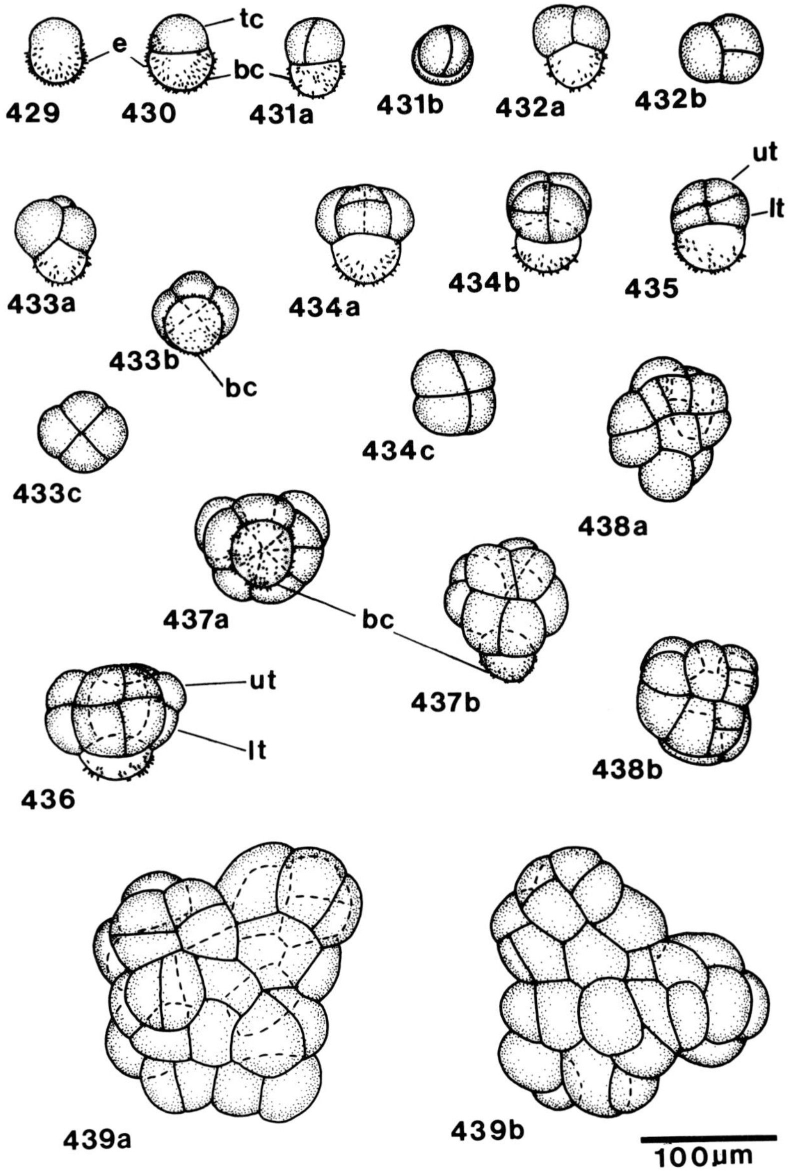

429
430
431a
431b
432a
432b
433a
433b
433c
434a
434b
434c
435
436
437a
437b
438a
438b
439a
439b
e
tc
bc
ut
lt
100µm

FIGURES 440-447. Protonemata of _Haplomitrium blumii_ (Nees) Schust. -440. Initial transverse division delimiting a terminal cell and a basal cell. -441. First longitudinal division of the terminal cell. Lateral view. -442. Apical view showing four quadrants. -443. Initiation of tier production. Lateral view. -444a-b. Two views of the same protonema bearing two tiers of cells. 443a. Lateral view. 443b. Basal view. -445a-b. Different views of the same protonema showing generalized divisions. -446a-b. Two different lateral views of the same massive, generalized protonema. -447. Lateral view of a massive protonema. Note the slime papillae.

Key: bc, basal cell; e, exospore; lt, lower tier; sp, slime papilla; tc, terminal cell; ut, upper tier.

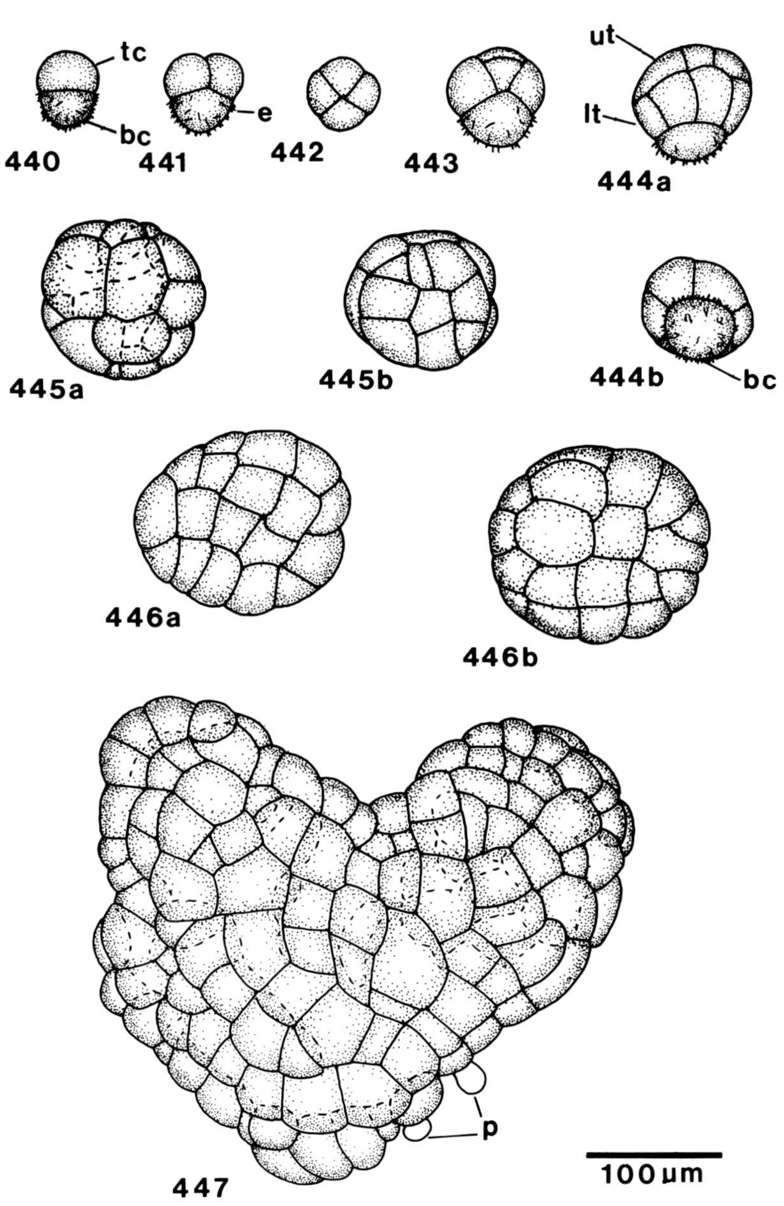

tc
bc
e
ut
lt
bc
p
100 µm
440
441
442
443
444a
445a
445b
444b
446a
446b
447

FIGURES 448-454. Scanning electron micrographs of the protonemata of *Haplomitrium mnioides* (Lindb.) Schust. -448. Young cylindrical shoots arising from all surfaces of a massive protonema. -449. Protonema with one shoot elongating faster than the others. Note the similar sized cells produced through multiple tier formation. -450. Four quadrants built from a protonemal cell at the initiation of the shoot generation. Note the early production of a slime papilla. -451. Portion of a protonema showing the cylindrical axes, each with a central cylindrical shoot initial surrounded by four peripheral cells. -452a-b. Two views of a shoot tip showing a cylindrical shoot initial and the four surrounding cells. 452a. Apical view. 452b. Lateral view. -453. Apical tier of a young cylinder composed of six peripheral cells surrounding a single, swollen central cell. -454. Portion of protonema with cylindrical shoots at different stages of elongation. Note the two daughter cells formed by the first division of a cylindrical shoot initial, and four apical quadrants.

Key: ccl, central cell; csi, cylindrical shoot initial; dc, daughter cell; q, quadrant of four cells; sp, slime papilla.

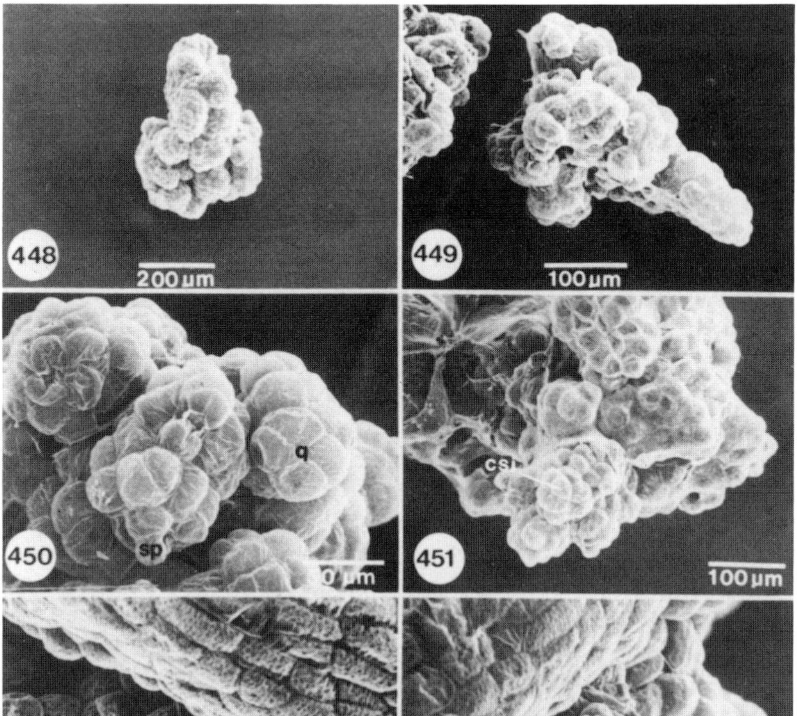
448
200 µm
449
100 µm
q
450
sp
csl
451
100 µm
csl
452a
csl
452b
ccl
453
dc
q
454

FIGURES 455-460. Protonemata of *Haplomitrium mnioides* (Lindb.) Schust. -455a-d. Light micrographs showing transverse serial sections made through a leafless, cylindrical shoot from the apical to the basal tiers of the shoot showing the fundamental cell arrangements of each tier. 455a. Apical tier composed of 6 epidermal cells and 1 inner cell. 455b. Tier composed of 8 epidermal cells and 4 inner cells. Approximately 28-32 µm below the apical tier. 455c. Tier composed of 10 epidermal cells and 8 inner cells. Approximately 44-56 µm below apical tier. 455d. Tier composed of 12 epidermal cells and 8 inner cells. Approximately 60-72 µm below the apical tier. Note the slime cell. -456-457. Scanning electron micrographs. 456. Clavate slime papilla of a cylindrical shoot. 457. Enlarged protonemal epidermal cells which will become active unicellular slime cells. -458-460. Light micrographs. 458. Protonemal cells with spherical oil bodies. 459. Median leaf cells of a juvenile sporeling showing numerous adult-like oil bodies. 460. Longitudinal section through the apex of a sporeling shoot. Note the apical cell, derivatives and protective slime cell.

Key: ccl, central cell; ep, epidermal cell; ic, inner cell; ob, oil body; p, plastid; sp, slime papilla; x, apical cell.

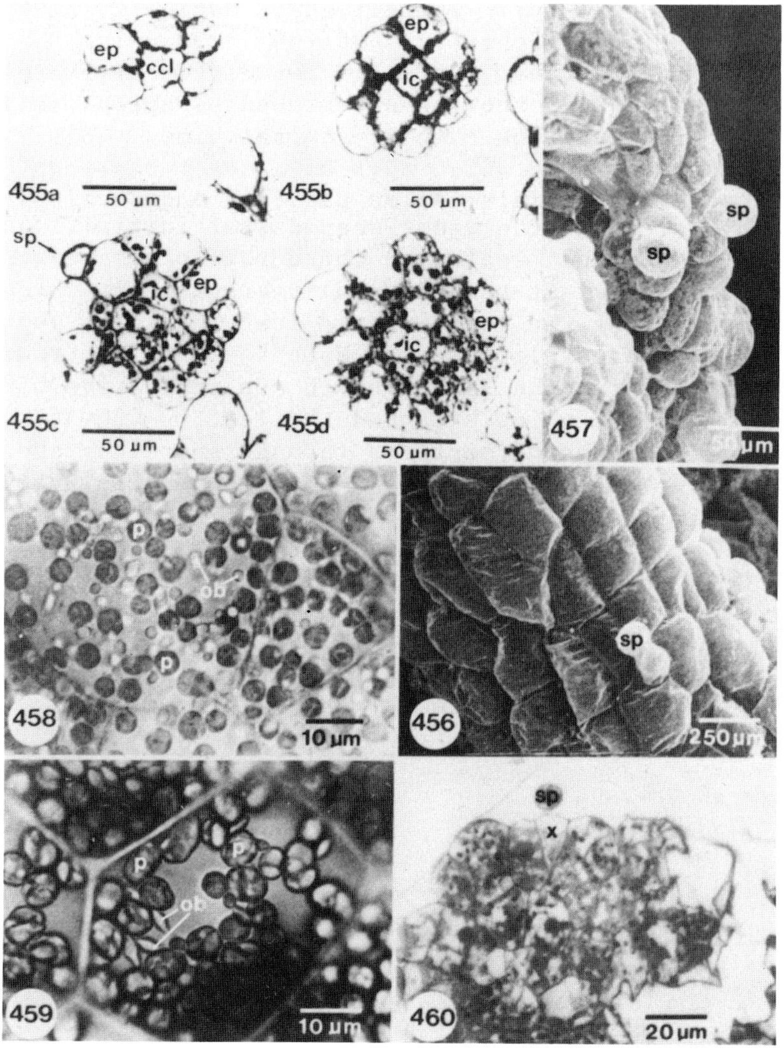
ep
ccl
455a
50 µm
ep
ic
455b
50 µm
sp
ic
ep
455c
50 µm
ic
ep
455d
50 µm
sp
sp
457
50 µm
p
ob
p
458
10 µm
sp
456
250 µm
p
p
ob
459
10 µm
sp
x
460
20 µm

FIGURES 461-467. Sporelings of Haplomitrium mnioides (Lindb.) Schust. -461-465. Scanning electron micrographs. 461a-b. Two different views of the same shoot tip showing a tetrahedral apical cell and young derivatives. 461a. Apical view. 461b. Near lateral view. 462. Apex of a shoot bearing protective, unicellular slime papillae and developing merophytes. Note the wedge-shaped leaf initial. 463. Leafy shoot with a scale and three rows of leaves. 464. Young leaf at the triangular stage built through segmentation of the wedge-shaped leaf initial. 465. Fully formed sporeling leaf. Note the straight rows of cells comprising the lamina which are indicative of the basal divisions which expanded the leaf. -466-467. Light micrographs of sections through sporeling shoots showing the central strands. 466. Transection. 467. Longitudinal section.

Key: b, branch; c, cortex; ep, epidermis; h, hydroid; le, leaf; li, leaf apical cell; mr, developing merophyte; sa, scale; sp, slime papilla; x, shoot apical cell.

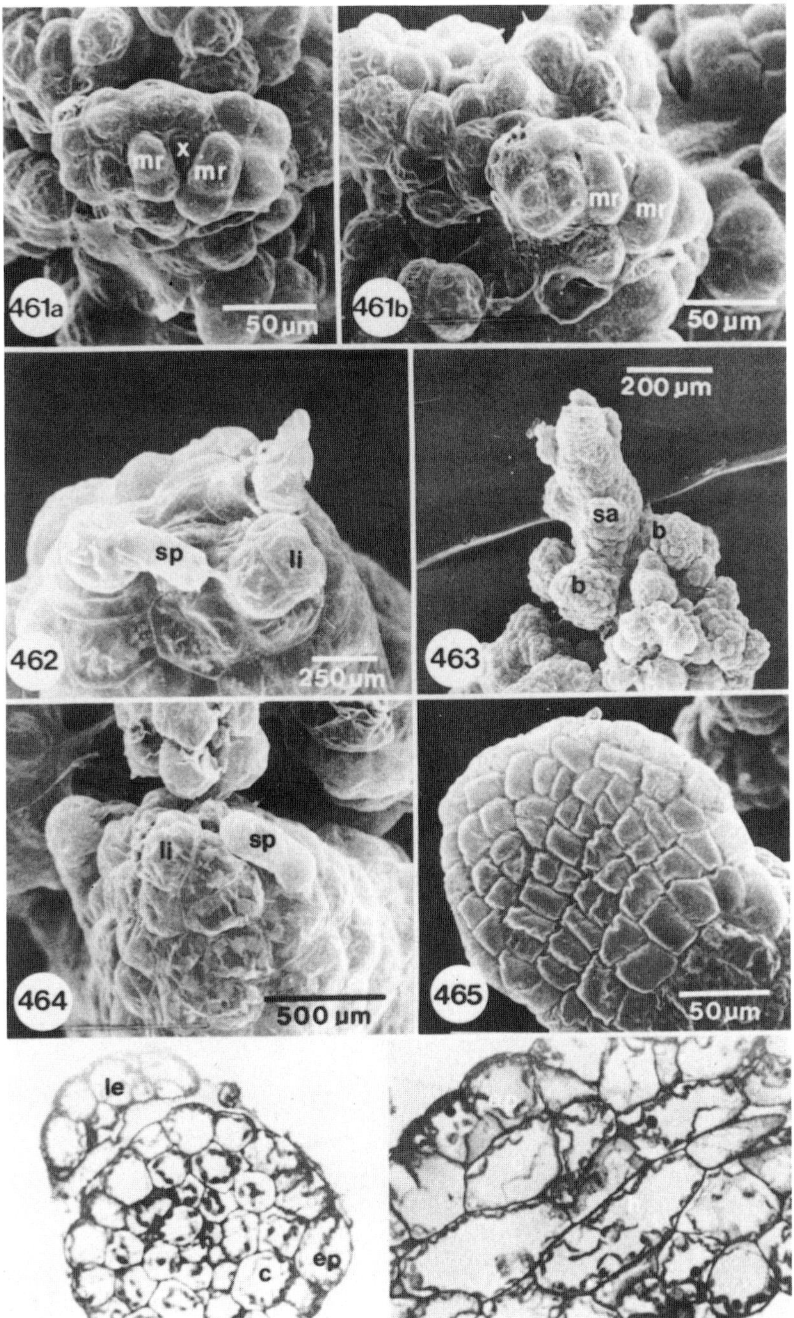
461a
mr
x
mr
50 µm
461b
mr
mr
50 µm
462
sp
li
250 µm
463
200 µm
sa
b
b
464
li
sp
500 µm
465
50 µm
466
le
ep
c
50 µm
467

FIGURES 468-471. Sporelings of _Haplomitrium mnioides_ (Lindb.) Schust. -468-470. Scanning electron micrographs. 468. Branch initiation (arrow) from a leafy shoot. 469. Elongating branch of a leafy shoot. 470. Portion of a sporeling showing multiple branches on a leafy shoot. -471. Light micrograph of a juvenile plant.

Key: b, branch.

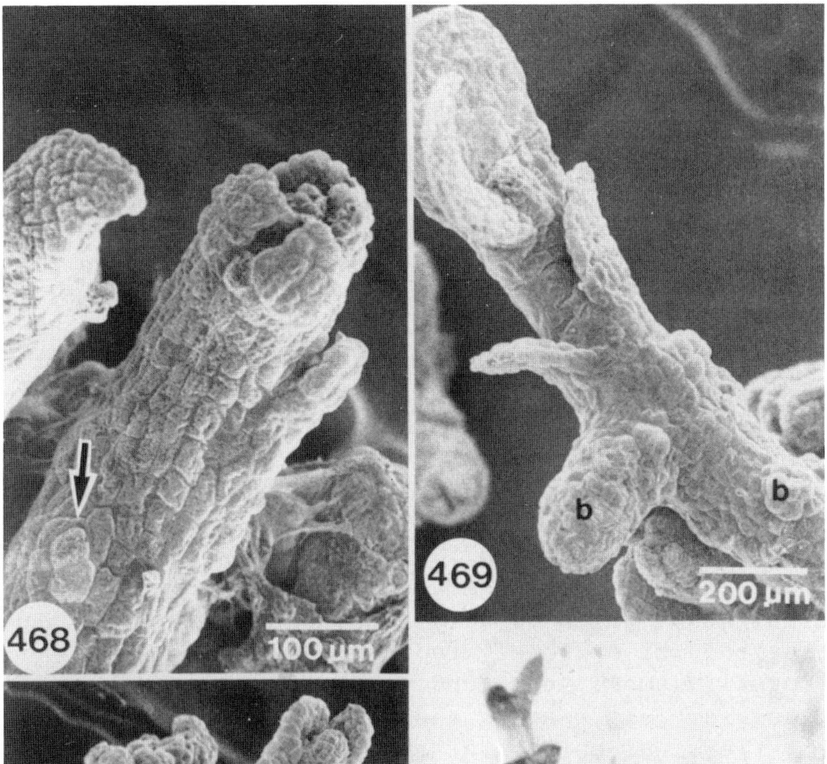

468
100 µm
469
b
b
200 µm
470
b
b
200 µm
471
1 mm

FIGURES 472-478. Media-induced growth forms of juvenile *Haplomitrium mnioides* (Lindb.) Schust. plants grown on Hatcher, Beneke, and Hatcher + 10^{-4} M L-proline media. All other environmental variables were held constant. All juveniles are 26 months old. The plants on Hatcher medium constitute the control group. -472. Leafy shoot produced by plants grown from spores to the cylindrical shoot stage on F^1 medium, then transferred to Hatcher medium to complete their development. The shoots display the morphology typical of plants grown from spores on Hatcher medium. -473. Leafy shoots from plants which were grown from spores on Hatcher medium. The shoots are 3-ranked, slightly anisophyllous, with entire, subrotund leaves which are moderately spaced (= the "normal" morphology). -474. Leafy shoot bearing a terminal antheridial rosette. Plants were grown from spores on Hatcher medium. -475-477. Plants grown from spores on Beneke medium. Note the aberrant morphology. 475a-b. Aerial shoot which is very long, unbranched, and bears three rows of extremely reduced leaves. 475b. Abnormally small subrotund leaf and a 5-celled triangular leaf bearing a single terminal slime cell. 476. Terminal portion of an aerial axis bearing a repressed branch and distantly spaced, reduced leaves. 477. Rudimentary "scales" (= 2-celled slime papillae) located at the base of a shoot. -478. Plants grown on Hatcher + 10^{-4} M L-proline medium. The leaves of all three rows are transversely inserted and are fairly similar to one another in size.

Key: an, antheridium; br, bract; b, branch; rs, rudimentary scale.

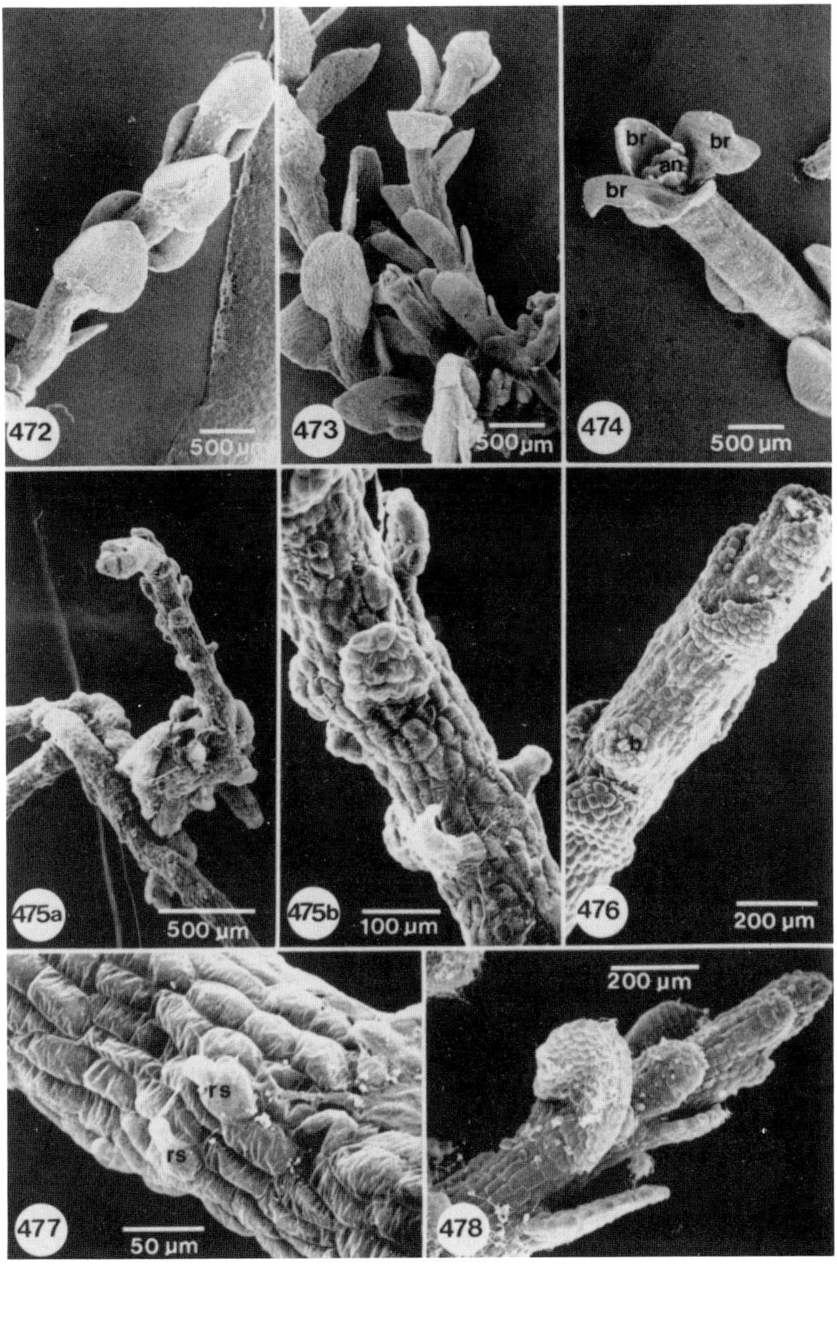
472
500 µm
473
500 µm
br
br
an
br
474
500 µm
475a
500 µm
475b
100 µm
b
476
200 µm
rs
rs
477
50 µm
200 µm
478

FIGURES 479-485. Light induced phenovariations in juvenile *Haplomitrium mnioides* (Lindb.) Schust. plants. All plants were grown on Hatcher medium under uniform temperature and photoperiod regimes, and are 26 months old. -479-480. Morphologies induced by different light intensities. The control group received 1929.9 lux. 479. Plants grown under a light intensity of 861.1 lux. 480a-b. Plants grown under a light intensity of 3002.2 lux. The leaves of all three rows are transversely inserted, but the leaves of one row are reduced in size. 480a. Habit. 480b. Apex of shoot showing distinct anisophylly. -481-484. Morphologies induced by differences in light quality. The control group received full-spectrum light (1.2 $\mu W/cm^2/nm$ in the blue spectrum, 0.8 $\mu W/cm^2/nm$ in the red spectrum, 3216.0 lux). 481. Shoot of a plant grown under full-spectrum light. The shoot is anisophyllous with transversely inserted subrotund leaves. 482. Shoots of plants grown under light high in the red spectrum (0.69 $\mu W/cm^2/nm$ in the red spectrum, 0 $\mu W/cm^2/nm$ in the blue spectrum, 534.6 lux). The shoots are stunted and the leaves are absent. 483-484. Plants grown under light high in the blue spectrum (0.56 $\mu W/cm^2/nm$ in the blue spectrum, 0 $\mu W/cm^2/nm$ in the red spectrum, 740.2 lux). 483. The shoots are stunted and the leaves are absent. 484. Slime papillae on the leafless axis. -485a-b. Plants grown under incandescent light (11.0 $\mu W/cm^2/nm$ in the far-red spectrum, 2.1 $\mu W/cm^2/nm$ in the blue spectrum, 8.5 $\mu W/cm^2/nm$ in the red spectrum, 3910.4 lux). 485a. Some shoots are obliquely ascending, while others are erect. 485b. Two shoots lacking leaves, but bearing slime hairs and slime papillae.

Key: sh, slime hair; sp, slime papilla.

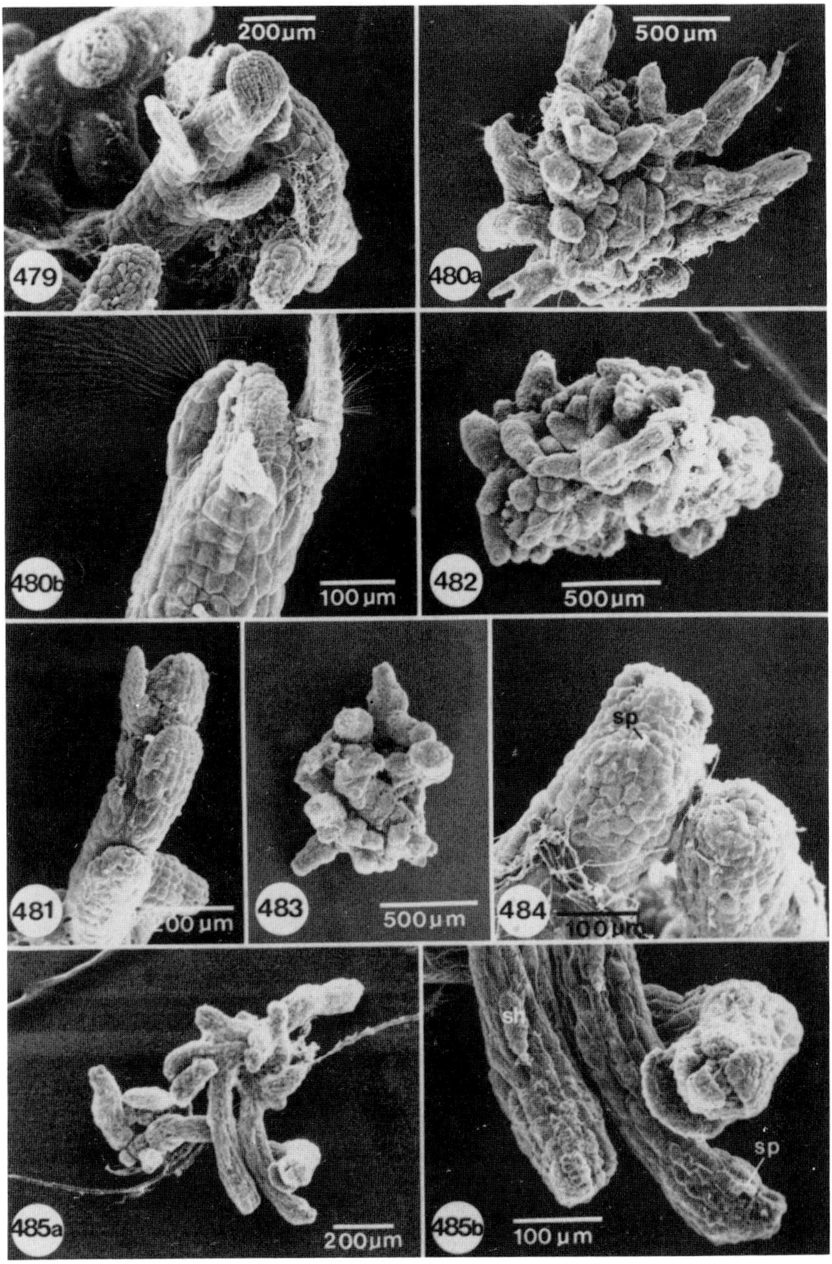
479
200 µm
480a
500 µm
480b
100 µm
482
500 µm
481
200 µm
483
500 µm
484
sp
100 µm
485a
200 µm
485b
sh
sp
100 µm

FIGURES 486-490. Culture-induced phenovariations of adult *Haplomitrium gibbsiae* (Steph.) Schust. plants. -486. Light micrograph of a tall, slender plant bearing distantly spaced, transversely inserted leaves and a single, long, slender stolon. -487-489. Scanning electron micrograph. 487. Distal portion of a shoot with bifid and deeply bilobed leaves. 488. Crescent-shaped leaf. 489. Slime hair along the stem. -490. Light micrograph of a distinctly anisophyllous shoot bearing one row of reduced, transversely inserted leaves and two rows of succubously inserted lateral leaves.

Key: le, leaf; sn, stolon; sh, slime hair. (Figs. 486-490: *Fife 8453* [ABSH].)

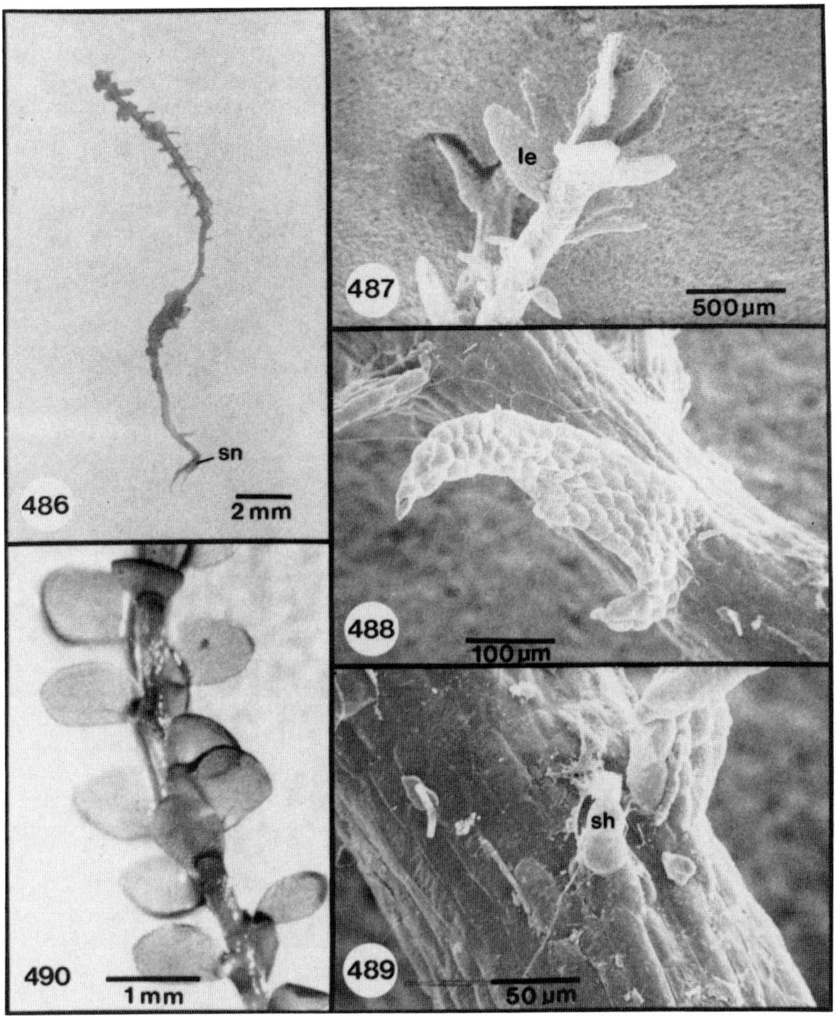
sn
486
2 mm
le
487
500 µm
488
100 µm
sh
489
50 µm
490
1 mm

FIGURES 491-496. Culture-induced phenovariations of adult *Haplomitrium mnioides* (Lindb.) Schust. plants. -491. Erect, nearly isophyllous vegetative shoot bearing highly dissected leaves. -492. Leaf bearing numerous lateral projections. -493. Terminal antheridial rosette. The perigonial bracts bear numerous lateral projections. -494. Aberrant growths along the stem and on the ventral surface of the basal lamina of a leaf. -495. Slime hairs and slime papilla on the stem. -496. Irregularly shaped scales.

Key: an, antheridium; ag, aberrant growth; br, perigonial bract; lp, lateral projection; mc, mucilage cell; sh, slime hair; sp, slime papilla. (Figs. 491-496: *Doei* *s*.*n*. [ABSH].)

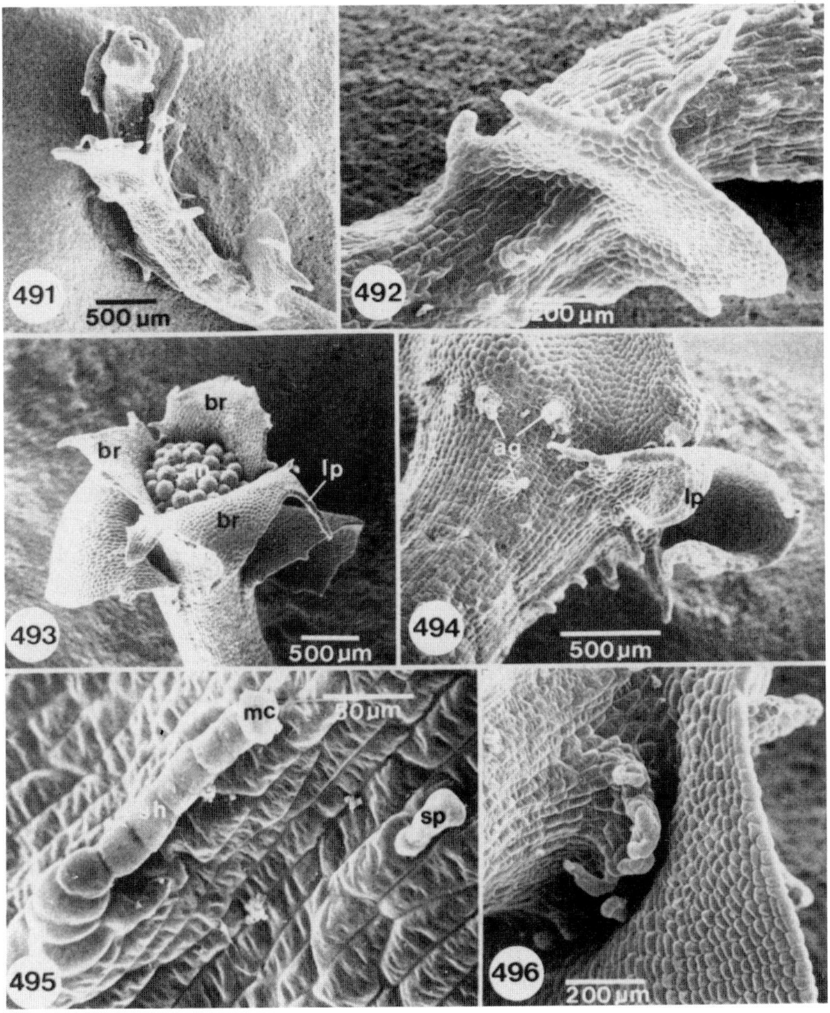
491
500 µm
492
493
br
br
br
lp
500 µm
494
ag
lp
500 µm
495
mc
sp
496
200 µm

FIGURES 497-502. Culture-induced aberrant growths on the shoots of *Haplomitrium mnioides* (Lindb.) Schust. and culture-induced branching in *H. gibbsiae* (Steph.) Schust. and *H. hookeri* (Smith) Nees. -497-499. Scanning electron micrographs of *H. mnioides*. 497. Multicellular outgrowth on the ventral surface of the basal lamina of a leaf. 498-499. Multicellular outgrowths on the stem. 498. Terminal slime papillae absent. 499. Terminal slime papillae present. -500-502. Culture-induced branching. 500-501. Elongate terminal branches lending a dichotomously forked appearance to the plant. 500. *H. gibbsiae*. 501. *H. hookeri*. 502a-b. Branching in *H. hookeri* plants maintained on Hatcher medium supplemented with kinetin. 502a. Distal portion of a shoot with a leafless branch in the axil of at least one leaf of virtually every series of merophytes. 502b. Leafless branch covered with mucilage hairs and minute scales terminated by slime papillae.

Key: ag, aberrant growth; b, branch; sa, scale; sp, slime papilla. (Figs. 497-499: *Doei s.n.* [ABSH]; 500: *Fife 8453* [ABSH]; 501, 502: *Lorbeer H56* [PR].)

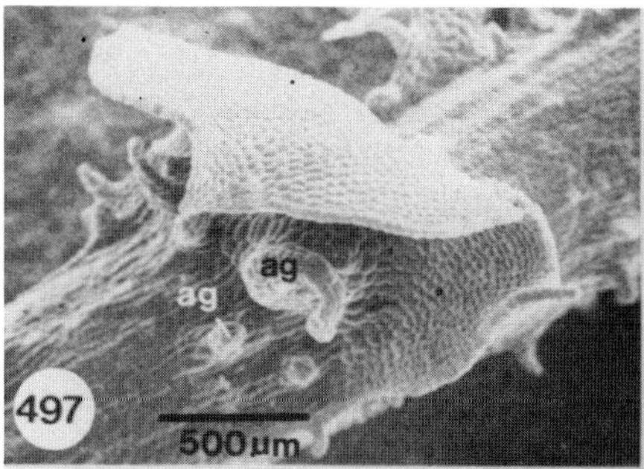
ag
ag
497
500 µm

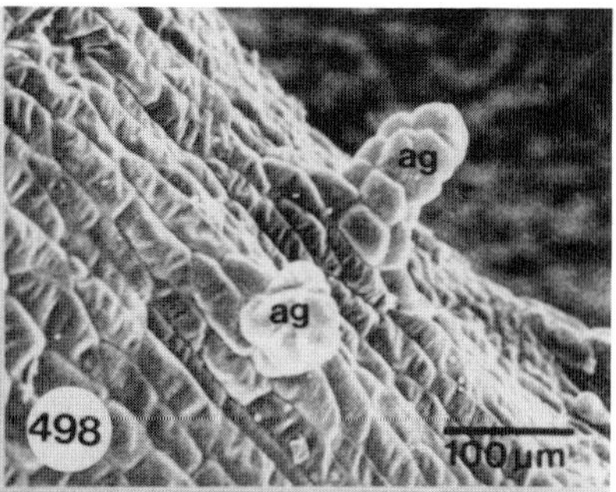
ag
ag
498
100 µm

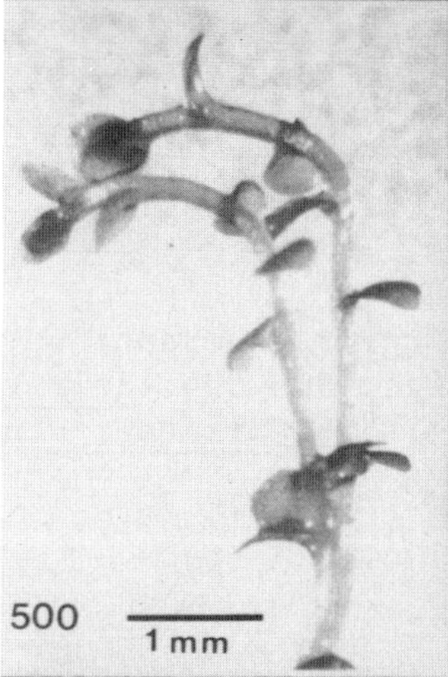
500
1 mm

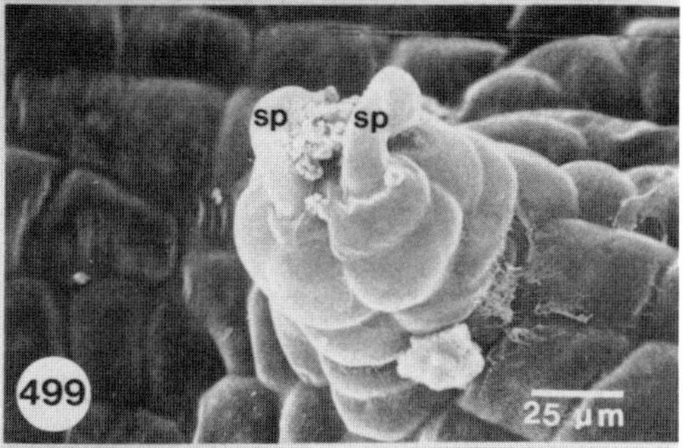
sp
sp
499
25 µm

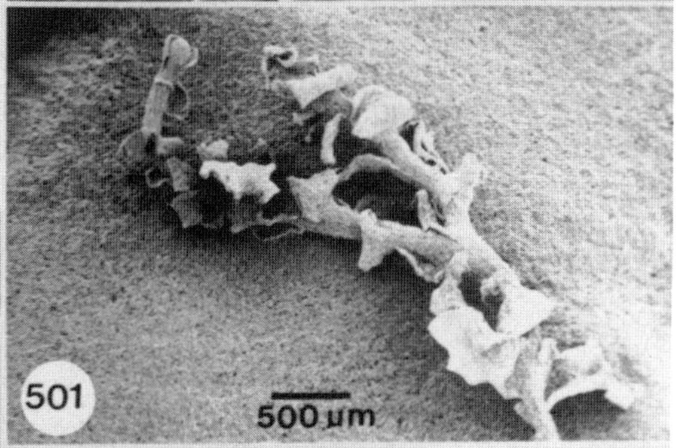
501
500 µm

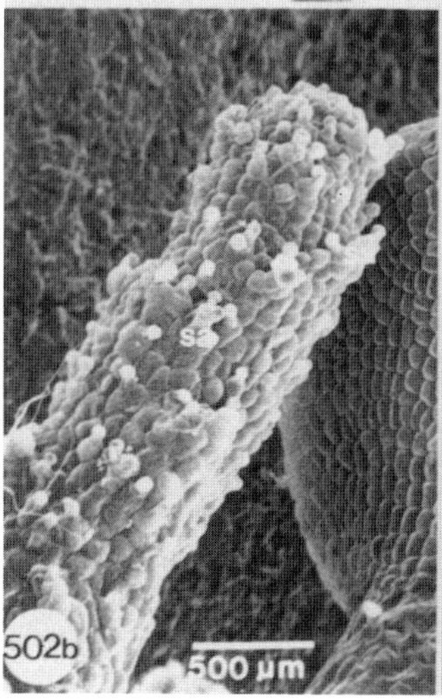
sa
502b
500 µm

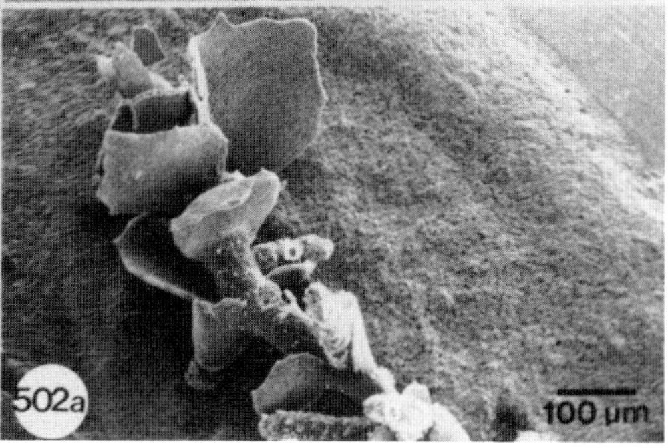
502a
100 µm

FIGURES 503-506. Culture-induced regeneration in adult *Haplomitrium* Nees plants. -503-504. *H. intermedium* Berrie. 503. Early stage in regenerant development. The regenerant is a sessile, multicellular globose mass of cells formed on the leaf lamina. 504a-b. Slender, filamentous regenerant. 504a. Distal portion of a shoot bearing a regenerant originating from the lamina of a leaf. 504b. Apical view showing the more or less triangular-shaped cylinder initial. -505. Light micrograph of a stalked, cylindrical regenerant originating from the leaf lamina of a *H. hookeri* (Smith) Nees plant. Note the terminal slime papillae. -506. Scanning electron micrograph showing a *H. mnioides* regenerant originating from the leaf margin. Note the tetrahedral apical cell.

Key: csi, cylinder initial; rg, regenerant; sp, slime papilla; x, apical cell. (Figs. 503-504: *Selkirk s.n.* [ABSH]; 505: *Lorbeer H56* [PR]; 506: *Mizutani s.n.* [ABSH].)

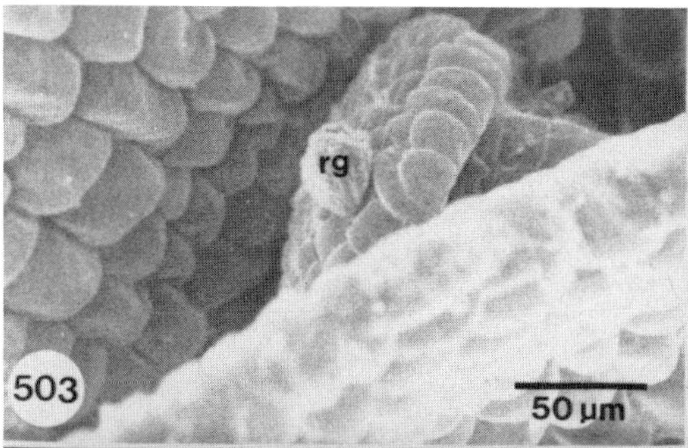
rg
503
50 µm

rg
504a
500 µm

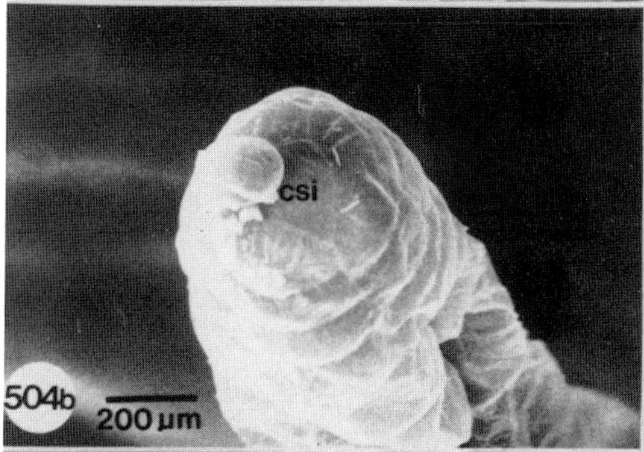
csi
504b
200 µm

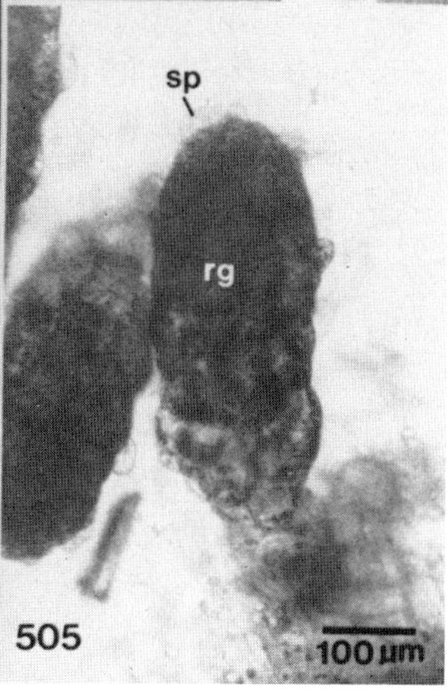
sp
rg
505
100 µm

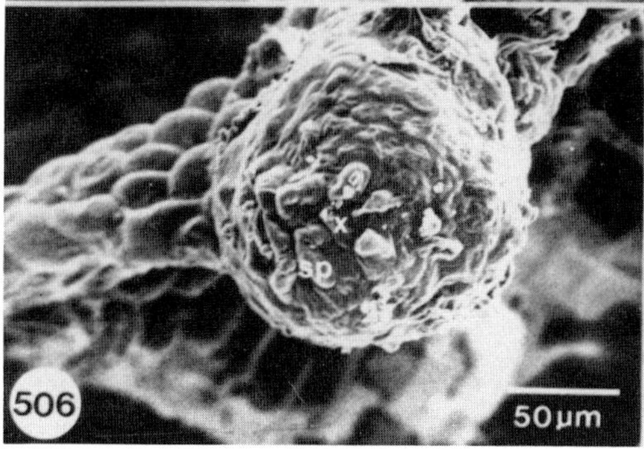
x
sp
506
50 µm

FIGURE 507. Male Haplomitrium gibbsiae (Steph.) Schust. shoot bearing culture-induced scattered antheridia.

Key: an, antheridium. (Fig. 507: Fife 8453 [ABSH].)

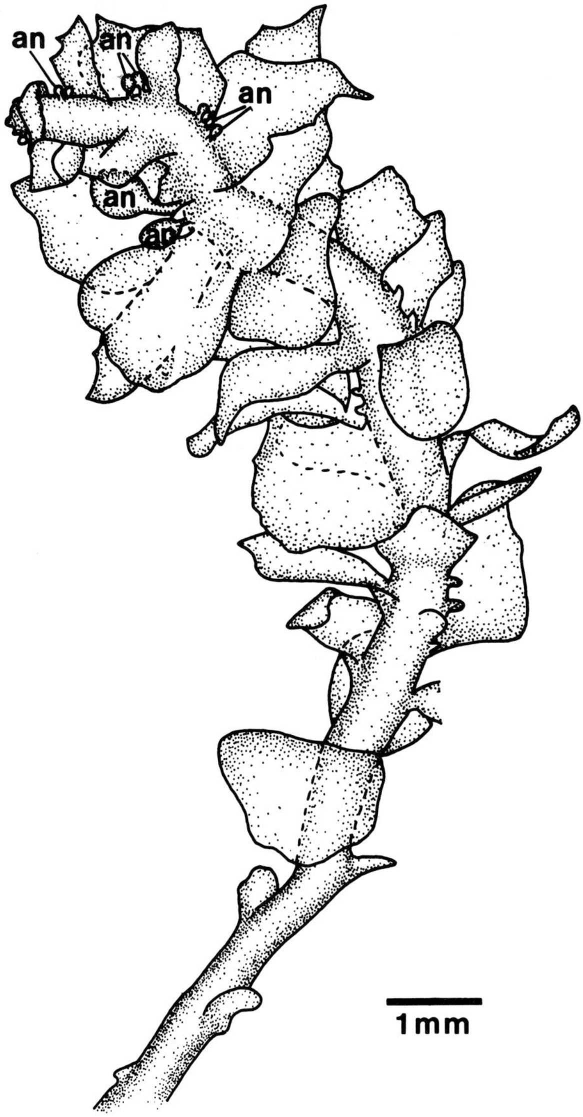

507

FIGURE 508. World distribution of *Haplomitrium* Nees

Key:

▲ *H. blumii* (Nees) Schust.

✱ *H. gibbsiae* (Steph.) Schust.

● *H. hookeri* (Smith) Nees

✶ *H. intermedium* Berrie

□ *H. mnioides* (Lindb.) Schust.

✖ *H. monoicum* Engel

▼ *H. ovalifolium* Schust.

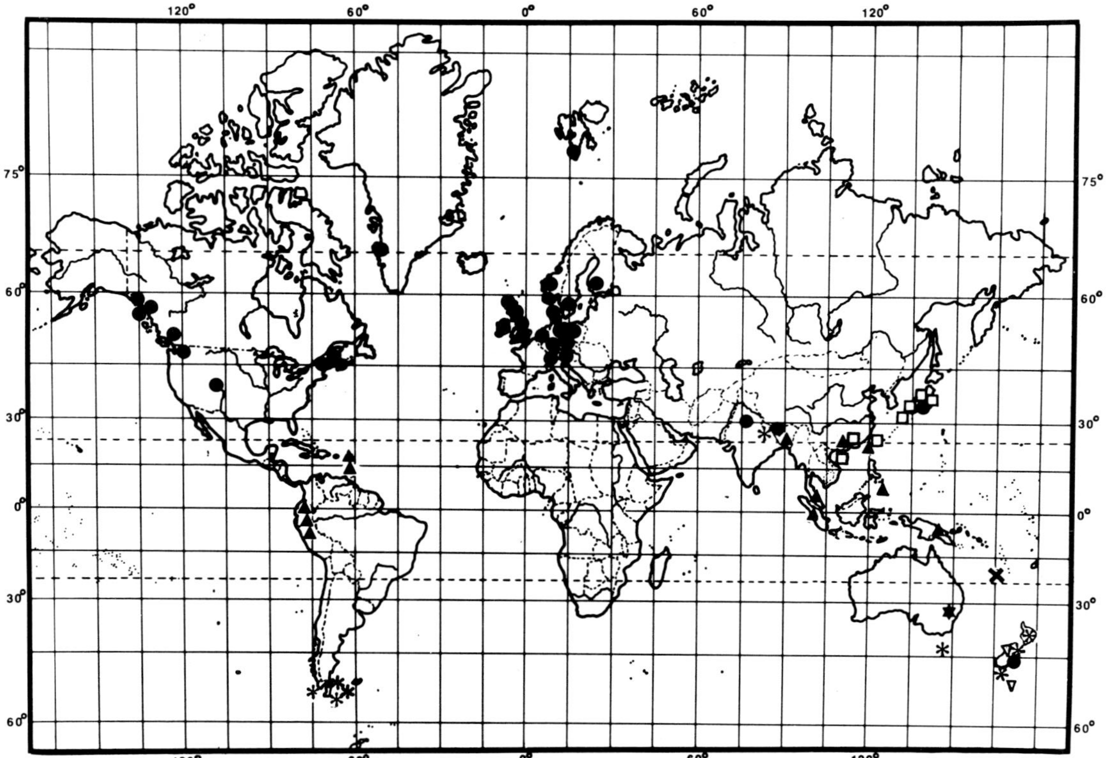
120°
60°
0°
60°
120°
75°
60°
30°
0°
30°
60°